FOURIER SERIES

FOURIER SERIES

GEORGI P. TOLSTOV

Professor of Mathematics
Moscow State University

Translated from the Russian by
Richard A. Silverman

DOVER PUBLICATIONS, INC.
NEW YORK

This Dover edition, first published in 1976, is an unabridged republication, with slight corrections, of the work originally published by Prentice-Hall, Inc., Englewood Cliffs, New Jersey in 1962.

International Standard Book Number

ISBN-13: 978-0-486-63317-6
ISBN-10: 0-486-63317-9

Library of Congress Catalog Card Number: 75-41883

Manufactured in the United States by Courier Corporation
63317923 2013
www.doverpublications.com

AUTHOR'S PREFACE

My book on Fourier series, which originally appeared in Russian, has already been translated into Chinese, German, Polish and Rumanian. I am very grateful to Dr. R. A. Silverman and to the Prentice-Hall Publishing Company for undertaking to prepare and publish an English version of the second Russian edition. It would be most gratifying to me if the book were to serve the needs of American readers.

I would like to thank V. Y. Kozlov, L. A. Tumarkin and A. I. Plesner for the helpful advice they gave me while I was writing this book.

G. P. T.

TRANSLATOR'S PREFACE

The present volume is the second in a new series of translations of outstanding Russian textbooks and monographs in the fields of mathematics, physics and engineering, under my editorship. It is hoped that Professor Tolstov's book will constitute a valuable addition to the English-language literature on Fourier series.

The following two changes, made with Professor Tolstov's consent, are worth mentioning:

1. To enhance the value of the English-language edition, a large number of extra problems have been added by myself and Professor Allen L. Shields of the University of Michigan. We have consulted a variety of sources, in particular, *A Collection of Problems in Mathematical Physics* by N. N. Lebedev, I. P. Skalskaya, and Y. S. Uflyand (Moscow, 1955), from which most of the problems appearing at the end of Chapter 9 have been taken.

2. To keep the number of cross references to a minimum, four chapters (8 and 9, 10 and 11) of the Russian original have been combined to make two chapters (8 and 9) of the present edition.

I have also added a Bibliography, containing suggestions for collateral and supplementary reading. Finally, it should be noted that sections marked with asterisks contain material of a more advanced nature, which can be omitted without loss of continuity.

R. A. S.

CONTENTS

FOURIER SERIES

1

TRIGONOMETRIC FOURIER SERIES

1. Periodic Functions

A function $f(x)$ is called *periodic* if there exists a constant $T > 0$ for which

$$f(x + T) = f(x), \tag{1.1}$$

for any x in the domain of definition of $f(x)$. (It is understood that both x and $x + T$ lie in this domain.) Such a constant T is called a *period* of the function $f(x)$. The most familiar periodic functions are $\sin x$, $\cos x$, $\tan x$, etc. Periodic functions arise in many applications of mathematics to problems of physics and engineering. It is clear that the sum, difference, product, or quotient of two functions of period T is again a function of period T.

If we plot a periodic function $y = f(x)$ on any closed interval $a \leqslant x \leqslant a + T$, we can obtain the entire graph of $f(x)$ by periodic repetition of the portion of the graph corresponding to $a \leqslant x \leqslant a + T$ (see Fig. 1).

If T is a period of the function $f(x)$, then the numbers $2T, 3T, 4T, \ldots$ are also periods. This follows immediately by inspecting the graph of a periodic function or from the series of equalities[1]

$$f(x) = f(x + T) = f(x + 2T) = f(x + 3T) = \cdots$$

[1] We suggest that the reader prove the validity not only of these equalities but also of the following equalities:

$$f(x) = f(x - T) = f(x - 2T) = f(x - 3T) = \cdots$$

which are obtained by repeated use of the condition (1.1). Thus, if T is a period, so is kT, where k is any positive integer, i.e., if a period exists, it is *not unique.*

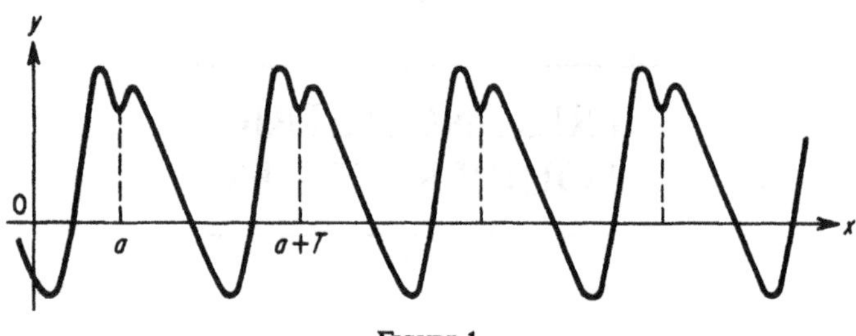

FIGURE 1

Next, we note the following property of any function $f(x)$ of period T:

If $f(x)$ is integrable on any interval of length T, then it is integrable on any other interval of the same length, and the value of the integral is the same, i.e.,

$$\int_a^{a+T} f(x)\,dx = \int_b^{b+T} f(x)\,dx, \tag{1.2}$$

for any a and b.

This property is an immediate consequence of the interpretation of an integral as an area. In fact, each integral (1.2) equals the area included between the curve $y = f(x)$, the x-axis and the ordinates drawn at the end points of the interval, where areas lying above the x-axis are regarded as positive and areas lying below the x-axis are regarded as negative. In the present case, the areas represented by the two integrals are the same, because of the periodicity of $f(x)$ (see Fig. 2).

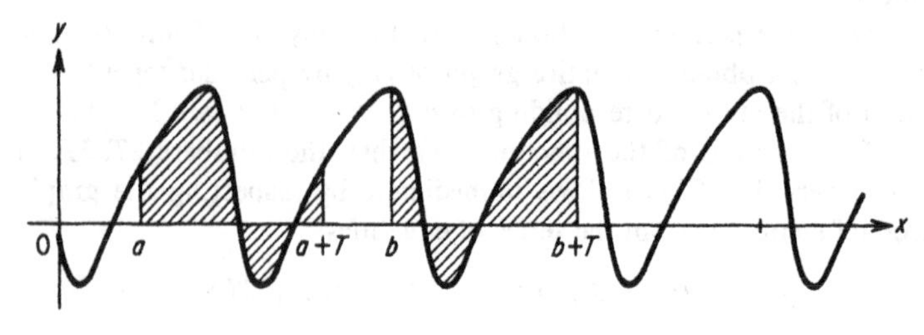

FIGURE 2

Hereafter, when we say that a function $f(x)$ of period T is *integrable*, we shall mean that it is integrable on an interval of length T. It follows from the property just proved that $f(x)$ is also integrable on any interval of finite length.

2. Harmonics

The simplest periodic function, and the one of greatest importance for the applications, is

$$y = A \sin (\omega x + \varphi),$$

where A, ω, and φ are constants. This function is called a *harmonic of amplitude* $|A|$, *(angular) frequency* ω, *and initial phase* φ. The period of such a harmonic is $T = 2\pi/\omega$, since for any x

$$A \sin \left[\omega\left(x + \frac{2\pi}{\omega}\right) + \varphi\right] = A \sin [(\omega x + \varphi) + 2\pi] = A \sin (\omega x + \varphi).$$

The terms "amplitude," "frequency," and "initial phase" stem from the following mechanical problem involving the simplest kind of oscillatory motion, i.e., *simple harmonic motion*: Suppose that a point mass M, of mass m, moves along a straight line under the action of a *restoring force* F which is proportional to the distance of M from a fixed origin O and which is directed towards O (see Fig. 3). Regarding s as positive if M lies to the right of O and

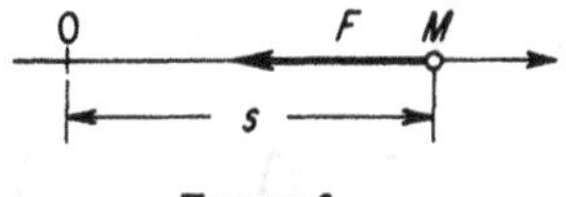

FIGURE 3

negative if M lies to the left of O, i.e., assigning the usual positive direction to the line, we find that $F = -ks$, where $k > 0$ is a constant of proportionality. Therefore

$$m\frac{d^2s}{dt^2} = -ks$$

or

$$\frac{d^2s}{dt^2} + \omega^2 s = 0,$$

where we have written $\omega^2 = k/m$, so that $\omega = \sqrt{k/m}$.

It is easily verified that the solution of this differential equation is the function $s = A \sin (\omega t + \varphi)$, where A and φ are constants, which can be calculated from a knowledge of the position and velocity of the point M at the initial time $t = 0$. This function s is a harmonic, and in fact is a periodic function of time with period $T = 2\pi/\omega$. Thus, under the action of the restoring force F, the point M undergoes oscillatory motion. The amplitude $|A|$ is the maximum deviation of the point M from O, and the quantity $1/T$ is the number of oscillations in an interval containing 2π units of time (e.g., seconds). This explains the term "frequency". The quantity φ is the initial phase and characterizes the initial position of the point, since for $t = 0$ we have $s_0 = \sin \varphi$.

We now examine the appearance of the curve $y = A \sin (\omega x + \varphi)$. We assume that $\omega > 0$, since otherwise $\sin (-\omega x + \varphi)$ is merely replaced by $-\sin (\omega x - \varphi)$. The simplest case is obtained when $A = 1, \omega = 1, \varphi = 0$; this gives the familiar *sine curve* $y = \sin x$ [see Fig. 4(a)]. For $A = 1$, $\omega = 1$, $\varphi = \pi/2$, we obtain the *cosine curve* $y = \cos x$, whose graph is the same as that of $y = \sin x$ shifted to the left by an amount $\pi/2$.

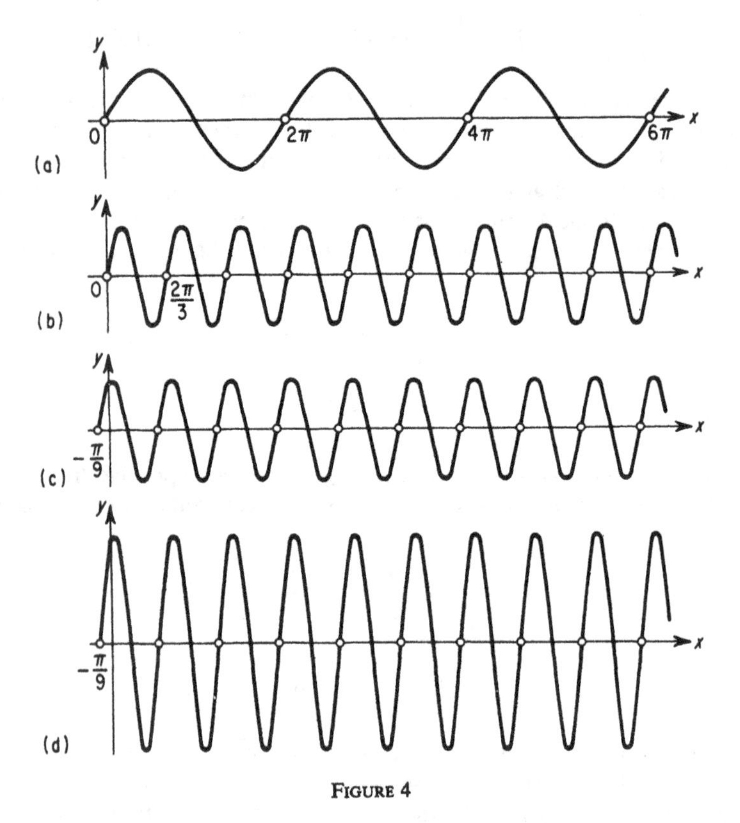

FIGURE 4

Next, consider the harmonic $y = \sin \omega x$, and set $\omega x = z$, thereby obtaining $y = \sin x$, an ordinary sine curve. Thus, the graph of $y = \sin \omega x$ is obtained by deforming the graph of a sine curve: This deformation reduces to a uniform compression along the x-axis by a factor ω if $\omega > 1$, and to a uniform expansion along the x-axis by a factor $1/\omega$ if $\omega < 1$. Figure 4(b) shows the harmonic $y = \sin 3x$, of period $T = 2\pi/3$.

Now, consider the harmonic $y = \sin(\omega x + \varphi)$, and set $\omega x + \varphi = \omega z$, so that $x = z - \varphi/\omega$. We already know the graph of $\sin \omega z$. Therefore, the graph of $y = \sin(\omega x + \varphi)$ is obtained by shifting the graph of $y = \sin \omega x$ along the x-axis by the amount $-\varphi/\omega$. Figure 4(c) represents the harmonic

$$y = \sin\left(3x + \frac{\pi}{3}\right)$$

with period $2\pi/3$ and initial phase $\pi/3$.

Finally, the graph of the harmonic $y = A\sin(\omega x + \varphi)$ is obtained from that of the harmonic $y = \sin(\omega x + \varphi)$ by multiplying all ordinates by the number A. Figure 4(d) shows the harmonic

$$y = 2\sin\left(3x + \frac{\pi}{3}\right).$$

These results may be summarized as follows:

The graph of the harmonic $y = A\sin(\omega x + \varphi)$ is obtained from the graph of the familiar sine curve by uniform compression (or expansion) along the coordinate axes plus a shift along the x-axis.

Using a well-known formula from trigonometry, we write

$$A\sin(\omega x + \varphi) = A(\cos\omega x \sin\varphi + \sin\omega x \cos\varphi).$$

Then, setting

$$a = A\sin\varphi, \qquad b = A\cos\varphi, \tag{2.1}$$

we convince ourselves that every harmonic can be represented in the form

$$a\cos\omega x + b\sin\omega x. \tag{2.2}$$

Conversely, every function of the form (2.2) is a harmonic. To prove this, it is sufficient to solve (2.1) for A and B. The result is

$$A = \sqrt{a^2 + b^2}, \quad \sin\varphi = \frac{a}{A} = \frac{a}{\sqrt{a^2 + b^2}}, \quad \cos\varphi = \frac{b}{A} = \frac{b}{\sqrt{a^2 + b^2}},$$

from which φ is easily found.

From now on, we shall write harmonics in the form (2.2). For example, for the harmonic shown in Fig. 4(d), this form is

$$2\sin\left(3x + \frac{\pi}{3}\right) = \sqrt{3}\cos 3x + \sin 3x$$

It will also be convenient to explicitly introduce the period T in (2.2). If we set $T = 2l$, then, since $T = 2\pi/\omega$, we have

$$\omega = \frac{2\pi}{T} = \frac{\pi}{l},$$

and therefore, the harmonic with period $T = 2l$ can be written as

$$a \cos \frac{\pi x}{l} + b \sin \frac{\pi x}{l}. \tag{2.3}$$

3. Trigonometric Polynomials and Series

Given the period $T = 2l$, consider the harmonics

$$a_k \cos \frac{\pi k x}{l} + b_k \sin \frac{\pi k x}{l} \qquad (k = 1, 2, \ldots) \tag{3.1}$$

with frequencies $\omega_k = \pi k/l$ and periods $T_k = 2\pi/\omega_k = 2l/k$. Since

$$T = 2l = kT_k,$$

the number $T = 2l$ is simultaneously a period of all the harmonics (3.1), for an integral multiple of a period is again a period (see Sec. 1). Therefore, every sum of the form

$$s_n(x) = A + \sum_{k=1} \left(a_k \cos \frac{\pi k x}{l} + b_k \sin \frac{\pi k x}{l}\right),$$

where A is a constant, is a function of period $2l$, since it is a sum of functions of period $2l$. (The addition of a constant obviously does not destroy periodicity; in fact, a constant can be regarded as a function for which *any* number is a period.) The function $s_n(x)$ is called a *trigonometric polynomial of order n* (and period $2l$).

Even though it is a sum of various harmonics, a trigonometric polynomial in general represents a function of a much more complicated nature than a simple harmonic. By suitably choosing the constants $A, a_1, b_1, a_2, b_2, \ldots$ we can form functions $y = s_n(x)$ with graphs quite unlike the smooth and symmetric graph of a simple harmonic. For example, Fig. 5 shows the trigonometric polynomial

$$y = \sin x + \tfrac{1}{2} \sin 2x + \tfrac{1}{4} \sin 3x.$$

The *infinite trigonometric series*

$$A + \sum_{k=1}^{\infty} \left(a_k \cos \frac{\pi k x}{l} + b_k \sin \frac{\pi k x}{l}\right)$$

(if it converges) also represents a function of period $2l$. The nature of functions which are sums of such infinite trigonometric series is even more diverse. Thus, the following question arises naturally: Can any given

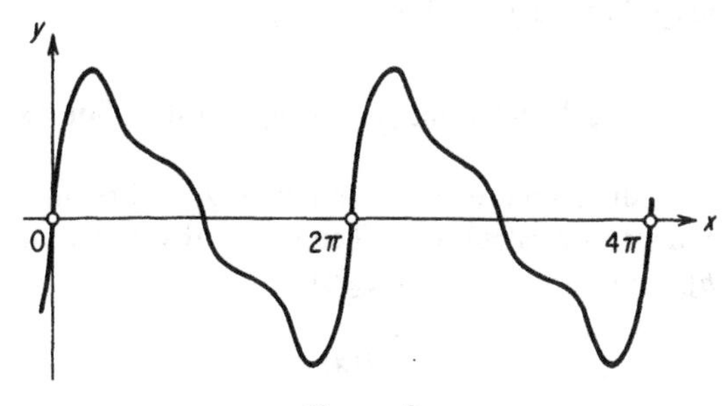

FIGURE 5

function of period $T = 2l$ be represented as the sum of a trigonometric series? We shall see later that such a representation is in fact possible for a very wide class of functions.

For the time being, suppose that $f(x)$ belongs to this class. This means that $f(x)$ can be expanded as a sum of harmonics, i.e., as a sum of functions with a very simple structure. The graph of the function $y = f(x)$ is obtained as a "superposition" of the graphs of these harmonics. Thus, to give a mechanical interpretation, we can represent a complicated oscillatory motion $f(x)$ as a sum of individual oscillations which are particularly simple. However, one must not imagine that trigonometric series are applicable only to oscillation phenomena. This is far from being the case. In fact, the concept of a trigonometric series is also very useful in studying many phenomena of a quite different nature.

If

$$f(x) = A + \sum_{k=1}^{\infty} \left(a_k \cos \frac{\pi k x}{l} + b_k \sin \frac{\pi k x}{l} \right), \tag{3.2}$$

then, setting $\pi x/l = t$ or $x = tl/\pi$, we find that

$$\varphi(t) = f\left(\frac{tl}{\pi}\right) = A + \sum_{k=1}^{\infty} (a_k \cos kt + b_k \sin kt), \tag{3.3}$$

where the harmonics in this series all have period 2π. This means that if a function $f(x)$ of period $2l$ has the expansion (3.2), then the function $\varphi(t) = f(tl/\pi)$ is of period 2π and has the expansion (3.3). Obviously, the converse is also true, i.e., if a function $\varphi(t)$ of period 2π has the expansion (3.3), then the function $f(x) = \varphi(\pi x/l)$ is of period $2l$ and has the expansion (3.2).

Thus, it is enough to know how to solve the problem of expansion in trigonometric series for functions of the "standard" period 2π. Moreover, in this case, the series has a simpler appearance. Therefore, we shall develop the theory for series of the form (3.3), and only the final results will be converted to the "language" of the general series (3.2).

4. A More Precise Terminology. Integrability. Series of Functions

We now introduce a more precise terminology and recall some facts from differential and integral calculus. When we say that $f(x)$ is integrable on the interval $[a, b]$, we mean that the integral

$$\int_a^b f(x)\,dx \tag{4.1}$$

(which may be improper) exists in the elementary sense. Thus, our integrable functions $f(x)$ will always be either continuous or have a finite number of points of discontinuity in the interval $[a, b]$, at which the function can be either bounded or unbounded.

In courses on integral calculus, it is proved that if a function has a finite number of discontinuities, then if the integral

$$\int_a^b |f(x)|\,dx$$

exists, so does the integral (4.1). (The converse is not always true.) In this case, the function $f(x)$ is said to be *absolutely integrable*. If $f(x)$ is absolutely integrable and $\varphi(x)$ is a bounded integrable function, then the product $f(x)\varphi(x)$ is absolutely integrable. The following rule for integration by parts holds:

Let $f(x)$ and $\varphi(x)$ be continuous on $[a, b]$, but perhaps non-differentiable at a finite number of points. Then, if $f'(x)$ and $\varphi'(x)$ are absolutely integrable,[2] *we have*

$$\int_a^b f(x)\varphi'(x)\,dx = \Big[f(x)\varphi(x)\Big]_{x=a}^{x=b} - \int_a^b f'(x)\varphi(x)\,dx. \tag{4.2}$$

Another familiar result is the fact that if the functions $f_1(x), f_2(x), \ldots, f_n(x)$ are integrable on $[a, b]$, then their sum is also integrable, and

$$\int_a^b \left[\sum_{k=1}^n f_k(x)\right] dx = \sum_{k=1}^n \int_a^b f_k(x)\,dx. \tag{4.3}$$

[2] Instead of absolute integrability of both derivatives, we can weaken this requirement to absolute integrability of just one of the derivatives. However, the stronger form of the requirement is sufficient for what follows.

We now consider an *infinite* series of functions

$$f_1(x) + f_2(x) + \cdots + f_k(x) + \cdots = \sum_{k=1}^{\infty} f_k(x). \tag{4.4}$$

Such a series is said to be *convergent* for a given value of x if its *partial sums*

$$s_n(x) = \sum_{k=1}^{n} f_k(x) \qquad (n = 1, 2, \ldots)$$

have a finite limit

$$s(x) = \lim_{n\to\infty} s_n(x).$$

The quantity $s(x)$ is said to be the *sum* of the series, and is obviously a function of x. If the series converges for all x in the interval $[a, b]$, then its sum $s(x)$ is defined on the whole interval $[a, b]$.

We now ask whether the formula (4.3) can be extended to the case of a convergent series of functions which are integrable on the interval $[a, b]$, i.e., is the formula

$$\int_a^b \left[\sum_{k=1}^{\infty} f_k(x)\right] dx = \int_a^b s(x)\, dx = \sum_{k=1}^{\infty} \int_a^b f_k(x)\, dx \tag{4.5}$$

valid? In other words, can the series be integrated *term by term*? It turns out that (4.5) is not always valid, if for no other reason than that a series of integrable or even continuous functions may not even have an integrable sum. A similar problem arises in connection with the possibility of term by term differentiation of series. We now single out an important class of series of functions to which these operations can be applied.

The series (4.4) is said to be *uniformly convergent* on the interval $[a, b]$ if for any positive number ε, there exists a number N such that the inequality

$$|s(x) - s_n(x)| \leqslant \varepsilon \tag{4.6}$$

holds for all $n \geqslant N$ and *for all x in the interval* $[a, b]$. Thus, if we examine the graph of the sum of the series $s(x)$ and of the partial sum $s_n(x)$, uniform convergence means that for all sufficiently large indices n and for all x, the curve representing $s(x)$ and the curve representing $s_n(x)$ are less than ε apart, where ε is any preassigned number, so that the two curves are *uniformly*[3] close (see Fig. 6).

Not every series which converges on an interval $[a, b]$ converges uniformly there. The following is a very useful and simple test for the uniform convergence of a series of functions (Weierstrass' M-test):

---

[3] I.e., for all x in $[a, b]$.

If the series of positive numbers

$$M_1 + M_2 + \cdots + M_k + \cdots$$

converges and if for any x in the interval $[a, b]$ we have $|f_k(x)| \leqslant M_k$ from a certain k on, then the series (4.3) converges uniformly (and absolutely) on $[a, b]$.

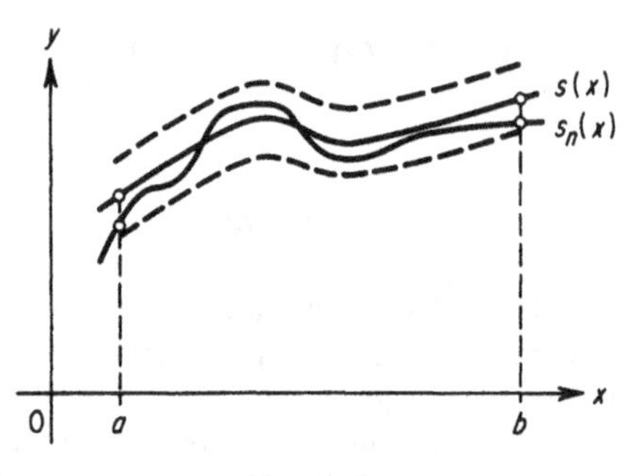

FIGURE 6

The following important theorems are valid:

THEOREM 1. *If the terms of the series (4.4) are continuous on $[a, b]$ and if the series is uniformly convergent on $[a, b]$, then*

a) *The sum of the series is continuous;*

b) *The sum can be integrated term by term, i.e., (4.5) holds.*

THEOREM 2. *If the series (4.4) converges, if its terms are differentiable and if the series*

$$f_1'(x) + f_2'(x) + \cdots + f_k'(x) + \cdots = \sum_{k=1}^{\infty} f_k'(x)$$

is uniformly convergent on $[a, b]$, then

$$\left(\sum_{k=1}^{\infty} f_k(x)\right)' = s'(x) = \sum_{k=1}^{\infty} f_k'(x),$$

i.e., the series (4.4) can be differentiated term by term.[4]

5. The Basic Trigonometric System. The Orthogonality of Sines and Cosines

By the *basic trigonometric system* we mean the system of functions

$$1, \cos x, \sin x, \cos 2x, \sin 2x, \ldots, \cos nx, \sin nx, \ldots \tag{5.1}$$

[4] In courses on analysis, it is usually assumed also that the derivatives are continuous, in order to simplify the proof.

All these functions have the common period 2π (although $\cos nx$ and $\sin nx$ also have the smaller period $2\pi/n$). We now prove some auxiliary formulas.

For any integer $n \neq 0$, we have

$$\int_{-\pi}^{\pi} \cos nx \, dx = \left[\frac{\sin nx}{n}\right]_{x=-\pi}^{x=\pi} = 0,$$
$$\int_{-\pi}^{\pi} \sin nx \, dx = \left[-\frac{\cos nx}{n}\right]_{x=-\pi}^{x=\pi} = 0, \tag{5.2}$$

and

$$\int_{-\pi}^{\pi} \cos^2 nx \, dx = \int_{-\pi}^{\pi} \frac{1 + \cos 2nx}{2} \, dx = \pi,$$
$$\int_{-\pi}^{\pi} \sin^2 nx \, dx = \int_{-\pi}^{\pi} \frac{1 - \cos 2nx}{2} \, dx = \pi. \tag{5.3}$$

Using the familiar trigonometric formulas

$$\cos \alpha \cos \beta = \tfrac{1}{2}[\cos(\alpha + \beta) + \cos(\alpha - \beta)],$$
$$\sin \alpha \sin \beta = \tfrac{1}{2}[\cos(\alpha - \beta) - \cos(\alpha + \beta)]$$

we find that

$$\int_{-\pi}^{\pi} \cos nx \cos mx \, dx = \tfrac{1}{2}\int_{-\pi}^{\pi} [\cos(n + m)x + \cos(n - m)x] \, dx = 0,$$
$$\int_{-\pi}^{\pi} \sin nx \sin mx \, dx = \tfrac{1}{2}\int_{-\pi}^{\pi} [\cos(n - m)x - \cos(n + m)x] \, dx = 0 \tag{5.4}$$

for any integers n and m ($n \neq m$). Finally, using the formula

$$\sin \alpha \cos \beta = \tfrac{1}{2}[\sin(\alpha + \beta) + \sin(\alpha - \beta)],$$

we find that

$$\int_{-\pi}^{\pi} \sin nx \cos mx \, dx = \tfrac{1}{2}\int_{-\pi}^{\pi} [\sin(n + m)x + \sin(n - m)x] \, dx = 0 \tag{5.5}$$

for any n and m. The formulas (5.2), (5.4), and (5.5) show that the integral over the interval $[-\pi, \pi]$ of the product of any two *different* functions of the system (5.1) vanishes.

We shall agree to call two functions $\varphi(x)$ and $\psi(x)$ *orthogonal*[5] on the interval $[a, b]$ if

$$\int_a^b \varphi(x)\psi(x)\,dx = 0.$$

With this definition, we can say that the functions of the system (5.1) are pairwise orthogonal on the interval $[-\pi, \pi]$, or more briefly, that *the system* (5.1) *is orthogonal on* $[-\pi, \pi]$.

As we know, the integral of a periodic function is the same over any interval whose length equals the period (see Sec. 1). Therefore, the formulas (5.2) through (5.5) are valid not only for the interval $[-\pi, \pi]$ but also for any interval $[a, a + 2\pi]$, i.e., the system (5.1) is orthogonal on every such interval.

6. Fourier Series for Functions of Period 2π

Suppose the function $f(x)$ of period 2π has the expansion

$$f(x) = \frac{a_0}{2} + \sum_{k=1}^{\infty} (a_k \cos kx + b_k \sin kx), \tag{6.1}$$

where, to simplify the subsequent formulas, we denote the constant term by $a_0/2$. We now pose the problem of determining the coefficients a_0, a_k and b_k $(k = 1, 2, \ldots)$ from a knowledge of $f(x)$. To do this, we make the following *assumption*: It is assumed that the series (6.1), and the series to be written presently, can be integrated term by term, i.e., it is assumed that for all these series the integral of the sum equals the sum of the integrals. [It is thereby also assumed that the function $f(x)$ is integrable.] Then, integrating (6.1) from $-\pi$ to π, we obtain

$$\int_{-\pi}^{\pi} f(x)\,dx = \frac{a_0}{2}\int_{-\pi}^{\pi} dx + \sum_{k=1}^{\infty}\left(a_k \int_{-\pi}^{\pi} \cos kx\,dx + b_k \int_{-\pi}^{\pi} \sin kx\,dx\right).$$

By (5.2), all the integrals in the sum vanish, so that

$$\int_{-\pi}^{\pi} f(x)\,dx = \pi a_0. \tag{6.2}$$

Next, we multiply both sides of (6.1) by $\cos nx$ and integrate the result from $-\pi$ to π, as before, obtaining

[5] In geometry, the word *orthogonality* connotes perpendicularity. One must not think that the concept of orthogonality of two functions corresponds to anything like perpendicularity of their graphs, despite the fact that this concept is related to a suitably generalized notion of perpendicularity. In this regard, see Ch. 2, Sec. 10.

$$\int_{-\pi}^{\pi} f(x)\cos nx\,dx = \frac{a_0}{2}\int_{-\pi}^{\pi}\cos nx\,dx$$
$$+\sum_{k=1}^{\infty}\left(a_k\int_{-\pi}^{\pi}\cos kx\cos nx\,dx\right.$$
$$\left.+\,b_k\int_{-\pi}^{\pi}\sin kx\cos nx\,dx\right).$$

By (5.2), the first integral on the right vanishes. Since the functions of the system (5.1) are pairwise orthogonal, all the integrals in the sum also vanish, except one. The only integral that remains is the coefficient of a_n:

$$\int_{-\pi}^{\pi}\cos^2 nx\,dx = \pi$$

[see (5.3)]. Thus we have

$$\int_{-\pi}^{\pi} f(x)\cos nx\,dx = a_n\pi. \tag{6.3}$$

Similarly, we find that

$$\int_{-\pi}^{\pi} f(x)\sin nx\,dx = b_n\pi. \tag{6.4}$$

It follows from (6.2) to (6.4) that

$$a_n = \frac{1}{\pi}\int_{-\pi}^{\pi} f(x)\cos nx\,dx \qquad (n = 0, 1, 2, \ldots),$$
$$b_n = \frac{1}{\pi}\int_{-\pi}^{\pi} f(x)\sin nx\,dx \qquad (n = 1, 2, \ldots). \tag{6.5}$$

Thus, finally, if $f(x)$ is integrable and can be expanded in a trigonometric series, and if this series and the series obtained from it by multiplying by $\cos nx$ and $\sin nx$ $(n = 1, 2, \ldots)$ can be integrated term by term, then the coefficients a_n and b_n are given by the formulas (6.5).

Now, suppose we are given an integrable function $f(x)$ of period 2π, and we wish to represent $f(x)$ as the sum of a trigonometric series. If such a representation is possible at all (and if the requirement of term by term integrability is satisfied), then by what has been said, the coefficients a_n and b_n must be given by (6.5). Therefore, in looking for a trigonometric series whose sum is a given function $f(x)$, it is natural to examine first the series whose coefficients are given by (6.5), and to see whether this series has the required properties. As we shall see later, this will be the case for a large class of functions.

The coefficients a_n and b_n calculated by the formulas (6.5) are called the *Fourier coefficients* of the function $f(x)$, and the trigonometric series with these coefficients is called the *Fourier series* of $f(x)$. Incidentally, we note that the formulas (6.5) involve integrating a function of period 2π. Therefore, the interval of integration $[-\pi, \pi]$ can be replaced by any other interval of length 2π (see Sec. 1), so that together with the formulas (6.5), we have

$$a_n = \frac{1}{\pi}\int_a^{a+2\pi} f(x)\cos nx\,dx \qquad (n = 0, 1, 2, \ldots),$$
$$b_n = \frac{1}{\pi}\int_a^{a+2\pi} f(x)\sin nx\,dx \qquad (n = 1, 2, \ldots). \tag{6.6}$$

The above considerations make it natural to devote special attention to Fourier series. It we form the Fourier series of a function $f(x)$ without deciding in advance whether it converges to $f(x)$, we write

$$f(x) \sim \frac{a_0}{2} + \sum_{n=1}^{\infty}(a_n \cos nx + b_n \sin nx).$$

This notation means only that the Fourier series written on the right *corresponds* to the function $f(x)$. The sign $\sim$ can be replaced by the sign $=$ only if we succeed in proving that the series converges and that its sum equals $f(x)$. A simple consequence of these considerations is the following theorem, which is quite useful:

THEOREM 1. *If a function $f(x)$ of period 2π can be expanded in a trigonometric series which converges uniformly on the whole real axis,*[6] *then this series is the Fourier series of $f(x)$.*

Proof. Suppose that $f(x)$ satisfies (6.1), where the series is uniformly convergent. By Theorem 1 of Sec. 4, $f(x)$ is continuous and term by term integration of the series is possible. This gives the formula (6.2). Next, we consider the equality

$$f(x)\cos nx = \frac{a_0}{2}\cos nx + \sum_{k=1}^{\infty}(a_k \cos kx \cos nx + b_k \sin kx \cos nx), \tag{6.7}$$

and show that the series on the right is uniformly convergent. Set

$$s_m(x) = \frac{a_0}{2} + \sum_{k=1}^{m}(a_k \cos kx + b_k \sin kx),$$

and let ε be an arbitrary positive number. If the series (6.1) converges uniformly, then there exists a number N such that

$$|f(x) - s_m(x)| \leqslant \varepsilon$$

[6] By the periodicity of $f(x)$ we can require uniform convergence on $[-\pi, \pi]$, rather than on the whole real axis.

for all $m \geqslant N$. The product $s_m(x) \cos nx$ is obviously the mth partial sum of the series (6.7). Then, the inequality

$$|f(x) \cos nx - s_m(x) \cos nx| = |f(x) - s_m(x)| \, |\cos nx| \leqslant \varepsilon,$$

which holds for all $m \geqslant N$, implies the uniform convergence of the series (6.7). It follows that this series can be integrated term by term, and the result of the integration is the formula (6.3). Similarly, we prove the formula (6.4). Thus, finally, the formulas (6.5) hold for the coefficients a_n and b_n, which means that (6.1) is the Fourier series of $f(x)$.

The modern theory of Fourier series allows us to prove the following more general result, whose proof we cannot give because of its complexity:

THEOREM 2. *If an absolutely integrable function $f(x)$ of period 2π can be expanded in a trigonometric series which converges to $f(x)$ everywhere, except possibly at a finite number of points (within one period), then this series is the Fourier series of $f(x)$.*

This theorem confirms the assertion made above, that in looking for a trigonometric series which has a given function $f(x)$ as its sum, we should first consider the Fourier series of $f(x)$.

7. Fourier Series for Functions Defined on an Interval of Length 2π

A problem which arises quite often in the applications is that of expanding a function $f(x)$ in trigonometric series, when $f(x)$ is defined only on the interval $[-\pi, \pi]$. In this case, nothing at all is said about the periodicity of $f(x)$. Nevertheless, this does not prevent us from writing the Fourier series of $f(x)$, since the formulas (6.5) involve only the interval $[-\pi, \pi]$. Moreover, $f(x)$ can be extended by periodicity from $[-\pi, \pi]$ onto the whole x-axis. This leads to a periodic function which coincides with $f(x)$ on $[-\pi, \pi]$ and which has a Fourier series identical with that of $f(x)$. In fact, if the Fourier series of $f(x)$ turns out to converge to $f(x)$, then, since it is a periodic function, the sum of this Fourier series automatically gives us the required periodic extension of $f(x)$ from $[-\pi, \pi]$ onto the whole x-axis.

Thus, it does not matter whether we talk about the Fourier series of a function defined on $[-\pi, \pi]$, or whether we talk about the Fourier series of the function obtained from $f(x)$ by *periodic extension* along the x-axis. This implies that it is sufficient to formulate the tests for convergence of Fourier series for the case of periodic functions.

In connection with the problem of extending $f(x)$ by periodicity from the interval $[-\pi, \pi]$ onto the whole x-axis, the following remarks are in order: If $f(-\pi) = f(\pi)$, there is no difficulty in making the extension, since in this case, if $f(x)$ is continuous on $[-\pi, \pi]$, its extension will be continuous on the whole x-axis [see Fig. 7(a)]. However, if $f(-\pi) \neq f(\pi)$, we cannot accomplish the required extension without changing the values of $f(-\pi)$ and $f(\pi)$, since the periodicity requires that $f(-\pi)$ and $f(\pi)$ coincide. This difficulty can be avoided in two ways: (1) We can completely avoid considering the values of $f(x)$ at $x = -\pi$ and $x = \pi$, thereby making the function undefined at these points and hence making the periodic extension of $f(x)$ undefined at the points $x = (2k + 1)\pi$, $k = 0, \pm 1, \pm 2, \ldots$; (2) We can suitably modify the values of the function $f(x)$ at $x = -\pi$ and $x = \pi$ by making these values equal. It is important to note that in both cases, the Fourier coefficients will have the same values as before, since changing the values of a function at a finite number of points, or even failing to define it at a finite number of points, cannot affect the value of an integral, in particular, the values of the integrals (6.5) defining the Fourier coefficients. Thus, whether or not we carry out the indicated modification of the function $f(x)$, its Fourier series remains unchanged.

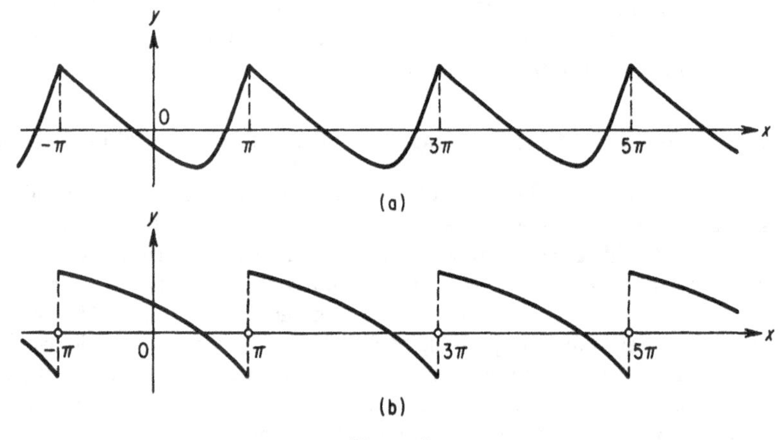

FIGURE 7

It should be observed that if $f(-\pi) \neq f(\pi)$ and if $f(x)$ is continuous on the interval $[-\pi, \pi]$, then the periodic extension of $f(x)$ onto the whole x-axis will have discontinuities at all the points $x = (2k + 1)\pi$, $k = 0, \pm 1, \pm 2, \ldots$, no matter how we change the values of the function at $x = -\pi$ and $x = \pi$ [see Fig. 7(b)]. The problem of finding the values to which the Fourier series of $f(x)$ may be expected to converge at $x = \pm\pi$, when $f(-\pi) \neq f(\pi)$, is a special one, and will be solved later.

Finally, suppose that $f(x)$ is defined on an arbitrary interval $[a, a + 2\pi]$ of length 2π, and that it is required to expand $f(x)$ in a trigonometric series.

As before, we arrive at the conclusion that it does not matter whether we talk about the Fourier series of $f(x)$ or about the Fourier series of the function obtained from $f(x)$ by extending it periodically onto the whole x-axis. If $f(x)$ is continuous on the interval $[a, a + 2\pi]$ but $f(a) \neq f(a + 2\pi)$, we obtain an extension which is discontinuous at the points $x = a + 2k\pi$ $(k = 0, \pm 1, \pm 2, \ldots)$.

8. Right-Hand and Left-Hand Limits. Jump Discontinuities

We introduce the notation

$$\lim_{\substack{x \to x_0 \\ x < x_0}} f(x) = f(x_0 - 0), \qquad \lim_{\substack{x \to x_0 \\ x > x_0}} f(x) = f(x_0 + 0),$$

provided these limits exist and are finite.[7] The first of these limits is called the *left-hand limit* of $f(x)$ at the point x_0, and the second is called the *right-hand limit* of $f(x)$ at x_0. These limits both exist at points of continuity (by the very definition of continuity), and we have

$$f(x_0 - 0) = f(x_0) = f(x_0 + 0) \tag{8.1}$$

at continuity points.

If x_0 is a point of discontinuity of the function $f(x)$, then the right-hand and left-hand limits (either or both of them) may exist in some cases and fail to exist in others. If both limits exist, we say that the point x_0 is a point of discontinuity of the first kind, or simply, a point of *jump discontinuity*. If at least one of these limits does not exist, then the point x_0 is called a *point of discontinuity of the second kind*. We shall be particularly interested in jump discontinuities. If x_0 is such a point, then the quantity

$$\delta = f(x_0 + 0) - f(x_0 - 0) \tag{8.2}$$

is called the *jump* of the function $f(x)$ at x_0.

The following example illustrates this situation. Suppose that

$$f(x) = \begin{cases} -x^3 & \text{for } x < 1, \\ 0 & \text{for } x = 1, \\ \sqrt{x} & \text{for } x > 1, \end{cases} \tag{8.3}$$

with the graph shown in Fig. 8. The value of the function at $x = 1$ is indicated by the little circle. At $x = 1$, the left-hand and right-hand limits are obviously

$$f(1 - 0) = -1, \qquad f(1 + 0) = 1.$$

---

[7] If $x_0 = 0$, we do not write $f(0 + 0)$ and $f(0 - 0)$, but simply $f(+0)$ and $f(-0)$.

Therefore, the jump of the function at $x = 1$ is

$$\delta = f(1 + 0) - f(1 - 0) = 2,$$

which is in complete agreement with the intuitive idea of a jump (see Fig. 8).

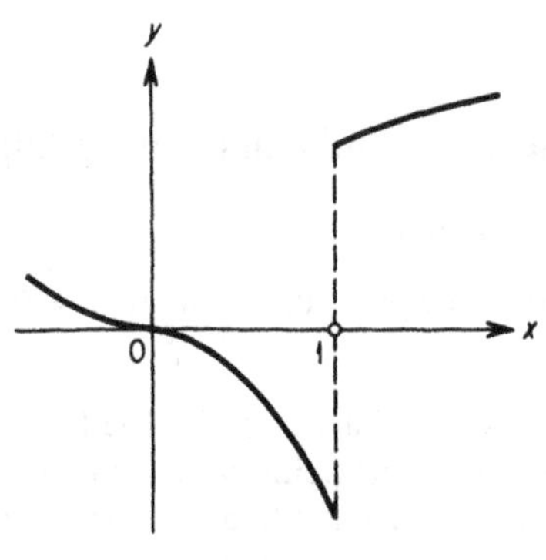

FIGURE 8

If $f(x)$ is a function which is continuous on the interval $[-\pi, \pi]$, then if $f(-\pi) \neq f(\pi)$, jump discontinuities appear in making the periodic extension of $f(x)$ from $[-\pi, \pi]$ onto the whole x-axis [see Fig. 7(b)], and all the jump discontinuities are equal to the number

$$\delta = f(-\pi) - f(\pi).$$

9. Smooth and Piecewise Smooth Functions

The function $f(x)$ is said to be *smooth* on the interval $[a, b]$ if it has a *continuous* derivative on $[a, b]$. In geometrical language, this means that the direction of the tangent changes *continuously*, without jumps, as it moves along the curve $y = f(x)$ [see Fig. 9(a)]. Thus, the graph of a smooth function is a smooth curve without any "corners."[8]

The function $f(x)$ is said to be *piecewise smooth* on the interval $[a, b]$ if either $f(x)$ and its derivative are both continuous on $[a, b]$, or they have only a finite number of jump discontinuities on $[a, b]$. It is easy to see that the graph of a piecewise smooth function is either a continuous curve or a discontinuous curve which can have a finite number of *corners* (at which the derivative has jumps). As we approach any discontinuity or corner (from one side or the other), the direction of the tangent approaches a definite limiting position, since the derivative can have only jump discontinuities.

---

[8] "Corner" = Russian "угловая точка," a point at which the curve has two distinct tangents. (*Translator*)

Figures 9(b) and 9(c) illustrate the graphs of continuous and discontinuous piecewise smooth functions. From now on, we shall regard smooth functions as a special case of piecewise smooth functions.

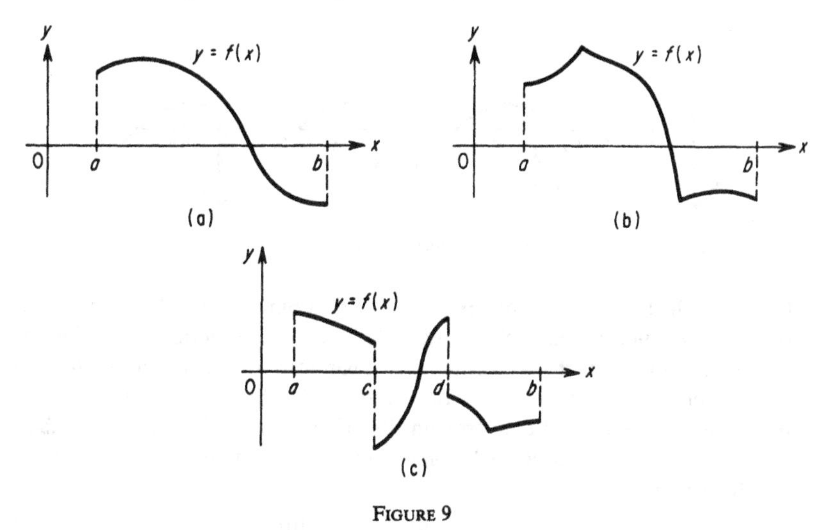

FIGURE 9

A continuous or discontinuous function $f(x)$ which is defined on the whole x-axis, is said to be piecewise smooth if it is piecewise smooth on every interval of finite length. In particular, this concept applies to periodic functions. Every piecewise smooth function $f(x)$ [whether continuous or discontinuous] is bounded and has a bounded derivative everywhere, except at its corners and points of discontinuity [at all these points, $f'(x)$ does not exist].

10. A Criterion for the Convergence of Fourier Series

We now give a more useful criterion for the convergence of a Fourier series, deferring the proof of this criterion until Ch. 3:

The Fourier series of a piecewise smooth (continuous or discontinuous) function $f(x)$ of period 2π converges for all values of x. The sum of the series equals $f(x)$ at every point of continuity and equals the number

$$\tfrac{1}{2}\,[f(x+0)+f(x-0)],$$

the arithmetic mean of the right-hand and left-hand limits, at every point of discontinuity (see Fig. 10). *If $f(x)$ is continuous everywhere, then the series converges absolutely and uniformly.*

Suppose the function $f(x)$ is defined only on $[-\pi, \pi]$, and is piecewise smooth on the interval $[-\pi, \pi]$ and continuous at its end points. As noted in Sec. 7, the Fourier series of $f(x)$ coincides with the Fourier series of the

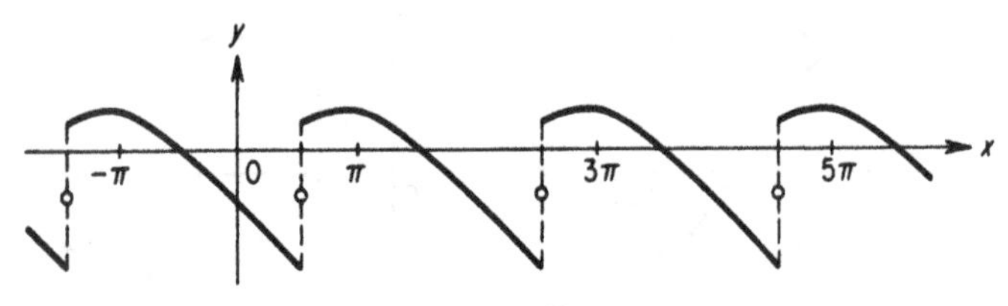

FIGURE 10

function which is the periodic extension of $f(x)$ onto the whole x-axis. But in this case, such an extension obviously leads to a function $f(x)$ which is piecewise smooth on the whole x-axis. Therefore, the criterion just formulated implies that the Fourier series of $f(x)$ will converge everywhere. In particular, the series will converge on the original interval $[-\pi, \pi]$. In fact, for $-\pi < x < \pi$, the series will converge to $f(x)$ at the points of continuity and to the value

$$\tfrac{1}{2}\,[f(x + 0) + f(x - 0)]$$

at the points of discontinuity. But what will happen at the end points of the interval $[-\pi, \pi]$?

At the end points, two cases are possible:

1) $f(-\pi) = f(\pi)$. In this case, the periodic extension obviously leads to a function which is continuous at the points $\pm\pi$ [and also at all the points $x = (2k + 1)\pi$, $k = 0, \pm 1, \pm 2, \ldots$]. Therefore, by our criterion, the Fourier series will also converge to $f(x)$ at the end points of $[-\pi, \pi]$.

2) $f(-\pi) \neq f(\pi)$. In this case, the periodic extension leads to a function which is discontinuous at the points $\pm\pi$ [and also at the points $x = (2k + 1)\pi$, $k = 0, \pm 1, \pm 2, \ldots$], where for the extension of $f(x)$ we obviously have

$$f(-\pi - 0) = f(\pi), \qquad f(-\pi + 0) = f(-\pi),$$
$$f(\pi + 0) = f(-\pi), \qquad f(\pi - 0) = f(\pi)$$

(see Fig. 11). Therefore, at $x = -\pi$, and $x = \pi$, the Fourier series will converge to the values

$$\left.\begin{matrix} \dfrac{f(-\pi + 0) + f(-\pi - 0)}{2}, \\ \dfrac{f(\pi + 0) + f(\pi - 0)}{2} \end{matrix}\right\} = \frac{f(-\pi) + f(\pi)}{2}.$$

Thus, the Fourier series of a function $f(x)$ defined on the interval $[-\pi, \pi]$ and continuous at $x = \pm\pi$ behaves at the points $x = \pm\pi$ just as it does at the other points of continuity, provided that $f(-\pi) = f(\pi)$. However, if

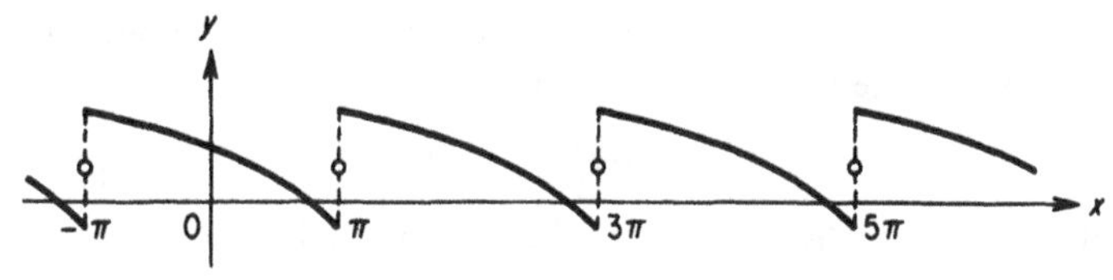

FIGURE 11

$f(-\pi) \neq f(\pi)$, the series obviously cannot converge to $f(x)$ at $x = \pi$, and in this case, it is meaningful to pose the problem of expanding $f(x)$ in Fourier series only for $-\pi < x < \pi$ and not for $-\pi \leqslant x \leqslant \pi$. A similar remark can be made concerning the Fourier series of a function specified in an interval of the type $[a, a + 2\pi]$, where a is any number.

In solving any concrete problem, if the reader draws a graph of the periodic extension of the function (this is always recommended!) and bears in mind the criterion just formulated, then the nature of the behavior of the Fourier series at the end points of the interval will be immediately apparent.

11. Even and Odd Functions

Let the function $f(x)$, defined either on the whole x-axis or on some interval, be symmetric with respect to the origin of coordinates. We say that $f(x)$ is an *even* function if

$$f(-x) = f(x)$$

for every x. This definition implies that the graph of any even function $y = f(x)$ is symmetric with respect to the y-axis [see Fig. 12(a)]. It follows

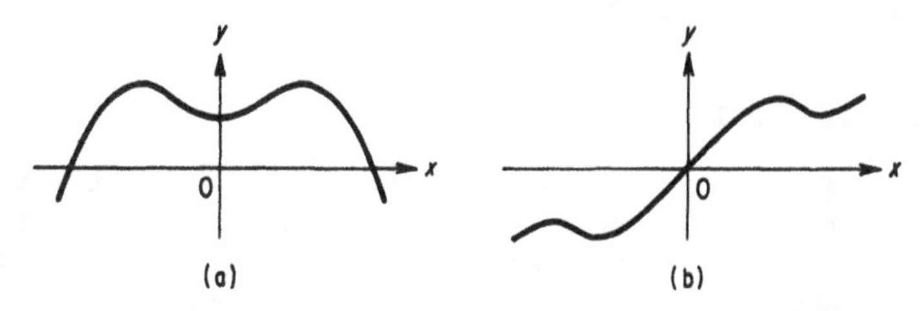

FIGURE 12

from the interpretation of the integral as an area that for even functions we have

$$\int_{-l}^{l} f(x)\,dx = 2\int_{0}^{l} f(x)\,dx \tag{11.1}$$

for any l, provided that $f(x)$ is defined and integrable on the interval $[-l, l]$.

We say that the function $f(x)$ is *odd* if

$$f(-x) = -f(x)$$

for every x. In particular, for an odd function we have

$$f(-0) = -f(0),$$

so that $f(0) = 0$. The graph of any odd function $y = f(x)$ is symmetric with respect to the point O [see Fig. 12(b)]. For odd functions

$$\int_{-l}^{l} f(x)\,dx = 0 \tag{11.2}$$

for any l, provided that $f(x)$ is defined and integrable on the interval $[-l, l]$.

The following properties are simple consequences of the definition of even and odd functions:

(a) The product of two even or odd functions is an even function;

(b) The production of an even and an odd function is an odd function.

In fact, if $\varphi(x)$ and $\psi(x)$ are even functions, then for $f(x) = \varphi(x)\psi(x)$, we have

$$f(-x) = \varphi(-x)\psi(-x) = \varphi(x)\psi(x) = f(x),$$

while if $\varphi(x)$ and $\psi(x)$ are odd, we have

$$f(-x) = \varphi(-x)\psi(-x) = [-\varphi(x)][-\psi(x)] = \varphi(x)\psi(x) = f(x).$$

This proves Property (a). On the other hand, if $\varphi(x)$ is even and $\psi(x)$ is odd, then

$$f(-x) = \varphi(-x)\psi(-x) = \varphi(x)[-\psi(x)] = -\varphi(x)\psi(x) = -f(x),$$

which proves Property (b).

12. Cosine and Sine Series

Let $f(x)$ be an *even* function defined on the interval $[-\pi, \pi]$, or else an even periodic function. Since $\cos nx$ ($n = 0, 1, 2, \ldots$) is obviously an even function, then by Property (a) of Sec. 11 the function $f(x)\cos nx$ is also even. On the other hand, the function $\sin nx$ ($n = 1, 2, \ldots$) is odd, so that the function $f(x)\sin nx$ is also odd, by Property (b) of Sec. 11. Then, using

(6.5), (11.1) and (11.2), we find that the Fourier coefficients of the even function $f(x)$ are

$$a_n = \frac{1}{\pi}\int_{-\pi}^{\pi} f(x)\cos nx\,dx = \frac{2}{\pi}\int_0^{\pi} f(x)\cos nx\,dx \qquad (n = 0, 1, 2, \ldots),$$
$$b_n = \frac{1}{\pi}\int_{-\pi}^{\pi} f(x)\sin nx\,dx = 0 \qquad (n = 1, 2, \ldots). \tag{12.1}$$

Therefore, the Fourier series of an even function contains only *cosines*, i.e.,

$$f(x) \sim \frac{a_0}{2} + \sum_{n=1}^{\infty} a_n \cos nx,$$

where the coefficients a_n are given by the formula (12.1).

Now, let $f(x)$ be an *odd* function, defined on the interval $[-\pi, \pi]$, or else an odd periodic function. Since $\cos nx$ $(n = 0, 1, 2, \ldots)$ is an even function, the function $f(x)\cos nx$ is odd, by Property (b) of Sec. 11, and since $\sin nx$ $(n = 1, 2, \ldots)$ is odd, the function $f(x)\sin nx$ is even, by Property (a) of Sec. 11. Then, using (6.5), (11.1), and (11.2), we find that the Fourier coefficients of the odd function $f(x)$ are

$$a_n = \frac{1}{\pi}\int_{-\pi}^{\pi} f(x)\cos nx\,dx = 0 \qquad (n = 0, 1, 2, \ldots),$$
$$b_n = \frac{1}{\pi}\int_{-\pi}^{\pi} f(x)\sin nx\,dx = \frac{2}{\pi}\int_0^{\pi} f(x)\sin nx\,dx \qquad (n = 1, 2, \ldots). \tag{12.2}$$

Therefore, the Fourier series of an odd function contains only *sines*, i.e.,

$$f(x) \sim \sum_{n=1}^{\infty} b_n \sin nx,$$

where the coefficients b_n are given by the formula (12.2). Since the Fourier series of an odd function contains only sines, it obviously vanishes for $x = -\pi$, $x = 0$, and $x = \pi$ (and in general for $x = k\pi$), regardless of the values of $f(x)$ at these points.

A problem which often arises is that of making an expansion in cosine series or sine series of an absolutely integrable function $f(x)$ defined on the interval $[0, \pi]$. To expand $f(x)$ in *cosine* series, we can reason as follows: Make the *even* extension of $f(x)$ from the interval $[0, \pi]$ onto the interval $[-\pi, 0]$ [see Fig. 13(a)]. Then all the previous considerations apply to the even extension of $f(x)$, so that its Fourier coefficients can be calculated by the formulas

$$a_n = \frac{2}{\pi}\int_0^{\pi} f(x)\cos nx\,dx \qquad (n = 0, 1, 2, \ldots),$$
$$b_n = 0 \qquad (n = 1, 2, \ldots), \tag{12.3}$$

which involve only the values of $f(x)$ in the interval $[0, \pi]$. Therefore, for computational purposes, there is no need to actually make the even extension of $f(x)$ from $[0, \pi]$ onto $[-\pi, 0]$.

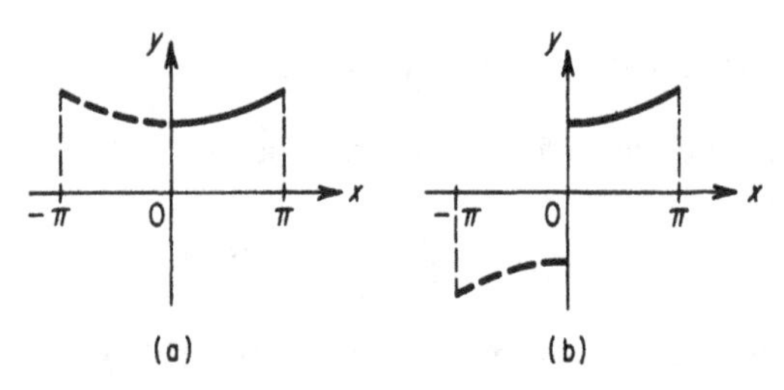

FIGURE 13

To expand $f(x)$ in *sine* series, we first make the *odd* extension of $f(x)$ from the interval $[0, \pi]$ onto the interval $[-\pi, 0]$ [see Fig. 13(b)]. In doing so, the oddness requires that we set $f(0) = 0$. Then, the previous considerations again apply to the odd extension of $f(x)$, so that its Fourier coefficients are given by the formulas

$$\begin{aligned} a_n &= 0 \qquad (n = 0, 1, 2, \ldots), \\ b_n &= \frac{2}{\pi} \int_0^\pi f(x) \sin nx\, dx \qquad (n = 1, 2, \ldots), \end{aligned} \tag{12.4}$$

which involve only the values of $f(x)$ in the interval $[0, \pi]$. Therefore, as in the case of cosine series, there is no need to actually make the odd extension of $f(x)$ from $[0, \pi]$ onto $[-\pi, 0]$.

However, in order to avoid mistakes in using the convergence criterion of Sec. 10, it is still recommended that a sketch be made of the function $f(x)$ and its even (or odd) extension onto the interval $[-\pi, 0]$, as well as of its periodic extension (with period 2π) onto the whole x-axis. This sketch will help in investigating the behavior of the "extended" function, which is the function to which the convergence criterion has to be applied.

13. Examples of Expansions in Fourier Series

Example 1. *Expand* $f(x) = x^2$ $(-\pi \leqslant x \leqslant \pi)$ *in Fourier series.* The function $f(x)$ is even; the graph of $f(x)$ together with its periodic extension is shown in Fig. 14. The extended function is continuous and piecewise smooth. Therefore, by the criterion of Sec. 10, its Fourier series converges to $f(x) = x^2$ everywhere in $[-\pi, \pi]$, and converges to the periodic extension of $f(x)$ outside $[-\pi, \pi]$. Moreover, the convergence is absolute and uniform.

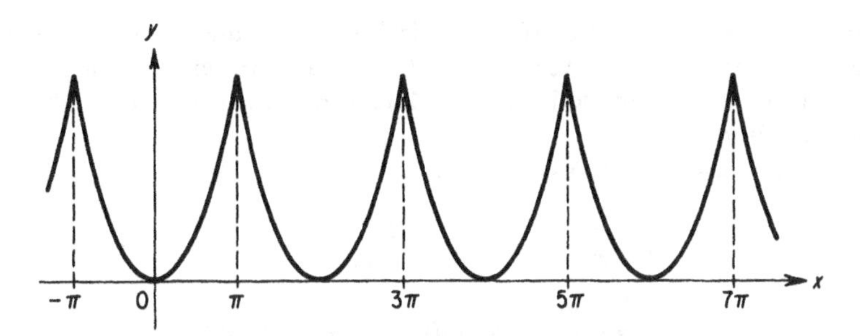

FIGURE 14

A calculation shows that

$$a_0 = \frac{2}{\pi}\int_0^{\pi} x^2\,dx = \frac{2}{\pi}\left[\frac{x^3}{3}\right]_{x=0}^{x=\pi} = \frac{2\pi^2}{3}.$$

Furthermore, integrating by parts, we find that

$$a_n = \frac{2}{\pi}\int_0^{\pi} x^2 \cos nx\,dx = -\frac{4}{\pi n}\int_0^{\pi} x \sin nx\,dx$$

$$= \frac{4}{\pi n^2}\left[x\cos nx\right]_{x=0}^{x=\pi} - \frac{4}{\pi n^2}\int_0^{\pi}\cos nx\,dx$$

$$= \frac{4}{n^2}\cos n\pi = (-1)^n\frac{4}{n^2},$$

while $b_n = 0$ $(n = 1, 2, \ldots)$, since $f(x)$ is even. Therefore, for $-\pi \leqslant x \leqslant \pi$, we have

$$x^2 = \frac{\pi^2}{3} - 4\left(\cos x - \frac{\cos 2x}{2^2} + \frac{\cos 3x}{3^2} - \cdots\right). \tag{13.1}$$

Example 2. *Expand $f(x) = |x|$ $(-\pi \leqslant x \leqslant \pi)$ in Fourier series.* The function $f(x)$ is even; Fig. 15 shows the graph of $f(x)$ together with its periodic extension. The extended function is continuous and piecewise smooth, so

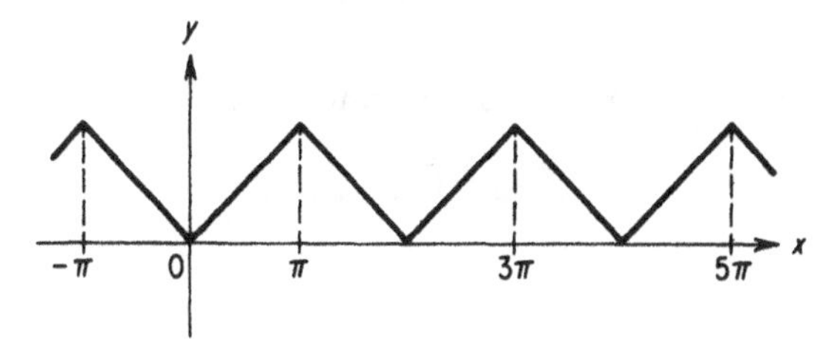

FIGURE 15

that the criterion of Sec. 10 is applicable. Therefore, its Fourier series converges to $f(x) = |x|$ everywhere in $[-\pi, \pi]$ and converges to the periodic extension of $f(x)$ outside $[-\pi, \pi]$. Moreover, the convergence is absolute and uniform.

Since $|x| = x$ for $x \geqslant 0$, we have

$$a_0 = \frac{2}{\pi}\int_0^\pi x\,dx = \frac{2}{\pi}\left[\frac{x^2}{2}\right]_{x=0}^{x=\pi} = \pi,$$

$$\begin{aligned} a_n &= \frac{2}{\pi}\int_0^\pi x\cos nx\,dx = -\frac{2}{\pi n}\int_0^\pi \sin nx\,dx \\ &= \frac{2}{\pi n^2}[\cos nx]_{x=0}^{x=\pi} = \frac{2}{\pi n^2}[\cos n\pi - 1] \\ &= \frac{2}{\pi n^2}[(-1)^n - 1]. \end{aligned}$$

It follows that $a_n = 0$ for even n, and that $a_n = -4/\pi n^2$ for odd n. Finally, $b_n = 0$ $(n = 1, 2, \ldots)$, since $f(x)$ is even. Thus, for $-\pi \leqslant x \leqslant \pi$, we have

$$|x| = \frac{\pi}{2} - \frac{4}{\pi}\left(\cos x + \frac{\cos 3x}{3^2} + \frac{\cos 5x}{5^2} + \cdots\right). \tag{13.2}$$

Example 3. *Expand $f(x) = |\sin x|$ in Fourier series.* This function is defined for all x, and represents a continuous, piecewise smooth, even function. Its graph is shown in Fig. 16. The criterion of Sec. 10 is applicable, and hence $f(x) = |\sin x|$ is everywhere equal to its Fourier series, which is absolutely and uniformly convergent.

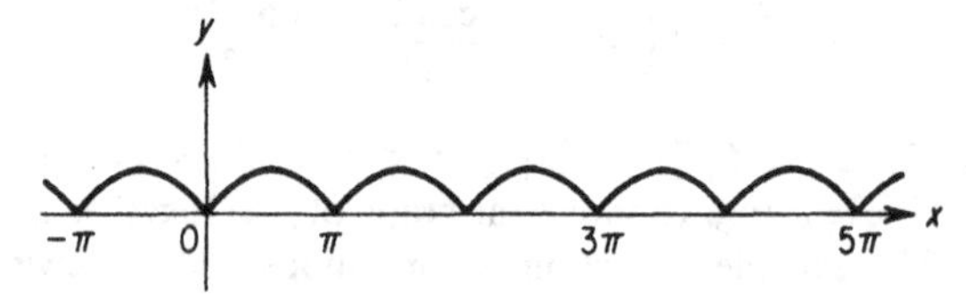

FIGURE 16

Since $|\sin x| = \sin x$ for $0 \leqslant x \leqslant \pi$, we have

$$a_0 = \frac{2}{\pi}\int_0^\pi \sin x\,dx = \frac{4}{\pi},$$

and

$$a_n = \frac{2}{\pi}\int_0^\pi \sin x\cos nx\,dx$$

$$= \frac{1}{\pi}\int_0^{\pi} [\sin (n + 1)x - \sin (n - 1)x]\, dx$$

$$= -\frac{1}{\pi}\left[\frac{\cos (n + 1)x}{n + 1} - \frac{\cos (n - 1)x}{n - 1}\right]_{x=0}^{x=\pi}$$

$$= -\frac{1}{\pi}\left[\frac{(-1)^{n+1}-1}{n + 1} - \frac{(-1)^{n-1}-1}{n - 1}\right] = -2\frac{(-1)^n+1}{\pi(n^2 - 1)},$$

for $n \neq 1$, while for $n = 1$

$$a_1 = \frac{2}{\pi}\int_0^{\pi} \sin x \cos x\, dx = \frac{1}{\pi}\int_0^{\pi} \sin 2x\, dx = 0.$$

Moreover, $b_n = 0$ $(n = 1, 2, \ldots)$, since $f(x)$ is even. Therefore, for all x we have

$$|\sin x| = \frac{2}{\pi} - \frac{4}{\pi}\left(\frac{\cos 2x}{3} + \frac{\cos 4x}{15} + \frac{\cos 6x}{35} + \cdots\right).$$

Example 4. *Expand* $f(x) = x$ $(-\pi < x < \pi)$ *in Fourier series.* The function $f(x)$ is odd; Fig. 17 shows the graph of $f(x)$ together with its periodic extension. The extended function is piecewise smooth and discontinuous at the points $x = (2k + 1)\pi$ $(k = 0, \pm 1, \pm 2, \ldots)$. The test of Sec. 10 is applicable, and the Fourier series of $f(x)$ converges to zero at the points of discontinuity.

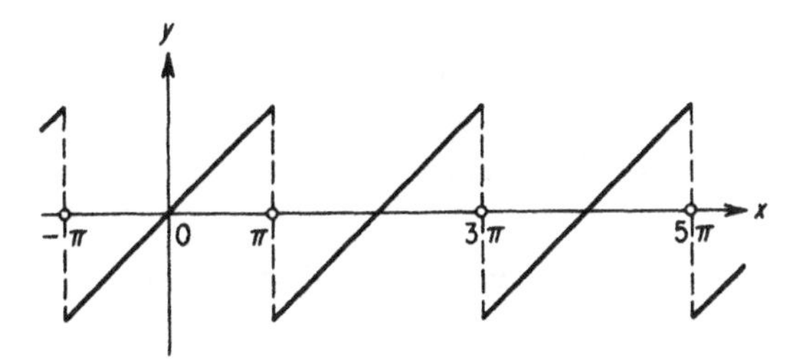

FIGURE 17

Since $f(x)$ is odd

$$a_n = 0 \qquad (n = 0, 1, 2, \ldots),$$

$$b_n = \frac{2}{\pi}\int_0^{\pi} x \sin nx\, dx$$

$$= -\frac{2}{\pi n}[x \cos nx]_{x=0}^{x=\pi} + \frac{2}{\pi n}\int_0^{\pi} \cos nx\, dx$$

$$= -\frac{2}{n}\cos n\pi = \frac{2}{n}(-1)^{n+1}.$$

Therefore, for $-\pi < x < \pi$, we have

$$x = 2\left(\sin x - \frac{\sin 2x}{2} + \frac{\sin 3x}{3} - \cdots\right). \tag{13.3}$$

Example 5. *Expand $f(x) = 1$ $(0 < x < \pi)$ in sine series.* Making the odd extension of $f(x)$ onto the interval $[-\pi, 0]$ produces a discontinuity at $x = 0$. Figure 18 shows the graph of $f(x)$ and its odd extension, together with its subsequent periodic extension (with period 2π) over the whole x-axis. The convergence criterion of Sec. 10 is applicable to this "extended" function. Therefore, its Fourier series converges to $f(x) = 1$ for $0 < x < \pi$. Outside the interval $0 < x < \pi$, it converges to the function shown in Fig. 18, with the sum of the series being equal to zero at the points $x = k\pi$ $(k = 0, \pm 1, \pm 2, \ldots)$.

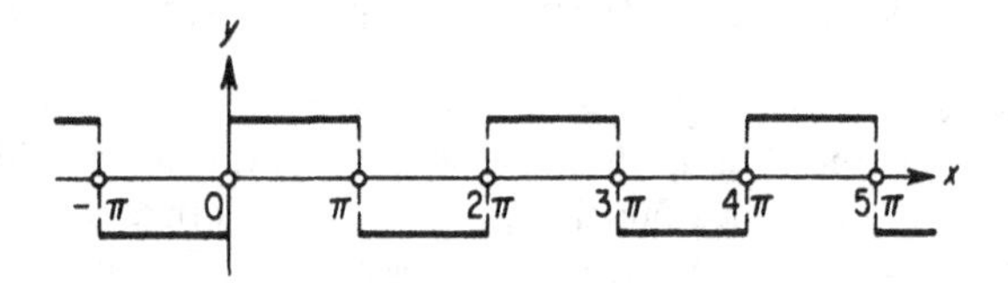

FIGURE 18

Since

$$a_n = 0 \qquad (n = 0, 1, 2, \ldots),$$

$$b_n = \frac{2}{\pi}\int_0^{\pi} \sin nx\, dx$$

$$= \frac{2}{\pi n}\left[-\cos nx\right]_{x=0}^{x=\pi} = \frac{2}{\pi n}\left[1 - (-1)^n\right],$$

we have

$$1 = \frac{4}{\pi}\left(\sin x + \frac{\sin 3x}{3} + \frac{\sin 5x}{5} + \cdots\right). \tag{13.4}$$

for $0 < x < \pi$.

Example 6. *Expand $f(x) = x$ $(0 < x < 2\pi)$ in Fourier series.* This example bears a superficial resemblance to Example 4, but the difference is immediately apparent if we construct the periodic extension of $f(x)$ (see Fig. 19). The criterion of Sec. 10 is applicable to this extended function. At the points of discontinuity, the Fourier series converges to the arithmetic mean of the right-hand and left-hand limits, i.e., to the value π. The function $f(x)$ is neither even nor odd.

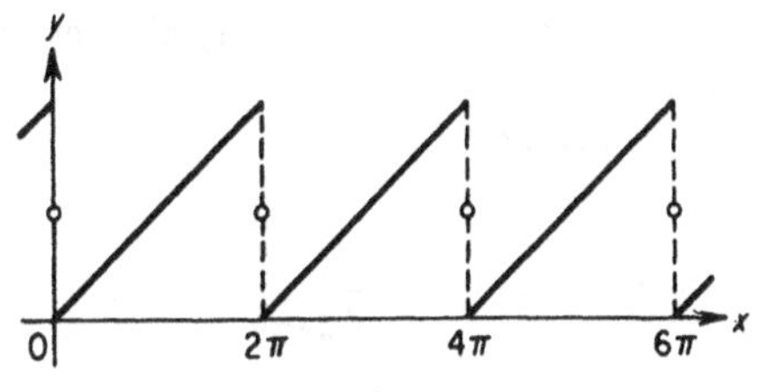

FIGURE 19

Since

$$a_0 = \frac{1}{\pi}\int_0^{2\pi} x\,dx = \frac{1}{\pi}\left[\frac{x^2}{2}\right]_{x=0}^{x=2\pi} = 2\pi,$$

$$\begin{aligned} a_n &= \frac{1}{\pi}\int_0^{2\pi} x \cos nx\,dx \\ &= \frac{1}{\pi n}[x \sin nx]_{x=0}^{x=2\pi} - \frac{1}{\pi n}\int_0^{2\pi} \sin nx\,dx = 0 \qquad (n = 1, 2, \ldots), \\ b_n &= \frac{1}{\pi}\int_0^{2\pi} x \sin nx\,dx \\ &= -\frac{1}{\pi n}[x \cos nx]_{x=0}^{x=2\pi} + \frac{1}{\pi n}\int_0^{2\pi} \cos nx\,dx = -\frac{2}{n}, \end{aligned}$$

we have

$$x = \pi - 2\left(\sin x + \frac{\sin 2x}{2} + \frac{\sin 3x}{3} + \cdots\right), \tag{13.5}$$

for $0 < x < 2\pi$.

Example 7. *Expand* $f(x) = x^2$ $(0 < x < 2\pi)$ *in Fourier series.* This example resembles Example 1, but the graph of the periodic extension of $f(x)$ immediately shows the difference (see Fig. 20). The criterion of Sec. 10 is

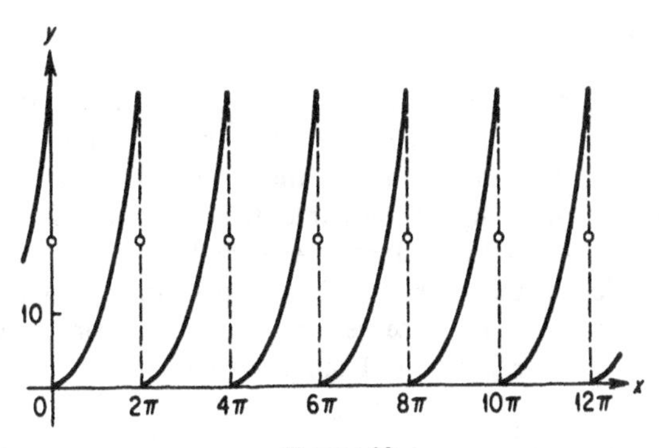

FIGURE 20

applicable, and at the points of discontinuity the series converges to the arithmetic mean of the right-hand and left-hand limits, i.e., to the value $2\pi^2$. The function $f(x)$ is neither even nor odd.

Since

$$a_0 = \frac{1}{\pi}\int_0^{2\pi} x^2\,dx = \frac{1}{\pi}\left[\frac{x^3}{3}\right]_{x=0}^{x=2\pi} = \frac{8\pi^2}{3},$$

$$a_n = \frac{1}{\pi}\int_0^{2\pi} x^2 \cos nx\,dx = -\frac{2}{\pi n}\int_0^{2\pi} x \sin nx\,dx$$

$$= \frac{2}{\pi n^2}\left[x\cos nx\right]_{x=0}^{x=2\pi} - \frac{2}{\pi n^2}\int_0^{2\pi}\cos nx\,dx = \frac{4}{n^2},$$

$$b_n = \frac{1}{\pi}\int_0^{2\pi} x^2 \sin nx\,dx$$

$$= -\frac{1}{\pi n}\left[x^2\cos nx\right]_{x=0}^{x=2\pi} + \frac{2}{\pi n}\int_0^{2\pi} x\cos nx\,dx$$

$$= -\frac{4\pi}{n} - \frac{2}{\pi n^2}\int_0^{2\pi}\sin nx\,dx = -\frac{4\pi}{n},$$

we have

$$x^2 = \frac{4\pi^2}{3} + 4\left(\cos x - \pi\sin x + \frac{\cos 2x}{2^2} - \frac{\pi\sin 2x}{2} + \cdots + \frac{\cos nx}{n^2} - \frac{\pi\sin nx}{n} + \cdots\right) \tag{13.6}$$

$$= \frac{4\pi^2}{3} + 4\sum_{n=1}^{\infty}\left(\frac{\cos nx}{n^2} - \frac{\pi\sin nx}{n}\right)$$

$$= \frac{4\pi^2}{3} + 4\sum_{n=1}^{\infty}\frac{\cos nx}{n^2} - 4\pi\sum_{n=1}^{\infty}\frac{\sin nx}{n},$$

for $0 < x < 2\pi$.

Example 8. *Expand* $f(x) = Ax^2 + Bx + C$ $(-\pi < x < \pi)$, *where* A, B, *and* C *are constants, in Fourier series.* The graph of $f(x)$ is a parabola. By periodic extension, we can obtain a continuous or a discontinuous function, depending on the choice of the constants A, B, and C. Figure 21 shows a possible extension for certain values of A, B, and C.

We could calculate the Fourier coefficients from the appropriate formulas, but there is no need to do so, since we can use the expansions for the functions x^2 and x $(-\pi < x < \pi)$, given in Examples 1 and 4. The result is

$$Ax^2 + Bx + C = \frac{A\pi^2}{3} + C + 4A\sum_{n=1}^{\infty}(-1)^n\frac{\cos nx}{n^2} - 2B\sum_{n=1}^{\infty}(-1)^n\frac{\sin nx}{n}.$$

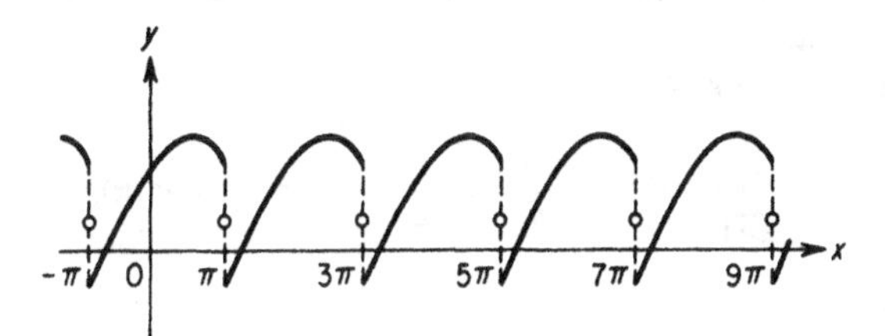

FIGURE 21

Example 9. *Expand* $f(x) = Ax^2 + Bx + C$ $(0 < x < 2\pi)$ *in Fourier series.* Figure 22 shows the periodic extension of $f(x)$ for a certain choice of the constants A, B, and C. Using the expansions of the functions x^2 and x $(0 < x < 2\pi)$, given in Examples 6 and 7, we find that

$$Ax^2 + Bx + C = \frac{4A\pi^2}{3} + B\pi + C + 4A\sum_{n=1}^{\infty}\frac{\cos nx}{n^2} - (4\pi A - 2B)\sum_{n=1}^{\infty}\frac{\sin nx}{n},$$

for $0 < x < 2\pi$.

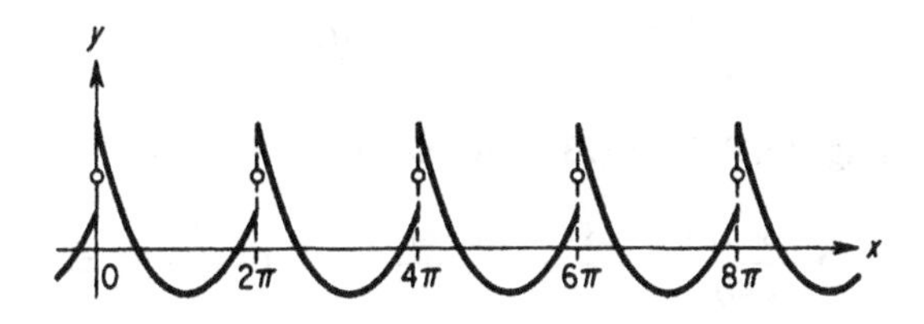

FIGURE 22

We can use these examples to calculate the sums of some important trigonometric series. For example, (13.5) immediately gives

$$\sum_{n=1}^{\infty}\frac{\sin nx}{n} = \frac{\pi - x}{2} \qquad (0 < x < 2\pi), \tag{13.7}$$

and from (13.5) and (13.6), we infer that

$$\sum_{n=1}^{\infty}\frac{\cos nx}{n^2} = \frac{3x^2 - 6\pi x + 2\pi^2}{12} \qquad (0 < x < 2\pi). \tag{13.8}$$

Since the terms of the series on the left do not exceed $1/n^2$ in absolute value, the series is uniformly convergent, which means that its sum is continuous for all x (see Sec. 4). Therefore, (13.8) is valid for $0 \leqslant x \leqslant 2\pi$, and not just for $0 < x < 2\pi$.

Similarly, (13.3) gives

$$\sum_{n=1}^{\infty}(-1)^{n+1}\frac{\sin nx}{n} = \frac{x}{2} \qquad (-\pi < x < \pi), \tag{13.9}$$

(13.1) gives

$$\sum_{n=1}^{\infty}(-1)^{n+1}\frac{\cos nx}{n^2} = \frac{\pi^2 - 3x^2}{12} \qquad (-\pi \leqslant x \leqslant \pi), \tag{13.10}$$

(13.4) gives

$$\sum_{n=0}^{\infty}\frac{\sin (2n+1)x}{2n+1} = \frac{\pi}{4} \qquad (0 < x < \pi), \tag{13.11}$$

and (13.2) gives

$$\sum_{n=0}^{\infty}\frac{\cos (2n+1)x}{(2n+1)^2} = \frac{\pi^2 - 2\pi x}{8} \qquad (0 \leqslant x \leqslant \pi). \tag{13.12}$$

Moreover, subtracting (13.11) from (13.7), we obtain

$$\sum_{n=1}^{\infty}\frac{\sin 2nx}{2n} = \frac{\pi - 2x}{4} \qquad (0 < x < \pi), \tag{13.13}$$

and subtracting (13.2) from (13.8), we obtain

$$\sum_{n=1}^{\infty}\frac{\cos 2nx}{(2n)^2} = \frac{6x^2 - 6\pi x + \pi^2}{24} \qquad (0 \leqslant x \leqslant \pi). \tag{13.14}$$

These formulas also allow us to calculate the sums of some *numerical* series. For example, if we set $x = 0$, (13.8) and (13.10) become

$$\frac{\pi^2}{6} = 1 + \frac{1}{2^2} + \frac{1}{3^2} + \frac{1}{4^2} + \cdots, \qquad \frac{\pi^2}{12} = 1 - \frac{1}{2^2} + \frac{1}{3^2} - \frac{1}{4^2} + \cdots$$

while if we set $x = \pi/2$, (13.11) becomes

$$\frac{\pi}{4} = 1 - \tfrac{1}{3} + \tfrac{1}{5} - \tfrac{1}{7} + \cdots$$

14. The Complex Form of a Fourier Series

Let the function $f(x)$ be integrable on the interval $[-\pi, \pi]$, and form its Fourier series

$$f(x) \sim \frac{a_0}{2} + \sum_{n=1}^{\infty} (a_n \cos nx + b_n \sin nx), \tag{14.1}$$

$$\begin{aligned} a_n &= \frac{1}{\pi}\int_{-\pi}^{\pi} f(x) \cos nx\, dx \qquad (n = 0, 1, 2, \ldots), \\ b_n &= \frac{1}{\pi}\int_{-\pi}^{\pi} f(x) \sin nx\, dx \qquad (n = 1, 2, \ldots). \end{aligned} \tag{14.2}$$

We shall make use of Euler's well-known formula, relating the trigonometric and exponential functions:

$$e^{i\varphi} = \cos\varphi + i \sin\varphi.$$

It is an immediate consequence of this formula that

$$\cos\varphi = \frac{e^{i\varphi} + e^{-i\varphi}}{2}, \quad \sin\varphi = \frac{e^{i\varphi} - e^{-i\varphi}}{2i}.$$

Therefore, we can write

$$\cos nx = \frac{e^{inx} + e^{-inx}}{2},$$

$$\sin nx = \frac{e^{inx} - e^{-inx}}{2i} = i\,\frac{-e^{inx} + e^{-inx}}{2}.$$

Substituting these expressions in (14.1), we obtain

$$f(x) \sim \frac{a_0}{2} + \sum_{n=1}^{\infty} \left(\frac{a_n - ib_n}{2} e^{inx} + \frac{a_n + ib_n}{2} e^{-inx}\right). \tag{14.3}$$

If we set

$$c_0 = \frac{a_0}{2}, \quad c_n = \frac{a_n - ib_n}{2}, \quad c_{-n} = \frac{a_n + ib_n}{2} \qquad (n = 1, 2, \ldots), \tag{14.4}$$

then the mth partial sum of the series (14.3), and hence of the series (14.1), can be written in the form

$$s_m(x) = c_0 + \sum_{n=1}^{m} (c_n e^{inx} + c_{-n} e^{-inx}) = \sum_{n=-m}^{m} c_n e^{inx}. \tag{14.5}$$

Therefore it is natural to write

$$f(x) \sim \sum_{n=-\infty}^{\infty} c_n e^{inx}. \tag{14.6}$$

This is the *complex form of the Fourier series of* $f(x)$. The convergence of the series (14.6) must be understood to mean the existence of the limit as $m \to \infty$ of the *symmetric* sums (14.5).

The coefficients c_n given by (14.4) are called the *complex Fourier coefficients* of the function $f(x)$. They satisfy the relations

$$c_n = \frac{1}{2\pi}\int_{-\pi}^{\pi} f(x)e^{-inx}\,dx \qquad (n = 0, \pm 1, \pm 2, \ldots). \tag{14.7}$$

In fact, by Euler's formula and (14.4), we have

$$\frac{1}{2\pi}\int_{-\pi}^{\pi} f(x)e^{-inx}\,dx = \frac{1}{2\pi}\left[\int_{-\pi}^{\pi} f(x)\cos nx\,dx - i\int_{-\pi}^{\pi} f(x)\sin nx\,dx\right]$$

$$= \tfrac{1}{2}(a_n - ib_n) = c_n$$

for positive indices and

$$\frac{1}{2\pi}\int_{-\pi}^{\pi} f(x)e^{inx}\,dx = \frac{1}{2\pi}\left[\int_{-\pi}^{\pi} f(x)\cos nx\,dx + i\int_{-\pi}^{\pi} f(x)\sin nx\,dx\right]$$

$$= \tfrac{1}{2}(a_n + ib_n) = c_{-n}$$

for negative indices. It is useful to bear in mind that if $f(x)$ is *real*, then the coefficients c_n and c_{-n} are *complex conjugates*. This is an immediate consequence of (14.4).

Incidentally, we note that the formula (14.7) can also be obtained directly, just as the formulas (14.2) were (see Sec. 6), if we assume that the sign $=$ appears in (14.6) instead of the sign $\sim$ and that term by term integration is legitimate. In fact, multiplying both sides of the equality

$$f(x) = \sum_{k=-\infty}^{\infty} c_k e^{ikx}$$

by e^{-inx} and integrating term by term over the interval $[-\pi, \pi]$, we obtain

$$\int_{-\pi}^{\pi} f(x)e^{-inx}\,dx = 2\pi c_n, \tag{14.8}$$

since for $k \neq n$ (see Sec. 5) we have

$$c_k \int_{-\pi}^{\pi} e^{i(k-n)x}\,dx$$

$$= c_k \int_{-\pi}^{\pi} [\cos (k - n)x + i \sin (k - n)x]\,dx = 0,$$

i.e., all the integrals on the right vanish except the one corresponding to the index $k = n$, while for $k = n$, we obtain the number $2\pi c_n$. The formula (14.7) is an immediate consequence of (14.8).

15. Functions of Period $2l$

If it is required to expand a function $f(x)$ of period $2l$ in Fourier series, we set $x = lt/\pi$, thereby obtaining the function $\varphi(t) = f(lt/\pi)$ of period 2π (see Sec. 3). For $\varphi(t)$ we can form the Fourier series

$$\varphi(t) \sim \frac{a_0}{2} + \sum_{n=1}^{\infty} (a_n \cos nt + b_n \sin nt), \tag{15.1}$$

where

$$a_n = \frac{1}{\pi}\int_{-\pi}^{\pi} \varphi(t) \cos nt\, dt = \frac{1}{\pi}\int_{-\pi}^{\pi} f\left(\frac{lt}{\pi}\right) \cos nt\, dt \qquad (n = 0, 1, 2, \ldots),$$

$$b_n = \frac{1}{\pi}\int_{-\pi}^{\pi} \varphi(t) \sin nt\, dt = \frac{1}{\pi}\int_{-\pi}^{\pi} f\left(\frac{lt}{\pi}\right) \sin nt\, dt \qquad (n = 1, 2, \ldots).$$

Returning to the original variable x by setting $t = \pi x/l$, we obtain

$$f(x) \sim \frac{a_0}{2} + \sum_{n=1}^{\infty}\left(a_n \cos\frac{\pi n x}{l} + b_n \sin\frac{\pi n x}{l}\right), \tag{15.2}$$

where

$$\begin{aligned} a_n &= \frac{1}{l}\int_{-l}^{l} f(x) \cos\frac{\pi n x}{l}\, dx \qquad (n = 0, 1, 2, \ldots), \\ b_n &= \frac{1}{l}\int_{-l}^{l} f(x) \sin\frac{\pi n x}{l}\, dx \qquad (n = 1, 2, \ldots). \end{aligned} \tag{15.3}$$

The coefficients (15.3) are still called the *Fourier coefficients* of $f(x)$, and the series (15.2) is still called the *Fourier series* of $f(x)$. If the equality holds in (15.1), then the equality holds in (15.2), and conversely.

We could have constructed a theory of series of the form (15.2) directly, by starting from a trigonometric system of the form

$$1, \cos\frac{\pi x}{l}, \sin\frac{\pi x}{l}, \ldots, \cos\frac{\pi n x}{l}, \sin\frac{\pi n x}{l}, \ldots, \tag{15.4}$$

just as we did in the case of the basic trigonometric system (5.1). The system (15.4) consists of functions with the common period $2l$, and it is easily verified that these functions are orthogonal on every interval of length $2l$. The considerations of Secs. 6, 7, 10, 12, and 14 can be repeated as applied to the system (15.4), and the result is a formulation analogous to that given in these sections, except that π is replaced by l. In particular, instead of a function $f(x)$ of period $2l$, we can consider a function defined only on the interval $[-l, l]$ [or on any other interval of length $2l$, provided we appropriately change the limits of integration in (15.3)]. The Fourier series of such a function is identical with that of its periodic extension onto the whole

x-axis. The convergence criterion of Sec. 10 continues to "work," if we replace the period 2π by the period $2l$.

If $f(x)$ is even, the formulas (15.3) become

$$a_n = \frac{2}{l}\int_0^l f(x)\cos\frac{\pi nx}{l}\,dx \qquad (n = 0, 1, 2, \ldots),$$
$$b_n = 0 \qquad (n = 1, 2, \ldots), \tag{15.5}$$

while if $f(x)$ is odd, they become

$$a_n = 0 \qquad (n = 0, 1, 2, \ldots)$$
$$b_n = \frac{2}{l}\int_0^l f(x)\sin\frac{\pi nx}{l}\,dx \qquad (n = 1, 2, \ldots). \tag{15.6}$$

As in Sec. 12, we can use this fact to expand a function $f(x)$ defined only on the interval $[0, l]$ in cosine series or in sine series (making the even or the odd extension of $f(x)$ onto the interval $[-l, 0]$).

The complex form of the series (15.2) is

$$f(x) \sim \sum_{n=-\infty}^{+\infty} c_n e^{i\pi nx/l},$$

where

$$c_n = \frac{1}{2l}\int_{-l}^{l} f(x)e^{-i\pi nx/l}\,dx \qquad (n = 0, \pm 1, \pm 2, \ldots),$$

or

$$c_0 = \frac{a_0}{2}, \quad c_n = \frac{a_n - ib_n}{2}, \quad c_{-n} = \frac{a_n + ib_n}{2} \qquad (n = 1, 2, \ldots).$$

Example 1. *Expand the function $f(x)$, defined by*

$$f(x) = \begin{cases} \cos\frac{\pi x}{l} & \text{for } 0 \leqslant x \leqslant \frac{l}{2}, \\ 0 & \text{for } \frac{l}{2} < x \leqslant l \end{cases}$$

in cosine series. Figure 23 shows the graph of $f(x)$ and its even extension onto the interval $[-l, 0]$, together with its subsequent periodic extension

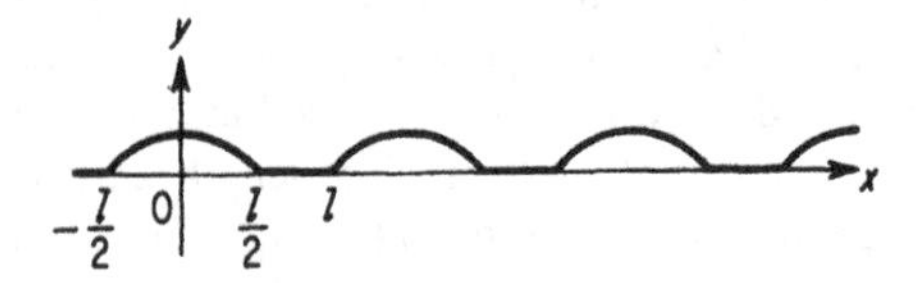

FIGURE 23

(with period $2l$) onto the whole x-axis. The convergence criterion can obviously be applied everywhere.

For $l/2 < x \leqslant l$, we have $f(x) = 0$, so that

$$a_0 = \frac{2}{l}\int_0^l f(x)\,dx = \frac{2}{l}\int_0^{l/2} \cos\frac{\pi x}{l}\,dx = \frac{2}{\pi},$$

$$a_n = \frac{2}{l}\int_0^l f(x)\cos\frac{\pi n x}{l}\,dx = \frac{2}{l}\int_0^{l/2}\cos\frac{\pi x}{l}\cos\frac{\pi n x}{l}\,dx.$$

Making the substitution $\pi x/l = t$, we obtain

$$a_n = \frac{2}{\pi}\int_0^{\pi/2}\cos t\cos nt\,dt = \frac{1}{\pi}\int_0^{\pi/2}[\cos(n+1)t + \cos(n-1)t]\,dt,$$

whence

$$a_1 = \frac{1}{\pi}\int_0^{\pi/2}(\cos 2t + 1)\,dt = \frac{1}{\pi}\left[\frac{\sin 2t}{2} + t\right]_{t=0}^{t=\pi/2} = \frac{1}{2},$$

$$a_n = \frac{1}{\pi}\left[\frac{\sin(n+1)t}{n+1} + \frac{\sin(n-1)t}{n-1}\right]_{t=0}^{t=\pi/2} \qquad (n > 1).$$

Therefore, for odd $n > 1$

$$a_n = 0,$$

while, for even n

$$a_n = -\frac{2(-1)^{n/2}}{\pi(n^2-1)}, \quad b_n = 0 \qquad (n = 1, 2, \ldots).$$

Thus we have

$$\frac{1}{\pi} + \frac{1}{2}\cos\frac{\pi x}{l} - \frac{2}{\pi}\sum_{n=1}^{\infty}\frac{(-1)^n}{4n^2-1}\cos\frac{2\pi n x}{l} = \begin{cases} \cos\dfrac{\pi x}{l} & \text{for } 0 \leqslant x \leqslant \dfrac{l}{2}, \\ 0 & \text{for } \dfrac{l}{2} < x \leqslant l. \end{cases}$$

This series converges on the whole x-axis to the function shown in Fig. 23.

Example 2. *Expand the function $f(x)$, defined by*

$$f(x) = \begin{cases} x & \text{for } 0 \leqslant x \leqslant \dfrac{l}{2}, \\ l - x & \text{for } \dfrac{l}{2} < x \leqslant l, \end{cases}$$

in sine series. Figure 24 shows the graph of $f(x)$ and its odd extension onto the interval $[-l, 0]$, together with its subsequent periodic extension (with period $2l$) onto the whole x-axis. The convergence criterion can be applied everywhere.

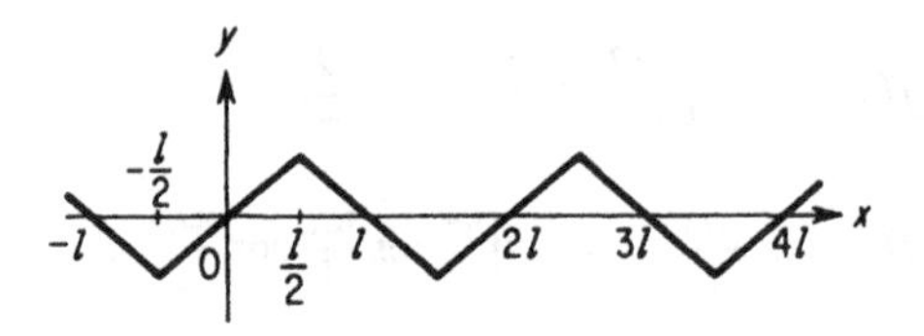

FIGURE 24

In this case, we have

$$a_n = 0 \qquad (n = 0, 1, 2, \ldots),$$

$$b_n = \frac{2}{l}\int_0^l f(x)\sin\frac{\pi n x}{l}\,dx$$

$$= \frac{2}{l}\int_0^{l/2} x\sin\frac{\pi n x}{l}\,dx + \frac{2}{l}\int_{l/2}^{l}(l-x)\sin\frac{\pi n x}{l}\,dx \qquad (n = 1, 2, \ldots).$$

Setting $\pi x/l = t$, we obtain

$$b_n = \frac{2l}{\pi^2}\int_0^{\pi/2} t\sin nt\,dt + \frac{2l}{\pi^2}\int_{\pi/2}^{\pi}(\pi - t)\sin nt\,dt$$

$$= \frac{2l}{\pi^2}\left[-\frac{t\cos nt}{n}\right]_{t=0}^{t=\pi/2} + \frac{2l}{\pi^2 n}\int_0^{\pi/2}\cos nt\,dt$$

$$+ \frac{2l}{\pi^2}\left[-\frac{(\pi - t)\cos nt}{n}\right]_{t=\pi/2}^{t=\pi} - \frac{2l}{\pi^2 n}\int_{\pi/2}^{\pi}\cos nt\,dt$$

$$= \frac{4l}{\pi^2 n^2}\sin\frac{\pi n}{2}.$$

Therefore

$$\frac{4l}{\pi^2}\left(\sin\frac{\pi x}{l} - \frac{1}{3^2}\sin\frac{3\pi x}{l} + \frac{1}{5^2}\sin\frac{5\pi x}{l} - \cdots\right) = \begin{cases} x & \text{for } 0 \leqslant x \leqslant \dfrac{l}{2}, \\ l - x & \text{for } \dfrac{l}{2} < x \leqslant l. \end{cases}$$

This series converges on the whole x-axis to the function shown in Fig. 24.

PROBLEMS

1. Expand the following functions in Fourier series:

a) $f(x) = e^{ax}$ $(-\pi < x < \pi)$, where $a \neq 0$ is a constant;
b) $f(x) = \cos ax$ $(-\pi \leqslant x \leqslant \pi)$, where a is not an integer;

c) $f(x) = \sin ax$ $(-\pi < x < \pi)$, where a is not an integer;

d) $f(x) = \begin{cases} 0 & \text{for} \quad -\pi < x < 0, \\ x & \text{for} \quad 0 \leqslant x < \pi. \end{cases}$

2. Using the expansion of Prob. 1b, show that

$$\frac{1}{\sin z} = \frac{1}{z} + \sum_{n=1}^{\infty} (-1)^n \left[\frac{1}{z - n\pi} + \frac{1}{z + n\pi}\right],$$

$$\cot z = \frac{1}{z} + \sum_{n=1}^{\infty} \left[\frac{1}{z - n\pi} + \frac{1}{z + n\pi}\right],$$

where z is any number which is not a multiple of π.

3. Using the expansion of Prob. 1a, expand the following functions in Fourier series:

a) The hyperbolic cosine

$$\cosh x = \frac{e^{ax} + e^{-ax}}{2} \qquad (-\pi \leqslant x \leqslant \pi);$$

b) The hyperbolic sine

$$\sinh x = \frac{e^{ax} - e^{-ax}}{2} \qquad (-\pi < x < \pi).$$

4. Expand the following functions in Fourier cosine series:

a) $f(x) = \sin ax$ $(0 \leqslant x \leqslant \pi)$, where a is not an integer;

b) $f(x) = \begin{cases} 1 & \text{for} \quad 0 \leqslant x \leqslant h, \\ 0 & \text{for} \quad h < x \leqslant \pi; \end{cases}$

c) $f(x) = \begin{cases} 1 - \dfrac{x}{2h} & \text{for} \quad 0 \leqslant x \leqslant 2h, \\ 0 & \text{for} \quad 2h < x \leqslant \pi. \end{cases}$

5. Expand the following functions in Fourier sine series:

a) $f(x) = \begin{cases} \sin \dfrac{\pi x}{l} & \text{for} \quad 0 \leqslant x < \dfrac{l}{2}, \\ 0 & \text{for} \quad \dfrac{l}{2} < x \leqslant l; \end{cases}$

b) $f(x) = \begin{cases} \sin \dfrac{\pi x}{l} & \text{for} \quad 0 \leqslant x < \dfrac{l}{2}, \\ -\sin \dfrac{\pi x}{l} & \text{for} \quad \dfrac{l}{2} < x \leqslant l. \end{cases}$

6. Expand the periodic function

$$f(x) = \left|\cos \frac{\pi x}{l}\right|, \quad l = \text{const}, \ l > 0$$

in Fourier series.

7. Let $f(x)$ have period 2π and let $|f(x) - f(y)| \leqslant c|x - y|^{\alpha}$, for some constants

$c > 0$, $\alpha > 0$, and for all x and y. [If $f(x)$ obeys this inequality, it is said to satisfy a *Hölder (or Lipschitz) condition of order* α.] Show that

$$|a_n| \leqslant \frac{c\pi^\alpha}{n^\alpha}, \quad |b_n| \leqslant \frac{c\pi^\alpha}{n^\alpha},$$

where a_n and b_n are the Fourier coefficients of $f(x)$.

8. Expand the following functions in Fourier sine series:

a) $f(x) = \cos x \quad (0 < x < \pi)$;
b) $f(x) = x^3 \quad (0 \leqslant x < \pi)$.

9. Let $f(x)$ be a function of period 2π defined for $-\pi < x < \pi$. Let $f(x)$ have the Fourier series

$$\frac{a_0}{2} + \sum_{n=1}^{\infty} (a_n \cos nx + b_n \sin nx),$$

and let

$$f_e(x) = \frac{f(x) + f(-x)}{2}, \quad f_o(x) = \frac{f(x) - f(-x)}{2}.$$

Show that $f_e(x)$ is an even function and $f_o(x)$ an odd function, with Fourier series

$$\frac{a_0}{2} + \sum_{n=1}^{\infty} a_n \cos nx, \quad \sum_{n=1}^{\infty} b_n \sin nx,$$

respectively. Show that the function $f(x - \pi)$ has the Fourier series

$$\frac{a_0}{2} + \sum_{n=1}^{\infty} (-1)^n(a_n \cos nx + b_n \sin nx).$$

10. Sum the series

a) $\displaystyle\sum_{n=1}^{\infty} \frac{\sin nx}{n}$; b) $\displaystyle\sum_{n=1}^{\infty} (-1)^n \frac{\sin nx}{n}$;

c) $\displaystyle\sum_{n=1}^{\infty} \frac{\cos nx}{n^2}$; d) $\displaystyle\sum_{n=1}^{\infty} (-1)^n \frac{\cos nx}{n^2}$

by using Example 8 of Sec. 13 and the results of the preceding problem.

11. Find the sum of each of the following numerical series by evaluating at a suitable point a Fourier series given in the text or in the problems:

a) $\displaystyle\sum_{n=1}^{\infty} \frac{1}{(2n-1)^2}$; b) $\displaystyle\frac{1}{2a} + \sum_{n=1}^{\infty} (-1)^n \frac{1}{n^2 + a^2}$;

c) $\displaystyle\frac{1}{2} + \sum_{n=1}^{\infty} \frac{\sin nh}{nh}$; d) $\displaystyle\frac{1}{2} + \sum_{n=1}^{\infty} \left(\frac{\sin nh}{nh}\right)^2$.

12. Show that the Fourier series for the function $f(x) = x$ on the interval $-\pi < x < \pi$ does not converge uniformly, but that the Fourier series for the function $f(x) = x^2$ does converge uniformly. Find the Fourier series for $f(x) = x^4$ by integrating the Fourier series for $f(x) = x^2$ between the limits 0 and x.

2

ORTHOGONAL SYSTEMS

1. Definitions

An infinite system of real functions

$$\varphi_0(x),\ \varphi_1(x),\ \varphi_2(x), \ldots, \varphi_n(x), \ldots \tag{1.1}$$

is said to be *orthogonal* on the interval $[a, b]$ if

$$\int_a^b \varphi_n(x)\varphi_m(x)\,dx = 0 \qquad (n \neq m,\ n, m = 0, 1, 2, \ldots). \tag{1.2}$$

We shall always assume also that

$$\int_a^b \varphi_n^2(x)\,dx \neq 0 \qquad (n = 0, 1, 2, \ldots). \tag{1.3}$$

The condition (1.2) says that every pair of functions of the system (1.1) is orthogonal (see Ch. 1, Sec. 5), while the condition (1.3) says that none of the functions of the system is identically zero.

We have already encountered special cases of orthogonal systems, i.e., the basic trigonometric system

$$1, \cos x, \sin x, \ldots, \cos nx, \sin nx, \ldots, \tag{1.4}$$

which is orthogonal on any interval of length 2π, and the more general trigonometric system

$$1, \cos\frac{\pi x}{l}, \sin\frac{\pi x}{l}, \ldots, \cos\frac{n\pi x}{l}, \sin\frac{n\pi x}{l}, \ldots, \tag{1.5}$$

which is orthogonal on any interval of length $2l$ (see Ch. 1, Secs. 5 and 15).

The system (1.1) is said to be *normalized* if

$$\int_a^b \varphi_n^2(x)\,dx = 1 \qquad (n = 0, 1, 2, \ldots).$$

Every orthogonal system can be normalized, i.e., one can always choose the constants $\mu_0, \mu_1, \ldots, \mu_n, \ldots$ in such a way that the system

$$\mu_0\varphi_0(x), \mu_1\varphi_1(x), \ldots, \mu_n\varphi_n(x), \ldots,$$

which is obviously still orthogonal, is now also normalized. Such a system is said to be *orthonormal.* In fact, the condition

$$\int_a^b \mu_n^2\varphi_n^2(x)\,dx = \mu_n^2 \int_a^b \varphi_n^2(x)\,dx = 1 \qquad (n = 0, 1, 2, \ldots)$$

implies that

$$\mu_n = \frac{1}{\sqrt{\int_a^b \varphi_n^2(x)\,dx}}.$$

Finally, we introduce the notation

$$\|\varphi_n\| = \sqrt{\int_a^b \varphi_n^2(x)\,dx} \qquad (n = 0, 1, 2, \ldots),$$

and call this number the *norm* of the function $\varphi_n(x)$. If the system (1.1) is normalized, then obviously

$$\|\varphi_n\| = 1 \qquad (n = 0, 1, 2, \ldots).$$

2. Fourier Series with Respect to an Orthogonal System

We now essentially repeat the considerations of Ch. 1, Sec. 6, in a more general context. Let $f(x)$ be a function defined on the interval $[a, b]$. Suppose that $f(x)$ can be represented as the sum of a series involving the functions of the orthogonal system (1.1), i.e., suppose that everywhere on $[a, b]$

$$f(x) = c_0\varphi_0(x) + c_1\varphi_1(x) + \cdots + c_n\varphi_n(x) + \cdots, \tag{2.1}$$

where $c_0, c_1, \ldots, c_n, \ldots$ are constants. To calculate these constants, we assume that the series

$$\begin{aligned} f(x)\varphi_n(x) = {} & c_0\varphi_0(x)\varphi_n(x) + c_1\varphi_1(x)\varphi_n(x) + \cdots + c_{n-1}\varphi_{n-1}(x)\varphi_n(x) \\ & + c_n\varphi_n^2(x) + c_{n+1}\varphi_{n+1}(x)\varphi_n(x) + \cdots \qquad (2.2) \\ & \qquad (n = 0, 1, 2, \ldots), \end{aligned}$$

obtained by multiplying the equations (2.1) by $\varphi_n(x)$, can be integrated term by term over the interval $[a, b]$. According to (1.2), this integration gives

$$\int_a^b f(x)\varphi_n(x)\,dx = c_n \int_a^b \varphi_n^2(x)\,dx \qquad (n = 0, 1, 2, \ldots).$$

Therefore we have

$$c_n = \frac{\int_a^b f(x)\varphi_n(x)\,dx}{\int_a^b \varphi_n^2(x)\,dx} = \frac{\int_a^b f(x)\varphi_n(x)\,dx}{\|\varphi_n\|^2} \qquad (n = 0, 1, 2, \ldots). \tag{2.3}$$

Now suppose that we are given a function $f(x)$ defined on the interval $[a, b]$, and we wish to make a series expansion of $f(x)$ with respect to the functions of the system (1.1), without knowing in advance whether or not such an expansion is possible. If such an expansion is actually possible (and if the required term by term integration is also possible), then, as we have just seen, we must arrive at the formulas (2.3). Therefore, in trying to find the required expansion of the function $f(x)$, it is natural to first examine the series with coefficients given by (2.3), and see whether this series might turn out to converge to $f(x)$. The coefficients given by (2.3) are called the *Fourier coefficients of $f(x)$ with respect to the system* (1.1), and the corresponding series is called the *Fourier series of $f(x)$ with respect to the system* (1.1). If the system (1.1) is normalized, then the formulas for the Fourier coefficients take a particularly simple form:

$$c_n = \int_a^b f(x)\varphi_n(x)\,dx \qquad (n = 0, 1, 2, \ldots). \tag{2.4}$$

Until it has been ascertained that the Fourier series of $f(x)$ actually converges to $f(x)$, we write

$$f(x) \sim c_0\varphi_0(x) + c_1\varphi_1(x) + \cdots + c_n\varphi_n(x) + \cdots$$

However, it should be noted that even in the case where the Fourier series turns out to be *divergent* (and this actually happens sometimes), the Fourier series still has various remarkable properties which will be discussed below.

If the functions of the system (1.1) are continuous, and if the series in the right-hand side of (2.1) is uniformly convergent, then it is easy to prove that the series (2.2) is also uniformly convergent and can therefore be integrated term by term (see the proof of Theorem 1 of Ch. 1, Sec. 6). This immediately implies the following

THEOREM. *If the functions of the system* (1.1) *are continuous and if the series expansion* (2.1) *of $f(x)$ is uniformly convergent, then* (2.1) *is the Fourier series of $f(x)$.*

3. Some Simple Orthogonal Systems

In addition to the orthogonal systems (1.4) and (1.5) just cited, we consider the following systems:

I. The system

$$1, \cos x, \cos 2x, \ldots, \cos nx, \ldots$$

is orthogonal on the interval $[0, \pi]$. In fact

$$\int_0^\pi \cos nx\,dx = \left[\frac{\sin nx}{n}\right]_{x=0}^{x=\pi} = 0 \qquad (n = 1, 2, \ldots), \tag{3.1}$$

which means that the functions $\cos nx$ and 1 are orthogonal. Moreover, we have

$$\int_0^\pi \cos nx \cos mx\,dx = \frac{1}{2}\int_0^\pi [\cos (n + m)x + \cos (n - m)x]\,dx$$

$$= \frac{1}{2}\int_0^\pi \cos (n + m)x\,dx + \frac{1}{2}\int_0^\pi \cos (n - m)x\,dx = 0 \qquad (n \neq m),$$

which follows from (3.1). This proves that the system I is orthogonal.

In writing Fourier series with respect to the system I, we continue to use the notation introduced in Ch. 1, i.e., we write

$$f(x) \sim \frac{a_0}{2} + a_1 \cos x + a_2 \cos 2x + \cdots + a_n \cos nx + \cdots.$$

With this notation for the Fourier coefficients, the formulas (2.3) give

$$\frac{a_0}{2} = \frac{\int_0^\pi f(x)\,dx}{\int_0^\pi 1 \cdot dx} = \frac{1}{\pi}\int_0^\pi f(x)\,dx,$$

and

$$a_n = \frac{\int_0^\pi f(x) \cos nx\,dx}{\int_0^\pi \cos^2 nx\,dx} \qquad (n = 1, 2, \ldots).$$

Since

$$\int_0^\pi \cos^2 nx\,dx = \int_0^\pi \frac{1 + \cos 2nx}{2}\,dx = \frac{\pi}{2},$$

we can write

$$a_n = \frac{2}{\pi}\int_0^\pi f(x) \cos nx\,dx \qquad (n = 0, 1, 2, \ldots).$$

Thus, as is to be expected, we arrive at the same formulas (12.3) as obtained in Ch. 1 for cosine series.

II. The system

$$\sin x, \sin 2x, \ldots, \sin nx, \ldots$$

is orthogonal on $[0, \pi]$. In fact

$$\int_0^\pi \sin nx \sin mx \, dx = \frac{1}{2}\int_0^\pi [\cos (n - m)x - \cos (n + m)x] \, dx = 0$$

for $n \neq m$ [see (3.1)]. As in the preceding case, in writing Fourier series with respect to the system II, we continue to use the notation adopted in Ch. 1:

$$f(x) \sim b_1 \sin x + b_2 \sin 2x + \cdots + b_n \sin nx + \cdots$$

Then by (2.3)

$$b_n = \frac{\int_0^\pi f(x) \sin nx \, dx}{\int_0^\pi \sin^2 nx \, dx} \qquad (n = 1, 2, \ldots).$$

Moreover, since

$$\int_0^\pi \sin^2 nx \, dx = \int_0^\pi \frac{1 - \cos 2nx}{2} \, dx = \frac{\pi}{2},$$

we have

$$b_n = \frac{2}{\pi}\int_0^\pi f(x) \sin nx \, dx \qquad (n = 1, 2, \ldots),$$

i.e., as is to be expected, we arrive at the formulas (12.4) obtained in Ch. 1 for sine series.

III. The system

$$\sin x, \sin 3x, \sin 5x, \ldots, \sin (2n + 1)x, \ldots$$

is orthogonal on $[0, \pi/2]$. In fact, for $n \neq m$ and $n, m = 0, 1, 2, \ldots$, we have

$$\int_0^{\pi/2} \sin (2n + 1)x \cdot \sin (2m + 1)x \, dx$$

$$= \frac{1}{2}\int_0^{\pi/2} [\cos 2(n - m)x - \cos 2(n + m + 1)x] \, dx$$

$$= \frac{1}{2}\left[\frac{\sin 2(n - m)x}{2(n - m)}\right]_{x=0}^{x=\pi/2} - \frac{1}{2}\left[\frac{\sin 2(n + m + 1)x}{2(n + m + 1)}\right]_{x=0}^{x=\pi/2} = 0.$$

The formulas (2.3) for the Fourier coefficients give

$$c_n = \frac{\int_0^{\pi/2} f(x) \sin (2n+1)x \, dx}{\int_0^{\pi/2} \sin^2 (2n+1)x \, dx} \qquad (n = 0, 1, 2, \ldots).$$

But

$$\int_0^{\pi/2} \sin^2 (2n+1)x \, dx = \int_0^{\pi/2} \frac{1 - \cos (4n+2)x}{2} \, dx = \frac{\pi}{4},$$

and therefore

$$c_n = \frac{4}{\pi} \int_0^{\pi/2} f(x) \sin (2n+1)x \, dx \qquad (n = 0, 1, 2, \ldots). \tag{3.2}$$

We now show that if $f(x)$ is a function defined on the interval $[0, \pi/2]$, then we can arrive at the expansion of $f(x)$ with respect to the system III by starting from the basic trigonometric system, just as in Ch. 1, Sec. 12, we arrived at the expansions in cosine series or sine series of a function defined on $[0, \pi]$ by using its even or odd extensions onto the interval $[-\pi, 0]$. To do this, we have to generalize the concepts of evenness and oddness of a function. Thus, let $f(x)$ be defined either on the whole x-axis or on some interval which is symmetric with respect to the point $x = l$. We shall say that $f(x)$ is *even with respect to* $x = l$ if

$$f(l - h) = f(l + h)$$

for every h. This means that the curve $y = f(x)$ is symmetric with respect to the *line* $x = l$ (see Fig. 25).

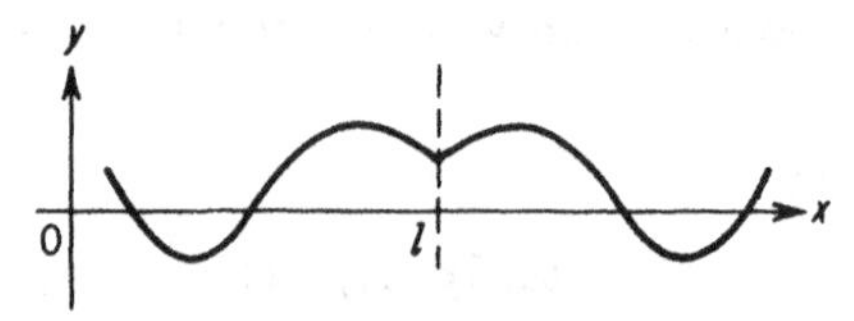

FIGURE 25

If a function $f(x)$ is even with respect to $x = l$, then obviously

$$\int_{l-a}^{l+a} f(x) \, dx = 2 \int_{l-a}^{l} f(x) \, dx,$$

and in particular (for $a = l$)

$$\int_0^{2l} f(x) \, dx = 2 \int_0^{l} f(x) \, dx. \tag{3.3}$$

Similarly, we say that $f(x)$ is *odd with respect to* $x = l$ if

$$f(l - h) = -f(l + h)$$

for every h. This means that the curve $y = f(x)$ is symmetric with respect to the *point* $(l, 0)$ (see Fig. 26). If a function $f(x)$ is odd with respect to $x = l$, then obviously

$$\int_{l-a}^{l+a} f(x)\, dx = 0,$$

and in particular

$$\int_{0}^{2l} f(x)\, dx = 0.$$

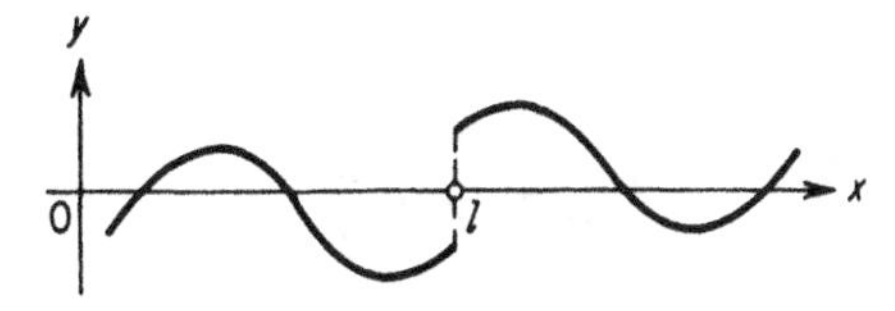

FIGURE 26

With this interpretation of evenness and oddness, we can assert that the product of two even or two odd functions is even, whereas the product of an even function and an odd function is odd. The proofs are essentially the same as those given in Ch. 1, Sec. 11.

Now let $f(x)$ be defined on the interval $[0, \pi/2]$, and make the even extension of $f(x)$ onto the interval $[\pi/2, \pi]$ (see Fig. 27). This gives a function

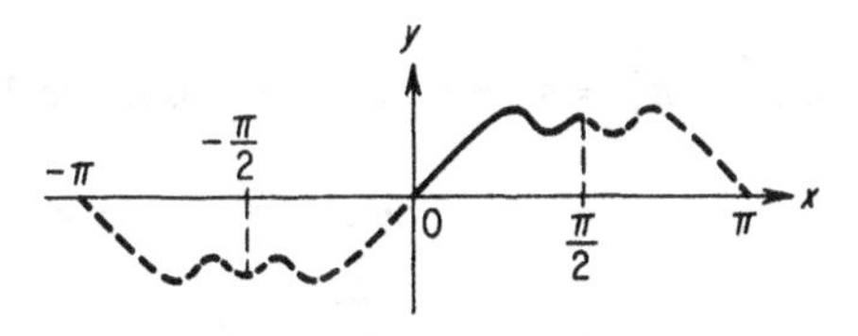

FIGURE 27

$g(x)$ defined on $[0, \pi]$ which coincides with $f(x)$ on $[0, \pi/2]$. Expand the function $g(x)$ in a cosine series, which is equivalent to making the *odd* extension of $g(x)$ onto the interval $[-\pi, 0]$ (see Ch. 1, Sec. 12); the result is

$$b_n = \frac{2}{\pi}\int_{0}^{\pi} g(x) \sin nx\, dx \qquad (n = 1, 2, \ldots). \tag{3.4}$$

Next, we observe that the functions of the system III are even with respect to $x = \pi/2$. In fact, for $n = 0, 1, 2, \ldots$, we have

$$\sin (2n + 1)\left(\frac{\pi}{2} - h\right)$$
$$= \sin (2n + 1)\frac{\pi}{2}\cdot\cos (2n + 1)h$$
$$- \cos (2n + 1)\frac{\pi}{2}\cdot\sin (2n + 1)h$$
$$= \sin (2n + 1)\frac{\pi}{2}\cdot\cos (2n + 1)h$$
$$+ \cos (2n + 1)\frac{\pi}{2}\cdot\sin (2n + 1)h = \sin (2n + 1)\left(\frac{\pi}{2} + h\right),$$

since

$$\cos (2n + 1)\frac{\pi}{2} = 0.$$

Therefore, since the functions $g(x)$ and $\sin (2n + 1)x$ are even with respect to $x = \pi/2$, it follows from (3.3) and (3.4) that

$$b_{2n+1} = \frac{2}{\pi}\int_0^\pi g(x)\sin (2n + 1)x\,dx$$
$$= \frac{4}{\pi}\int_0^{\pi/2} f(x)\sin (2n + 1)x\,dx \qquad (n = 0, 1, 2, \ldots).$$

On the other hand, the functions $\sin 2nx$ $(n = 1, 2, \ldots)$ are odd with respect to $x = \pi/2$, since

$$\sin 2n\left(\frac{\pi}{2} - h\right) = \sin \pi n \cos 2nh - \cos \pi n \sin 2nh$$
$$= -(\sin \pi n \cos 2nh + \cos \pi n \sin 2nh)$$
$$= -\sin 2n\left(\frac{\pi}{2} + h\right).$$

Therefore, the product $g(x)\sin 2nx$ $(n = 1, 2, \ldots)$ is odd with respect to $x = \pi/2$, and hence

$$b_{2n} = \frac{2}{\pi}\int_0^\pi g(x)\sin 2nx\,dx = 0 \qquad (n = 1, 2, \ldots).$$

Thus, finally, we have expanded the function $g(x)$, and consequently the function $f(x)$, in a sine series such that all the coefficients with even indices vanish. The coefficients with odd indices are given by the formulas (3.4), which coincide with (3.2).

We have gone through this rather lengthy argument in order to be able to apply the convergence criterion of Ch. 1, Sec. 10 (which was formulated for functions of period 2π) to the case of series expansions with respect to the system III. Our argument shows that this convergence criterion must be applied to the function obtained from $f(x)$ by first making an even extension of $f(x)$ onto $[\pi/2, \pi]$, then making an odd extension of the resulting function onto $[-\pi, 0]$,[1] and finally extending the result periodically (with period 2π) onto the whole real axis.

IV, V. The systems

$$1, \cos\frac{\pi x}{l}, \cos\frac{2\pi x}{l}, \ldots, \cos\frac{n\pi x}{l}, \ldots$$

and

$$\sin\frac{\pi x}{l}, \sin\frac{2\pi x}{l}, \ldots, \sin\frac{n\pi x}{l}, \ldots$$

are orthogonal on the interval $[0, l]$. In fact, if we make the substitution $\pi x/l = t$, the integrals of pairs of functions from each of these systems reduce to the corresponding integrals for the systems I and II.

VI. The system

$$\sin\frac{\pi x}{2l}, \sin\frac{3\pi x}{2l}, \sin\frac{5\pi x}{2l}, \ldots, \sin\frac{(2n+1)\pi x}{2l}, \ldots$$

is orthogonal on the interval $[0, l]$. In fact, making the substitution $\pi x/2l = t$, we obtain

$$\int_0^l \sin\frac{(2n+1)\pi x}{2l} \sin\frac{(2m+1)\pi x}{2l}\,dx$$

$$= \frac{2l}{\pi}\int_0^{\pi/2} \sin(2n+1)t\cdot\sin(2m+1)t\,dt = 0 \qquad (n \neq m),$$

since everything has been reduced to the orthogonality of functions of the system III. For the Fourier coefficients with respect to the system VI, we obtain

$$c_n = \frac{\displaystyle\int_0^l f(x)\sin\frac{(2n+1)\pi x}{2l}\,dx}{\displaystyle\int_0^l \sin^2\frac{(2n+1)\pi x}{2l}\,dx} = \frac{2}{l}\int_0^l f(x)\sin\frac{(2n+1)\pi x}{2l}\,dx.$$

If we want to expand a function $f(x)$ defined on $[0, l]$ in a Fourier series

[1] Figure 27 shows the graph of $y = f(x)$ together with the indicated twofold extension.

with respect to the system VI, we first make the substitution $\pi x/2l = t$. This reduces the problem to expanding the function $\varphi(t) = f(2lt/\pi)$, defined on $[0, \pi/2]$, with respect to the system III. Then, we return to the required expansion with respect to the system VI by transforming back to the variable x. It follows that we can apply the convergence criterion of Ch. 1, Sec. 10 to expansions with respect to the system VI (which are quite often encountered in the applications), since the criterion is applicable to the system III.

Later, we shall deal with orthogonal systems consisting of functions which are more complicated than trigonometric functions (e.g., Bessel functions).

4. Square Integrable Functions. The Schwarz Inequality

We shall say that the function $f(x)$ defined on the interval $[a, b]$ is *square integrable* if $f(x)$ and its square are integrable on $[a, b]$. Every bounded integrable function must be square integrable, but this is not always the case for unbounded integrable functions. For example, the integral

$$\int_0^1 \frac{dx}{\sqrt{x}}$$

exists, but the integral

$$\int_0^1 \frac{dx}{x}$$

does not exist.

Let $\varphi(x)$ and $\psi(x)$ be square integrable functions defined on $[a, b]$. First we note that it follows from the elementary inequality

$$|\varphi\psi| \leqslant \tfrac{1}{2}(\varphi^2 + \psi^2)$$

that the function $|\varphi\psi|$ is integrable.[2] Now consider the inequality

$$\int_a^b (\varphi + \lambda\psi)^2\, dx = \int_a^b \varphi^2\, dx + 2\lambda \int_a^b \varphi\psi\, dx + \lambda^2 \int_a^b \psi^2\, dx \geqslant 0,$$

which holds for an arbitrary constant λ, and set

$$\int_a^b \varphi^2\, dx = A, \quad \int_a^b \varphi\psi\, dx = B, \quad \int_a^b \psi^2\, dx = C.$$

Then

$$A + 2B\lambda + C\lambda^2 \geqslant 0$$

[2] Incidentally, this implies that a square integrable function is always absolutely integrable. (As we have seen, the converse is not always true.) To see this, it is sufficient to set $\psi(x) = 1$.

for any λ. Therefore, the graph of the polynomial

$$\mu = A + 2B\lambda + C\lambda^2$$

is a parabola lying above the x-axis or perhaps touching it (see Fig. 28).

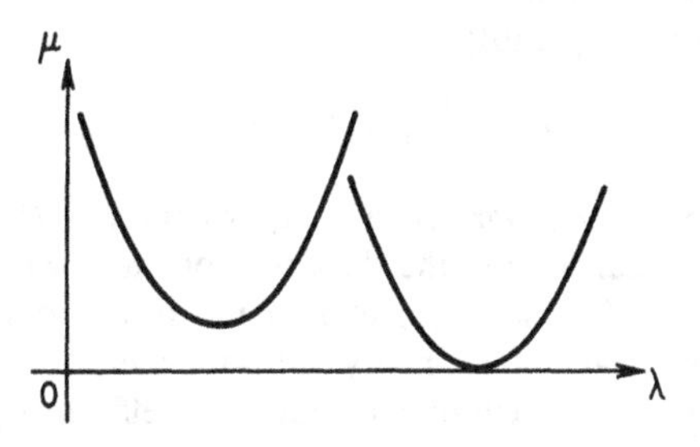

FIGURE 28

It follows that our polynomial cannot have *distinct* real zeros, since then the curve would intersect the x-axis in two points. Therefore, the discriminant of the polynomial must satisfy the inequality

$$B^2 - AC \leqslant 0,$$

i.e.,

$$B^2 \leqslant AC.$$

Recalling the meaning of A, B, and C, we obtain

$$\left(\int_a^b \varphi\psi \, dx\right)^2 \leqslant \int_a^b \varphi^2 \, dx \int_a^b \psi^2 \, dx. \tag{4.1}$$

This very useful inequality is called the *Schwarz inequality*.[3]

Using the Schwarz inequality, it is easy to show that the sum of a finite number of square integrable functions is also a square integrable function. In fact, for two functions we have

$$\int_a^b (\varphi + \psi)^2 \, dx = \int_a^b \varphi^2 \, dx + 2\int_a^b \varphi\psi \, dx + \int_a^b \psi^2 \, dx,$$

and the generalization to the sum of any number of functions is straightforward.

5. The Mean Square Error and its Minimum

Let $f(x)$ be an arbitrary square integrable function defined on the interval $[a, b]$, and let $\sigma_n(x)$ be a linear combination of the first $n + 1$ functions of the system (1.1), i.e.,

$$\sigma_n(x) = \gamma_0\varphi_0(x) + \gamma_1\varphi_1(x) + \cdots + \gamma_n\varphi_n(x), \tag{5.1}$$

[3] Called the *Buniakovski inequality* by Russian mathematicians. (*Translator*)

where $\gamma_0, \gamma_1, \ldots, \gamma_n$ are constants. By hypothesis, all the functions $\varphi_n(x)$ of the system (1.1) are square integrable [see (1.3)]. Therefore, the linear combination $\sigma_n(x)$ and the difference $f(x) - \sigma_n(x)$ $[n = 0, 1, 2, \ldots]$ are also square integrable functions.

Now consider the quantity

$$\delta_n = \int_a^b [f(x) - \sigma_n(x)]^2 \, dx, \tag{5.2}$$

which we call the *mean square error* in approximating $f(x)$ by $\sigma_n(x)$. There are many ways of calculating the deviation of the linear combination $\sigma_n(x)$ from the function $f(x)$; we have chosen the mean square error, which is particularly appropriate in the theory of Fourier series.

We now pose the problem of choosing the coefficients $\gamma_0, \gamma_1, \ldots, \gamma_n$ (for a given n) in such a way that the mean square error δ_n is a minimum. It follows from (5.2) that

$$\delta_n = \int_a^b f^2(x) \, dx - 2 \int_a^b f(x)\sigma_n(x) \, dx + \int_a^b \sigma_n^2(x) \, dx. \tag{5.3}$$

By (5.1) we have

$$\int_a^b f(x)\sigma_n(x) \, dx = \sum_{k=0}^{n} \gamma_k \int_a^b f(x)\varphi_k(x) \, dx.$$

But according to (2.3)

$$\int_a^b f(x)\varphi_k(x) \, dx = c_k \|\varphi_k\|^2 \qquad (k = 0, 1, 2, \ldots),$$

where the c_k are the Fourier coefficients of the function $f(x)$, and therefore

$$\int_a^b f(x)\sigma_n(x) \, dx = \sum_{k=0}^{n} \gamma_k c_k \|\varphi_k\|^2. \tag{5.4}$$

Furthermore, we have

$$\begin{aligned}
\int_a^b \sigma_n^2(x) \, dx &= \int_a^b \left(\sum_{k=0}^{n} \gamma_k \varphi_k(x) \right)^2 dx \\
&= \int_a^b \left(\sum_{k=0}^{n} \gamma_k^2 \varphi_k^2(x) + \sum_{p \neq q} \gamma_p \gamma_q \varphi_p(x) \varphi_q(x) \right) dx \\
&= \sum_{k=0}^{n} \gamma_k^2 \int_a^b \varphi_k^2(x) \, dx + \sum_{p \neq q} \gamma_p \gamma_q \int_a^b \varphi_p(x)\varphi_q(x) \, dx.
\end{aligned}$$

The last sum extends over all possible unequal indices p and q that do not exceed n. By the orthogonality of the system (1.1), this sum vanishes. Thus, we find that

$$\int_a^b \sigma_n^2(x)\,dx = \sum_{k=0}^{n} \gamma_k^2 \|\varphi_k\|^2. \tag{5.5}$$

Substituting (5.4) and (5.5) in (5.3), we obtain

$$\begin{aligned}\delta_n &= \int_a^b f^2(x)\,dx - 2\sum_{k=0}^{n} \gamma_k c_k \|\varphi_k\|^2 + \sum_{k=0}^{n} \gamma_k^2 \|\varphi_k\|^2 \\ &= \int_a^b f^2(x)\,dx + \sum_{k=0}^{n} (c_k - \gamma_k)^2 \|\varphi_k\|^2 - \sum_{k=0}^{n} c_k^2 \|\varphi_k\|^2.\end{aligned}$$

Here

$$\int_a^b f^2(x)\,dx = \text{const}, \quad \sum_{k=0}^{n} c_k^2 \|\varphi_k\|^2 = \text{const},$$

i.e., do not depend on $\gamma_0, \gamma_1, \ldots, \gamma_n$, and therefore the quantity δ_n is obviously a minimum when

$$\sum_{k=0}^{n} (c_k - \gamma_k)^2 \|\varphi_n\|^2 = 0,$$

which is equivalent to the conditions

$$\gamma_k = c_k \qquad (k = 0, 1, 2, \ldots, n).$$

Thus, finally, *the mean square error is a minimum when the coefficients in the linear combination* (5.1) *are the Fourier coefficients.* Denoting the minimum value of the mean square error by Δ_n, we have

$$\begin{aligned}\Delta_n &= \int_a^b \left[f(x) - \sum_{k=0}^{n} c_k\varphi_k(x)\right]^2 dx \\ &= \int_a^b f^2(x)\,dx - \sum_{k=0}^{n} c_k^2 \|\varphi_k\|^2.\end{aligned} \tag{5.6}$$

This expression shows that as n increases, the nonnegative quantity Δ_n can only decrease. Thus, as n increases, the partial sums of the Fourier series give a closer approximation to the function $f(x)$, i.e., an approximation with smaller mean square error.

6. Bessel's Inequality

Since $\Delta_n \geqslant 0$, it follows from (5.6) that

$$\int_a^b f^2(x)\,dx \geqslant \sum_{k=0}^{n} c_k^2 \|\varphi_k\|^2,$$

where n is arbitrary. The sum on the right can only increase as n increases.

Therefore, since it is bounded by a constant (the integral on the left), the sum has a finite limit as $n \to \infty$. Thus, the series

$$\sum_{k=0}^{\infty} c_k^2 \|\varphi_k\|^2$$

converges and

$$\int_a^b f^2(x)\,dx \geqslant \sum_{k=0}^{\infty} c_k^2 \|\varphi_k\|^2. \tag{6.1}$$

This very important result is called *Bessel's inequality.* It follows at once from the convergence of the series on the right that

$$\lim_{n\to\infty} c_n \|\varphi_n\| = 0. \tag{6.2}$$

If the system (1.1) is normalized, then Bessel's inequality takes the form

$$\int_a^b f^2(x)\,dx \geqslant \sum_{k=0}^{\infty} c_k^2,$$

and therefore, the sum of the squares of the Fourier coefficients is convergent. For a normalized system, (6.2) becomes

$$\lim_{n\to\infty} c_n = 0,$$

i.e., the Fourier coefficients approach zero as $n \to \infty$.

7. Complete Systems. Convergence in the Mean

The system (1.1) is said to be *complete* if for any square integrable function $f(x)$, the equalitv

$$\int_a^b f^2(x)\,dx = \sum_{k=0}^{\infty} c_k^2 \|\varphi_k\|^2 \tag{7.1}$$

holds (instead of Bessel's *inequality*). Here, as above, the c_k ($k = 0, 1, 2, \ldots$) are the Fourier coefficients of the function $f(x)$. The equality (7.1) is called the *completeness condition* for the system (1.1). The following simple result is an immediate consequence of the completeness condition:

THEOREM 1. *Let $f(x)$ and $F(x)$ be square integrable functions for which*

$$f(x) \sim c_0\varphi_0(x) + c_1\varphi_1(x) + \cdots,$$
$$F(x) \sim C_0\varphi_0(x) + C_1\varphi_1(x) + \cdots,$$

and let the system (1.1) *be complete. Then we have*

$$\int_a^b f(x)F(x)\,dx = \sum_{k=0}^{\infty} c_k C_k \|\varphi_k\|^2. \tag{7.2}$$

Proof. The sum $f(x) + F(x)$ and the difference $f(x) - F(x)$ are square integrable functions. Moreover, the sum has Fourier coefficients $c_k + C_k$ and the difference has Fourier coefficients $c_k - C_k$. By the completeness relation, we have

$$\int_a^b [f(x) + F(x)]^2 \, dx = \sum_{k=0}^{\infty} (c_k + C_k)^2 \|\varphi_k\|^2,$$

$$\int_a^b [f(x) - F(x)]^2 \, dx = \sum_{k=0}^{\infty} (c_k - C_k)^2 \|\varphi_k\|^2,$$

and then subtraction gives

$$4 \int_a^b f(x)F(x) \, dx = \sum_{k=0}^{\infty} 4c_k C_k \|\varphi_k\|^2.$$

The following result has important consequences:

THEOREM 2. *A necessary and sufficient condition for the system* (1.1) *to be complete is that the relation*

$$\lim_{n\to\infty} \int_a^b \left[f(x) - \sum_{k=0}^{n} c_k\varphi_k(x)\right]^2 dx = 0, \tag{7.3}$$

hold for any square integrable function $f(x)$, where the c_k ($k = 0, 1, 2, \ldots$) are the Fourier coefficients of $f(x)$ with respect to the system (1.1).

Proof. Use the relation (5.6) and the fact that the completeness condition is equivalent to

$$\lim_{n\to\infty} \left[\int_a^b f^2(x) \, dx - \sum_{k=0}^{n} c_k^2 \|\varphi_k\|^2\right] = 0.$$

If the relation (7.3) is satisfied, we say that the Fourier series *converges to $f(x)$ in the mean.* Therefore, Theorem 2 can also be formulated as follows:

A necessary and sufficient condition for the system (1.1) *to be complete is that the Fourier series of any square integrable function $f(x)$ converge to $f(x)$ in the mean.*

It should be noted that *ordinary convergence* of a Fourier series to the function from which it is formed does not always occur, even if the system (1.1) is complete. Nevertheless, as we have just shown, *convergence in the mean* always occurs for complete systems (it is understood that we are talking about square integrable functions). In particular, these remarks apply to the trigonometric system (whose completeness will be proved in Ch. 5, Sec. 2).

These remarks show the importance of the concept of convergence in the mean and suggest that this kind of convergence be regarded as a *generalization* of ordinary convergence. To completely justify this approach, we now show that a Fourier series can converge in the mean to *only one function* (with a certain stipulation). Thus suppose that in addition to (7.3), the relation

$$\lim_{n\to\infty} \int_a^b \left[F(x) - \sum_{k=0}^{n} c_k\varphi_k(x)\right]^2 dx = 0, \tag{7.4}$$

also holds. Then, using the elementary inequality

$$(a + b)^2 \leqslant 2(a^2 + b^2),$$

we find that

$$\begin{aligned} 0 &\leqslant \int_a^b [F(x) - f(x)]^2\, dx \\ &= \int_a^b \left[\left(F(x) - \sum_{k=0}^{n} c_k\varphi_k(x)\right) + \left(\sum_{k=0}^{n} c_k\varphi_k(x) - f(x)\right)\right]^2 dx \\ &\leqslant 2\int_a^b \left[F(x) - \sum_{k=0}^{n} c_k\varphi_k(x)\right]^2 dx + 2\int_a^b \left[f(x) - \sum_{k=0}^{n} c_k\varphi_k(x)\right]^2 dx. \end{aligned}$$

By (7.3) and (7.4), this implies

$$\int_a^b [F(x) - f(x)]^2\, dx = 0,$$

and since the integrand is positive, it follows that

$$F(x) = f(x)$$

at the points of continuity of the integrand. But the integrand has only a finite number of points of discontinuity. Hence, the functions $F(x)$ and $f(x)$ coincide everywhere, except possibly at a finite number of points. Two such functions should hardly be considered different in the theory of Fourier series, since the values of a function at individual points can have no influence at all on the behavior of its Fourier series (since its Fourier coefficients are expressed in terms of integrals, and integrals do not depend on the values of the integrand at a finite number of points!). This allows us to draw the following conclusion:

THEOREM 3. *If the system* (1.1) *is complete, then every square integrable function* $f(x)$ *is completely determined* (*except for its values at a finite number of points*) *by its Fourier series, whether or not this series converges.*

This means that no other function which is "essentially" different from a given function $f(x)$, i.e., differs from $f(x)$ at more than a finite number of points, can have the same Fourier series as $f(x)$.[4]

8. Important Properties of Complete Systems

We now establish some very important properties of complete systems:

THEOREM 1. *If the system* (1.1) *is complete, then any continuous function $f(x)$ which is orthogonal to all the functions of the system must be identically zero.*

Proof. If $f(x)$ is orthogonal to all the functions of the system, then all the Fourier coefficients of $f(x)$ vanish. Then the completeness relation (7.1) implies that

$$\int_a^b f^2(x)dx = 0,$$

so that

$$f(x) \equiv 0,$$

since $f(x)$ is continuous.

THEOREM 2. *If the system* (1.1) *is complete, if the functions of the system are continuous, and if the Fourier series of the continuous function $f(x)$ is uniformly convergent, then the sum of the series equals $f(x)$.*

Proof. Let

$$f(x) \sim c_0\varphi_0(x) + c_1\varphi_1(x) + \cdots + c_n\varphi_n(x) + \cdots,$$

and set

$$s(x) = c_0\varphi_0(x) + c_1\varphi_1(x) + \cdots + c_n\varphi_n(x) + \cdots. \tag{8.1}$$

Since the functions of the system (1.1) are continuous, and since the series (8.1) is uniformly convergent, the sum of the series is continuous. It follows from the theorem of Sec. 2 that (8.1) is the Fourier series of $s(x)$. Therefore, the continuous functions $f(x)$ and $s(x)$ have the same Fourier series. But then Theorem 3 of Sec. 7 implies that

$$f(x) \equiv s(x),$$

and by (8.1)

$$f(x) = c_0\varphi_0(x) + c_1\varphi_1(x) + \cdots + c_n\varphi_n(x) + \cdots.$$

---

[4] It should be kept in mind that in this book integrable functions, and hence square integrable functions, are always assumed to be continuous except possibly at a finite number of points (Ch. 1, Sec. 4). Otherwise Theorem 3 would not be true. (*Translator*)

THEOREM 3. *If the system* (1.1) *is complete, then the Fourier series of every square integrable function $f(x)$ can be integrated term by term, whether or not the series converges.*

In other words, if

$$f(x) \sim c_0\varphi_0(x) + c_1\varphi_1(x) + \cdots + c_n\varphi_n(x) + \cdots,$$

then

$$\int_{x_1}^{x_2} f(x)\,dx = c_0 \int_{x_1}^{x_2} \varphi_0(x)\,dx + c_1 \int_{x_1}^{x_2} \varphi_1(x)\,dx + \cdots + c_n \int_{x_1}^{x_2} \varphi_n(x)\,dx + \cdots, \tag{8.2}$$

where x_1 and x_2 are any points of the interval $[a, b]$.

Proof. Assuming for definiteness that $x_1 < x_2$, we have

$$\begin{aligned}\left| \int_{x_1}^{x_2} f(x)\,dx - \sum_{k=0}^{n} c_k \int_{x_1}^{x_2} \varphi_k(x)\,dx \right| &\leqslant \int_{x_1}^{x_2} \left| f(x) - \sum_{k=0}^{n} c_k\varphi_k(x) \right| dx, \\ &\leqslant \int_a^b \left| f(x) - \sum_{k=0}^{n} c_k\varphi_k(x) \right| dx \\ &\leqslant \sqrt{\int_a^b \left[f(x) - \sum_{k=0}^{n} c_k\varphi_k(x) \right]^2 dx \int_a^b 1 \cdot dx}\,,\end{aligned} \tag{8.3}$$

where we have used the Schwarz inequality (see Sec. 4). By Theorem 2 of Sec. 7, the last term in (8.3) approaches zero as $n \to \infty$. Therefore,

$$\lim_{n\to\infty} \left[\int_{x_1}^{x_2} f(x)\,dx - \sum_{k=0}^{n} c_k \int_{x_1}^{x_2} \varphi_k(x)\,dx \right] = 0,$$

which is equivalent to (8.2).

9. A Criterion for the Completeness of a System

In view of the importance of the concept of complete systems, it is appropriate to give a simple test for the completeness of a system. The following completeness criterion is very convenient:

If for every continuous function $F(x)$ on $[a, b]$ and any number $\varepsilon > 0$, there exists a linear combination

$$\sigma_n(x) = \gamma_0\varphi_0(x) + \gamma_1\varphi_1(x) + \cdots + \gamma_n\varphi_n(x) \tag{9.1}$$

for which

$$\int_a^b [F(x) - \sigma_n(x)]^2\,dx \leqslant \varepsilon,$$

then the system (1.1) *is complete.*

We begin by noting that given any square integrable function $f(x)$, there exists a continuous function $F(x)$ for which

$$\int_a^b [f(x) - F(x)]^2 \, dx \leqslant \varepsilon. \tag{9.2}$$

This fact is quite clear geometrically, but for the reader who does not regard it as completely obvious, we give the following proof:

The function $f(x)$ can have only a finite number of points of discontinuity. In particular, $f(x)$ can have only a finite number of points at which it becomes unbounded. Every such point can be included in an interval of such small length that the sum of the integrals of the function $f^2(x)$ over these intervals does not exceed $\varepsilon/4$. Define the auxiliary function $\Phi(x)$ as being equal to $f(x)$ outside these intervals and equal to zero inside them. $\Phi(x)$ is bounded and can have only a finite number of discontinuities, and obviously

$$\int_a^b [f(x) - \Phi(x)]^2 \, dx \leqslant \frac{\varepsilon}{4}. \tag{9.3}$$

Next, we include each point of discontinuity of $\Phi(x)$ in an interval of such small length that the total length l of all these new intervals satisfies the condition

$$4M^2 l \leqslant \frac{\varepsilon}{4},$$

where M is any number such that $|\Phi(x)| < M$ for $a \leqslant x \leqslant b$.

Finally, consider the *continuous* function $F(x)$ which equals $\Phi(x)$ outside the intervals just described and which is "linear" inside each of them (see Fig. 29). Obviously we have

$$\int_a^b [\Phi(x) - F(x)]^2 \, dx \leqslant 4M^2 l \leqslant \frac{\varepsilon}{4}. \tag{9.4}$$

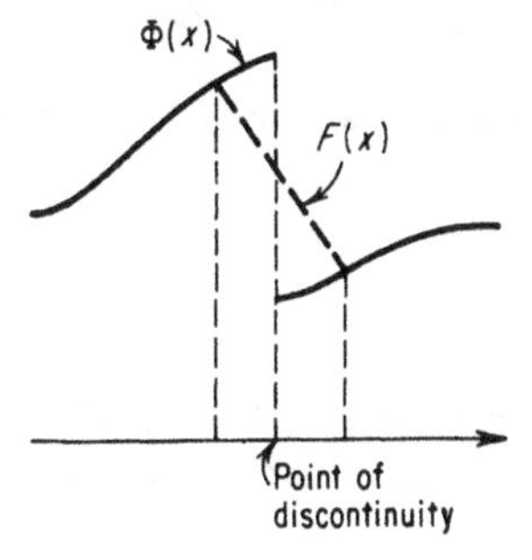

FIGURE 29

It follows from (9.3), (9.4) and the inequality

$$(a + b)^2 \leqslant 2(a^2 + b^2) \tag{9.5}$$

that

$$\int_a^b [f(x) - F(x)]^2 \, dx = \int_a^b [(f(x) - \Phi(x)) + (\Phi(x) - F(x))]^2 \, dx$$

$$\leqslant 2 \int_a^b [f(x) - \Phi(x)]^2 \, dx + 2 \int_a^b [\Phi(x) - F(x)]^2 \, dx \leqslant \varepsilon,$$

i.e., the function $F(x)$ satisfies the condition (9.2).

Returning to the proof of the completeness criterion, consider the linear combination $\sigma_n(x)$ which satisfies the inequality (9.1). Applying the inequality (9.5), we obtain

$$\begin{aligned} \int_a^b [f(x) - \sigma_n(x)]^2 \, dx &= \int_a^b [(f(x) - F(x)) + (F(x) - \sigma_n(x))]^2 \, dx \\ &\leqslant 2 \int_a^b [f(x) - F(x)]^2 \, dx + 2 \int_a^b [F(x) - \sigma_n(x)]^2 \, dx \leqslant 4\varepsilon, \end{aligned} \tag{9.6}$$

where $F(x)$ is the continuous function appearing in (9.2). We now recall that the linear combination whose coefficients are the Fourier coefficients c_k gives the least mean square error (see Sec. 5). Therefore (9.6) gives

$$\int_a^b [f(x) - \sum_{k=1}^{n} c_k \varphi_k(x)]^2 \, dx \leqslant 4\varepsilon,$$

whence by (5.6)

$$0 \leqslant \int_a^b f^2(x) \, dx - \sum_{k=1}^{n} c_k^2 \|\varphi_k\|^2 \leqslant 4\varepsilon,$$

which means that

$$0 \leqslant \int_a^b f^2(x) \, dx - \sum_{k=1}^{\infty} c_k^2 \|\varphi_k\|^2 \leqslant 4\varepsilon$$

as well. Since ε is arbitrary, this gives (7.1), i.e., the system (1.1) is complete as asserted.

*10. The Vector Analogy

Let **i**, **j**, and **k** be three space vectors of arbitrary length which are perpendicular (orthogonal) to each other. If we want to expand a given vector **r** as a sum of the form

$$\mathbf{r} = a\mathbf{i} + b\mathbf{j} + c\mathbf{k}, \tag{10.1}$$

then the scalar coefficients a, b, c are calculated as follows: Take the scalar product of both sides of the equation (10.1) first with $\mathbf{i}$, then with $\mathbf{j}$, and finally with $\mathbf{k}$. Then, since these vectors are orthogonal, we obtain

$$(\mathbf{r}, \mathbf{i}) = a|\mathbf{i}|^2,$$
$$(\mathbf{r}, \mathbf{j}) = b|\mathbf{j}|^2,$$
$$(\mathbf{r}, \mathbf{k}) = c|\mathbf{k}|^2,$$

so that

$$a = \frac{(\mathbf{r}, \mathbf{i})}{|\mathbf{i}|^2}, \qquad b = \frac{(\mathbf{r}, \mathbf{j})}{|\mathbf{j}|^2}, \qquad c = \frac{(\mathbf{r}, \mathbf{k})}{|\mathbf{k}|^2}. \tag{10.2}$$

(We use the absolute value sign to denote the length of a vector.)

Suppose now that we know the expansion (10.1) and we want to calculate the length of the vector $\mathbf{r}$. To do this, we take the scalar product of (10.1) with $\mathbf{r}$. The result is

$$|\mathbf{r}|^2 = a(\mathbf{r}, \mathbf{i}) + b(\mathbf{r}, \mathbf{j}) + c(\mathbf{r}, \mathbf{k}),$$

or, if we use the formula (10.2)

$$|\mathbf{r}|^2 = a^2|\mathbf{i}|^2 + b^2|\mathbf{j}|^2 + c^2|\mathbf{k}|^2. \tag{10.3}$$

The quantities $a|\mathbf{i}|$, $b|\mathbf{j}|$, $c|\mathbf{k}|$ are the projections of the vector $\mathbf{r}$ on the directions of the vectors $\mathbf{i}$, $\mathbf{j}$, $\mathbf{k}$. Therefore, the equation (10.3) is the familiar relation between the square of the length of a vector and its projections on three orthogonal directions. If besides the vector $\mathbf{r}$, we consider the vector

$$\mathbf{R} = A\mathbf{i} + B\mathbf{j} + C\mathbf{k} \tag{10.4}$$

and take the scalar product of $\mathbf{r}$ and $\mathbf{R}$, then (10.1) and (10.4) give

$$(\mathbf{r}, \mathbf{R}) = aA|\mathbf{i}|^2 + bB|\mathbf{j}|^2 + cC|\mathbf{k}|^2. \tag{10.5}$$

The reader who has familiarized himself with the contents of this chapter will now immediately perceive an analogy between these vector considerations and our treatment of Fourier series. In fact, let us regard every square integrable function defined on the interval $[a, b]$ as a generalized vector, and define the scalar product of two such generalized vectors by the formula

$$(\varphi, \psi) = \int_a^b \varphi(x)\psi(x)\,dx.$$

In particular, we have

$$\|\varphi\|^2 = \int_a^b \varphi^2(x)\,dx = (\varphi, \varphi).$$

Then, we regard the orthogonal system

$$\varphi_0(x), \varphi_1(x), \ldots, \varphi_n(x), \ldots \tag{10.6}$$

as a system of orthogonal vectors, in complete agreement with our definition of the scalar product. If we are given a square integrable function $f(x)$ and we want to represent $f(x)$ as a series with respect to the system (10.6), i.e.,

$$f(x) \sim c_0\varphi_0(x) + c_1\varphi_1(x) + \cdots + c_n\varphi_n(x) + \cdots,$$

then the argument which led to the relations (10.2) leads to the relations

$$c_n = \frac{(f, \varphi_n)}{\|\varphi_n\|^2} \qquad (n = 0, 1, 2, \ldots),$$

which we recognize as the formulas for the Fourier coefficients [see (2.3)]. In fact, we have gone through precisely this derivation in Sec. 2. The reader will recognize that the completeness condition (7.1) is a generalization of the formula (10.3), and that (7.2) is a generalization of (10.5).

We now make some remarks concerning the term "completeness." Since *any* three-dimensional vector $\mathbf{r}$ can be represented in the form (10.1), i.e., as a linear combination of the vectors $\mathbf{i}, \mathbf{j}, \mathbf{k}$, it is natural to call the system consisting of these three vectors *complete*. However, the situation is different if we try to represent an arbitrary three-dimensional vector $\mathbf{r}$ as a linear combination not of three orthogonal vectors but of just two vectors, say $\mathbf{i}$ and $\mathbf{j}$. Then, in general, we cannot write an equation of the form

$$\mathbf{r} = a\mathbf{i} + b\mathbf{j},$$

but the coefficients obtained from the formula (10.2) obviously satisfy the inequality

$$|\mathbf{r}|^2 \geqslant a^2|\mathbf{i}|^2 + b^2|\mathbf{j}|^2, \tag{10.7}$$

where the equality holds only if the vector $\mathbf{r}$ lies in the plane of the vectors $\mathbf{i}$ and $\mathbf{j}$. Thus, two orthogonal vectors are *not sufficient* to be used in this way to represent any space vector, and therefore we say that a system of two orthogonal vectors is *incomplete*.

Similar considerations apply to the case of expansions in Fourier series. If the system (10.6) satisfies the condition (7.1), then the system is "rich enough in functions" to permit every square integrable function to be represented by its Fourier series (in the sense of mean convergence). In this case, we say that (10.6) is a "complete system." However, if the system (10.6) does not satisfy the condition (7.1), then we say that it is *incomplete*. It can be shown that given any incomplete system, there are always square integrable functions $f(x), g(x), \ldots$ whose Fourier series *do not converge* to $f(x), g(x), \ldots$ in the mean. The reader will easily recognize that Bessel's inequality (7.1) is the analog of the inequality (10.7).

The case of a *normalized* system (10.6) corresponds to the case where the vectors $\mathbf{i}, \mathbf{j}, \mathbf{k}$ are *unit* vectors. In both cases, all the formulas simplify.

For example, the formulas (10.2) become

$$a = (\mathbf{r}, \mathbf{i}), \quad b = (\mathbf{r}, \mathbf{j}), \quad c = (\mathbf{r}, \mathbf{k})$$

(where a, b, and c now *coincide* with the projections of the vector $\mathbf{r}$ on $\mathbf{i}, \mathbf{j}, \mathbf{k}$, respectively), while the formulas for the Fourier coefficients become

$$c_n = (f, \varphi_n) \qquad (n = 0, 1, 2, \ldots).$$

The reader who is somewhat familiar with vectors in an n-dimensional space (in which case there are n pairwise orthogonal vectors) will find the analogy discussed here even more natural. However, no matter how large n is, the transition from n orthogonal vectors to an infinite number of orthogonal vectors (bear in mind that we are treating an orthogonal system of functions as a system of vectors) cannot be regarded as simply a quantitative change; in fact, a qualitative change in behavior occurs, since instead of ordinary sums, we have to deal with infinite series and convergence in the mean.

PROBLEMS

1. Give another proof of the Schwarz inequality (4.1) by considering the inequality

$$\int_a^b \int_a^b [f(x)g(y) - f(y)g(x)]^2 \, dx \, dy \geqslant 0.$$

2. Prove the inequality

$$\left(\sum_{i=1}^n a_i b_i\right)^2 \leqslant \sum_{i=1}^n a_i^2 \sum_{i=1}^n b_i^2,$$

where a_i and b_i are arbitrary real numbers.

Comment. This result, known as the *Cauchy inequality*, is the discrete analog of the Schwarz inequality.

3. Let the polynomial $P(x) = a_0 + a_1 x + \cdots + a_n x^n$ have coefficients satisfying the relation

$$\sum_{i=0}^n a_i^2 = 1.$$

Prove that

$$\int_0^1 |P(x)| \, dx \leqslant \frac{\pi}{2}.$$

Show that this inequality continues to hold if $\pi/2$ is replaced by $\pi/\sqrt{6}$.

4. Show that if $n_1, n_2, \ldots, n_k$ are distinct nonzero integers, then

$$\frac{1}{2\pi} \int_0^{2\pi} |1 + e^{in_1 x} + \cdots + e^{in_k x}| \, dx \leqslant \sqrt{k+1}.$$

Comment. It has been conjectured by the English mathematician J. E. Littlewood that there is a constant c such that the above integral $\geqslant c \log k$. This remains unproved, although the integral has been shown to be $\geqslant c (\log k)^{1/4}$.

5. Let f and g be square integrable functions. Prove that

$$\|f + g\| \leqslant \|f\| + \|g\|.$$

Comment. This result is often called the *triangle inequality*, since it is the generalization of the familiar geometrical fact that the length of any side of a triangle is $\leqslant$ the sum of the lengths of the other two sides.

6. Give an example of a sequence of functions which converges to 0 at each point of the interval $[0, 1]$, but which does not converge in the mean.

7. A system of functions $\varphi_0(x), \varphi_1(x), \ldots, \varphi_n(x), \ldots$ which is not necessarily orthogonal is said to be *complete* if every square integrable function can be approximated in the mean by a linear combination of the $\varphi_i(x)$, i.e., if given any square integrable function $g(x)$ and any $\varepsilon > 0$, there exist numbers $a_0, a_1, \ldots, a_n$ such that

$$\int_a^b [g(x) - (a_0\varphi_0(x) + a_1\varphi_1(x) + \cdots + a_n\varphi_n(x))]^2\,dx < \varepsilon.$$

Show that if the system $\{\varphi_i(x)\}$ is complete, then any continuous function which is orthogonal to all the functions of the system must be zero. (Cf. the corresponding result in Sec. 8 for the case of orthogonal systems.)

8. Let $\varphi_0, \varphi_1, \ldots, \varphi_n, \ldots$ be a complete orthonormal system of functions. For which of the following systems is there no nonzero continuous function orthogonal to every function in the system:

a) $\varphi_0 + \varphi_1, \varphi_0 + \varphi_2, \varphi_0 + \varphi_3, \ldots;$
b) $\varphi_0 + \varphi_1, \varphi_1 + \varphi_2, \varphi_2 + \varphi_3, \ldots;$
c) $\varphi_0 + 2\varphi_1, \varphi_1 + 2\varphi_2, \varphi_2 + 2\varphi_3, \ldots$?

In Part c) we assume that the functions φ_n are continuous and uniformly bounded, i.e., $|\varphi_n(x)| \leqslant M$ for $a \leqslant x \leqslant b$.

9. A system of functions $\varphi_0(x), \varphi_1(x), \ldots, \varphi_n(x), \ldots$ is said to be *linearly independent* if given any n, there is no set of numbers $a_0, a_1, \ldots, a_n$ which are not all zero such that the linear combination $a_0\varphi_0(x) + a_1\varphi_1(x) + \cdots + a_n\varphi_n(x)$ is identically zero. Show that

a) An orthogonal system of functions is linearly independent;
b) The functions $1, x, x^2, x^3, \ldots$ are linearly independent.

10. Given a linearly independent system of functions $f_0, f_1, \ldots, f_n, \ldots$ defined on the interval $[a, b]$, we define a new system $g_0, g_1, \ldots, g_n, \ldots$ as follows:

$$g_0 = f_0,$$

$$g_1 = f_1 - \frac{(f_1, g_0)}{\|g_0\|^2} g_0,$$

$$g_2 = f_2 - \frac{(f_2, g_1)}{\|g_1\|^2} g_1 - \frac{(f_2, g_0)}{\|g_0\|^2} g_0, \quad \text{etc.}$$

[Here (f, g) denotes the scalar product of the two functions f and g, i.e.,

$$(f, g) = \int_a^b f(x)g(x)\,dx,$$

as in Sec. 10.] This is the so-called *Gram-Schmidt orthogonalization process.* Interpret the process geometrically, and show that the new system $g_0, g_1, \ldots, g_n, \ldots$ is orthogonal and that $\|g_n\|^2 \neq 0$. Apply the process to the functions

$$1, x, x^2, x^3, \ldots \qquad (-1 \leqslant x \leqslant 1),$$

thereby generating the Legendre polynomials (except for numerical factors). Show that a nonzero function is orthogonal to all the f_i if and only if it is orthogonal to all the g_i, and show that the system $\{f_i\}$ is complete (see Prob. 7) if and only if the system $\{g_i\}$ is complete.

11. The *Legendre polynomials* are defined by the formula

$$P_n(x) = \frac{1}{2^n n!}\frac{d^n}{dx^n}(x^2 - 1)^n.$$

Show that

$$\int_{-1}^{1} P_n(x)P_m(x)\,dx = \begin{cases} 0 \text{ if } n \neq m, \\ \dfrac{2}{2n+1} \text{ if } n = m. \end{cases}$$

12. Expand the following functions in Legendre polynomials:

a) $f(x) = \begin{cases} 0 & \text{for } -1 < x < 0, \\ 1 & \text{for } 0 < x < 1; \end{cases}$

b) $f(x) = |x|$.

3

CONVERGENCE OF TRIGONOMETRIC FOURIER SERIES

1. A Consequence of Bessel's Inequality

For the basic trigonometric system

$$1, \cos x, \sin x, \ldots, \cos nx, \sin nx, \ldots \tag{1.1}$$

we have

$$\|1\| = \sqrt{\int_{-\pi}^{\pi} 1 \cdot dx} = \sqrt{2\pi},$$

$$\|\cos nx\| = \sqrt{\int_{-\pi}^{\pi} \cos^2 nx\, dx} = \sqrt{\pi} \qquad (n = 1, 2, \ldots),$$

$$\|\sin nx\| = \sqrt{\int_{-\pi}^{\pi} \sin^2 nx\, dx} = \sqrt{\pi} \qquad (n = 1, 2, \ldots).$$

Let $f(x)$ be a square integrable function defined on the interval $[-\pi, \pi]$. Then, as applied to the system (1.1), Bessel's inequality (see Ch. 2, Sec. 6) becomes

$$\int_{-\pi}^{\pi} f^2(x)\, dx \geqslant \left(\frac{a_0}{2}\right)^2 \|1\|^2 + \sum_{n=1}^{\infty} (a_n^2\|\cos nx\|^2 + b_n^2\| \sin nx\|^2)$$

or

$$\int_{-\pi}^{\pi} f^2(x)\, dx \geqslant \left(\frac{a_0}{2}\right)^2 2\pi + \sum_{n=1}^{\infty} (a_n^2 + b_n^2)\, \pi,$$

so that

$$\frac{1}{\pi}\int_{-\pi}^{\pi} f^2(x)\,dx \geqslant \frac{a_0^2}{2} + \sum_{n=1}^{\infty} (a_n^2 + b_n^2). \tag{1.2}$$

From now on, we shall write Bessel's inequality for the basic trigonometric system in this form. It will be shown later (see Ch. 5, Sec. 3) that the *equality* actually holds in (1.2). However, for our present purposes, the inequality in (1.2) is sufficient.

The inequality (1.2) implies the following result concerning the convergence of the series in the right-hand side:

THEOREM. *The sum of the squares of the Fourier coefficients of any square integrable function always converges.*

It is useful to note that this series *always diverges* for any other kind of function, i.e., for any function which is not square integrable. However, we shall not prove this fact.

2. The Limit as $n \to \infty$ of the Trigonometric Integrals $\int_a^b f(x) \cos nx\,dx$ and $\int_a^b f(x) \sin nx\,dx$

It is an immediate consequence of the preceding theorem that

$$\lim_{n\to\infty} a_n = \lim_{n\to\infty} b_n = 0 \tag{2.1}$$

for any square integrable function, since the general term of a convergent series must approach zero as $n \to \infty$. But

$$a_n = \frac{1}{\pi}\int_{-\pi}^{\pi} f(x) \cos nx\,dx,$$

$$b_n = \frac{1}{\pi}\int_{-\pi}^{\pi} f(x) \sin nx\,dx,$$

and therefore

$$\lim_{n\to\infty} \int_{-\pi}^{\pi} f(x) \cos nx\,dx = \lim_{n\to\infty} \int_{-\pi}^{\pi} f(x) \sin nx\,dx = 0. \tag{2.2}$$

It follows from (2.2) that

$$\lim_{n\to\infty} \int_a^b f(x) \cos nx\,dx = \lim_{n\to\infty} \int_a^b f(x) \sin nx\,dx = 0, \tag{2.3}$$

for *any interval* $[a, b]$ *whatsoever*. (We temporarily assume that $f(x)$ is square integrable, although we shall drop this requirement soon.) To show this, suppose first that $a < b \leqslant a + 2\pi$, i.e., $b - a \leqslant 2\pi$, and set $g(x) = f(x)$ for $a \leqslant x \leqslant b$ and $g(x) = 0$ for $b < x \leqslant a + 2\pi$. The function $g(x)$ is obviously square integrable on the interval $[a, a + 2\pi]$. We now extend $g(x)$ periodically (with period 2π) over the whole x-axis. Then, by a property of periodic functions (see Ch. 1, Sec. 1)

$$\int_a^{a+2\pi} g(x) \cos nx \, dx = \int_{-\pi}^{\pi} g(x) \cos nx \, dx,$$

so that by (2.2)

$$\lim_{n\to\infty} \int_a^{a+2\pi} g(x) \cos nx \, dx = \lim_{n\to\infty} \int_{-\pi}^{\pi} g(x) \cos nx \, dx = 0.$$

On the other hand, it follows from the definition of the function $g(x)$ that

$$\int_a^{a+2\pi} g(x) \cos nx \, dx = \int_a^b f(x) \cos nx \, dx,$$

and therefore

$$\lim_{n\to\infty} \int_a^b f(x) \cos nx \, dx = 0.$$

The same argument applies to the second integral in (2.3). Finally, if $b - a > 2\pi$, then the interval $[a, b]$ can be divided into a finite number of subintervals of length no greater than 2π, for each of which the property summarized by (2.3) has already been proved. But this implies that the property also holds for the whole interval.

We now get rid of the requirement that $f(x)$ be square integrable, and also of the requirement that n be an integer. To do this, we need two lemmas, which are rather obvious from a geometrical standpoint.

LEMMA 1. *Let $f(x)$ be continuous on the interval $[a, b]$. Then, for every $\varepsilon > 0$, there exists a continuous, piecewise smooth function $g(x)$ such that*

$$|f(x) - g(x)| \leqslant \varepsilon \tag{2.4}$$

for all x ($a \leqslant x \leqslant b$).

Proof. We divide the interval $[a, b]$ into subintervals by the points

$$a = x_0 < x_1 < x_2 < \cdots < x_m = b,$$

and for $g(x)$ we take the continuous function for which $g(x_k) = f(x_k)$ $(k = 0, 1, 2, \ldots, m)$ and which is linear on each interval $[x_{k-1}, x_k]$ $(k = 1, 2, \ldots, m)$. The graph of the function $y = g(x)$ is represented by a broken line with vertices on the curve $y = f(x)$ (see Fig. 30). Obviously $g(x)$ is a piecewise smooth function. Since $f(x)$ is continuous, the subintervals into which $[a, b]$ is subdivided can be chosen small enough to make (2.4) valid for all x in $[a, b]$.

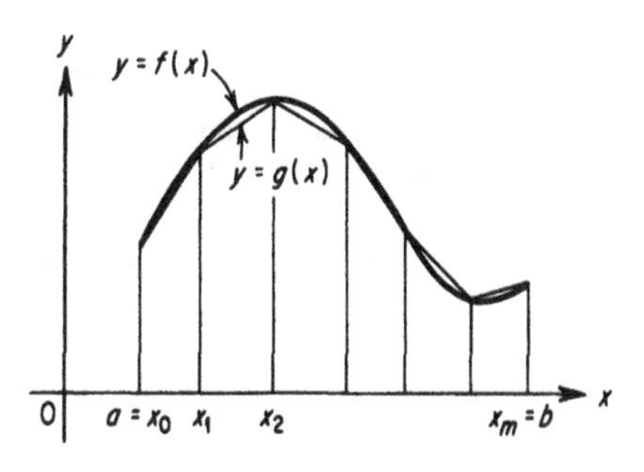

FIGURE 30

LEMMA 2. *Let $f(x)$ be absolutely integrable on the interval $[a, b]$. Then, for any $\varepsilon > 0$, there exists a continuous, piecewise smooth function $g(x)$ such that*

$$\int_a^b |f(x) - g(x)| \, dx \leqslant \varepsilon. \tag{2.5}$$

Proof.[1] The function $f(x)$ can have only a finite number of points of discontinuity, and in particular, only a finite number of points at which it becomes unbounded. We include every such point in an interval of such small length that the sum of the integrals of the function $|f(x)|$ over these intervals does not exceed $\varepsilon/3$. Next, define the auxiliary function $\Phi(x)$ as being equal to $f(x)$ outside these intervals and equal to zero inside them. $\Phi(x)$ is bounded and can have only a finite number of discontinuities, and obviously

$$\int_a^b |f(x) - \Phi(x)| \, dx \leqslant \frac{\varepsilon}{3}. \tag{2.6}$$

Next, we include each point of discontinuity of $\Phi(x)$ in an interval of such small length that the total length l of all these new intervals satisfies the condition

$$2Ml \leqslant \frac{\varepsilon}{3},$$

where M is any number such that $|\Phi(x)| < M$ for $a \leqslant x \leqslant b$.

Now, consider the *continuous* function $F(x)$ which equals $\Phi(x)$

[1] This proof is based on the same idea as the proof of the formula (9.2) of Ch. 2.

outside the intervals just described and which is "linear" inside each of them (see Fig. 29 of Ch. 2). Obviously we have

$$\int_a^b |\Phi(x) - F(x)|\, dx \leqslant 2Ml \leqslant \frac{\varepsilon}{3}. \tag{2.7}$$

Finally, according to Lemma 1, there exists a continuous, piecewise smooth function $g(x)$ for which

$$|F(x) - g(x)| \leqslant \frac{\varepsilon}{3(b-a)} \qquad (a \leqslant x \leqslant b),$$

so that

$$\int_a^b |F(x) - g(x)|\, dx \leqslant \frac{\varepsilon}{3}. \tag{2.8}$$

Then it follows from (2.6), (2.7), and (2.8) that

$$\int_a^b |f(x) - g(x)|\, dx = \int_a^b |[f(x) - \Phi(x)] + [\Phi(x) - F(x)]$$
$$+ [F(x) - g(x)]|\, dx \leqslant \int_a^b |f(x) - \Phi(x)|\, dx$$
$$+ \int_a^b |\Phi(x) - F(x)|\, dx + \int_a^b |F(x) - g(x)|\, dx \leqslant \varepsilon,$$

as was to be shown.

Remark. If $f(x)$ is an absolutely integrable *periodic* function, then $g(x)$ can be taken to be periodic.

THEOREM.[2] *For any absolutely integrable function $f(x)$, we have*

$$\lim_{m\to\infty} \int_a^b f(x) \cos mx\, dx = \lim_{m\to\infty} \int_a^b f(x) \sin mx\, dx = 0, \tag{2.9}$$

where m is not necessarily an integer.

Proof. Let ε be an arbitrarily small positive number. By Lemma 2, there exists a continuous, piecewise smooth function $g(x)$ such that

$$\int_a^b |f(x) - g(x)|\, dx \leqslant \frac{\varepsilon}{2}. \tag{2.10}$$

Consider the expression

$$\left|\int_a^b f(x) \cos mx\, dx\right| = \left|\int_a^b [f(x) - g(x)] \cos mx\, dx + \int_a^b g(x) \cos mx\, dx\right|$$
$$\leqslant \int_a^b |f(x) - g(x)|\, dx + \left|\int_a^b g(x) \cos mx\, dx\right|. \tag{2.11}$$

---

[2] This result is usually known as the *Riemann-Lebesgue lemma.* (*Translator*)

Integrating by parts, we obtain

$$\int_a^b g(x) \cos mx \, dx = \frac{1}{m} \Big[g(x) \sin mx \Big]_{x=a}^{x=b} - \frac{1}{m} \int_a^b g'(x) \sin mx \, dx.$$

The expression in brackets and the integral on the right are obviously bounded. Therefore, for sufficiently large m, we have

$$\left| \int_a^b g(x) \cos mx \, dx \right| \leqslant \frac{\varepsilon}{2}. \tag{2.12}$$

By (2.10) and (2.12), it follows from (2.11) that

$$\left| \int_a^b f(x) \cos mx \, dx \right| \leqslant \varepsilon$$

for all sufficiently large m, i.e.,

$$\lim_{m \to \infty} \int_a^b f(x) \cos mx \, dx = 0.$$

The same argument applies to the second integral in (2.9), and the theorem is proved.

Recalling the formulas for the Fourier coefficients, we can phrase this theorem as follows: *The Fourier coefficients of any absolutely integrable function approach zero as $n \to \infty$.* At the beginning of this section, we proved this property for any square integrable function, and we have now extended it to the case of any absolutely integrable function. It should be noted that if we drop the requirement that the function be absolutely integrable, its Fourier coefficients may not converge to zero as $n \to \infty$.

3. Formula for the Sum of Cosines. Auxiliary Integrals

We now show that

$$\tfrac{1}{2} + \cos u + \cos 2u + \cdots + \cos nu = \frac{\sin (n + \frac{1}{2})u}{2 \sin (u/2)}. \tag{3.1}$$

To prove this, we denote the sum on the left by S. Obviously we have

$$2S \sin \frac{u}{2} = \sin \frac{u}{2} + 2 \cos u \sin \frac{u}{2} + 2 \cos 2u \sin \frac{u}{2} + \cdots + 2 \cos nu \sin \frac{u}{2}.$$

Applying the formula

$$2 \cos \alpha \sin \beta = \sin (\alpha + \beta) - \sin (\alpha - \beta)$$

to every product on the right, we obtain

$$2S \sin \frac{u}{2} = \sin \frac{u}{2} + \left(\sin \frac{3}{2} u - \sin \frac{u}{2}\right) + \left(\sin \frac{5}{2} u - \sin \frac{3}{2} u\right) + \cdots$$
$$+ \left(\sin \left(n + \frac{1}{2}\right) u - \sin \left(n - \frac{1}{2}\right) u\right) = \sin \left(n + \frac{1}{2}\right) u.$$

Therefore

$$S = \frac{\sin (n + \frac{1}{2}) u}{2 \sin (u/2)},$$

as was to be proved.

Next, we prove two more auxiliary formulas. Integrating the equality (3.1) over the interval $[-\pi, \pi]$ and dividing the result by π, we obtain

$$1 = \frac{1}{\pi} \int_{-\pi}^{\pi} \frac{\sin (n + \frac{1}{2}) u}{2 \sin (u/2)} du, \tag{3.2}$$

for any n whatsoever (since the integrals of the cosines vanish). It is easy to see that the integrand in (3.2) is even (since changing the sign of u changes the sign of both the numerator and denominator and leaves their ratio unchanged). Therefore we have

$$\frac{1}{\pi} \int_{0}^{\pi} \frac{\sin (n + \frac{1}{2}) u}{2 \sin (u/2)} du = \frac{1}{\pi} \int_{-\pi}^{0} \frac{\sin (n + \frac{1}{2}) u}{2 \sin (u/2)} du = \frac{1}{2}. \tag{3.3}$$

4. The Integral Formula for the Partial Sum of a Fourier Series

Let $f(x)$ have period 2π and suppose that

$$f(x) \sim \frac{a_0}{2} + \sum_{k=1}^{\infty} (a_k \cos kx + b_k \sin kx).$$

Writing

$$s_n(x) = \frac{a_0}{2} + \sum_{k=1}^{n} (a_k \cos kx + b_k \sin kx),$$

and substituting the expressions for the Fourier coefficients, we obtain

$$s_n(x) = \frac{1}{2\pi} \int_{-\pi}^{\pi} f(t)\, dt + \frac{1}{\pi} \sum_{k=1}^{n} \left[\int_{-\pi}^{\pi} f(t) \cos kt\, dt \cdot \cos kx \right.$$
$$\left. + \int_{-\pi}^{\pi} f(t) \sin kt\, dt \cdot \sin kx \right]$$

$$= \frac{1}{\pi}\int_{-\pi}^{\pi} f(t)\left[\frac{1}{2} + \sum_{k=1}^{n} (\cos kt \cos kx + \sin kt \sin kx)\right] dt$$

$$= \frac{1}{\pi}\int_{-\pi}^{\pi} f(t)\left[\frac{1}{2} + \sum_{k=1}^{n} \cos k(t - x)\right] dt,$$

or using formula (3.1)

$$s_n(x) = \frac{1}{\pi}\int_{-\pi}^{\pi} f(t)\frac{\sin [(n + \frac{1}{2})(t - x)]}{2 \sin [\frac{1}{2}(t - x)]}\, dt.$$

We now change variables by setting $t - x = u$. The result is

$$s_n(x) = \frac{1}{\pi}\int_{-\pi-x}^{\pi-x} f(x + u)\frac{\sin (n + \frac{1}{2})u}{2 \sin (u/2)}\, du.$$

Now, the functions $f(x + u)$ and

$$\frac{\sin (n + \frac{1}{2})u}{2 \sin (u/2)}$$

are periodic in the variable u, with period 2π [see (3.1)], and the interval $[-\pi - x, \pi - x]$ is of length 2π. Therefore, the integral over this interval is the same as the integral over the interval $[-\pi, \pi]$ (see Ch. 1, Sec. 1) and we obtain

$$s_n(x) = \frac{1}{\pi}\int_{-\pi}^{\pi} f(x + u)\frac{\sin (n + \frac{1}{2})u}{2 \sin (u/2)}\, du. \tag{4.1}$$

This *integral formula* for the partial sum of a Fourier series allows us to establish conditions under which the convergence of the series to $f(x)$ can be guaranteed.

5. Right-Hand and Left-Hand Derivatives

Suppose that the function $f(x)$ is continuous *from the right* at x, i.e., $f(x + 0) = f(x)$. Then, we say that $f(x)$ has a *right-hand derivative* at the point x if the limit

$$\lim_{\substack{u\to 0\\ u>0}} \frac{f(x + u) - f(x)}{u} = f'_+(x) \tag{5.1}$$

exists and is finite. If $f(x)$ is continuous *from the left* at x, i.e., $f(x - 0) = f(x)$, and if the limit

$$\lim_{\substack{u\to 0\\ u<0}} \frac{f(x + u) - f(x)}{u} = f'_-(x) \tag{5.2}$$

exists and is finite, then we say that $f(x)$ has a *left-hand derivative* at x.

In the case where $f'_+(x) = f'_-(x)$, the function $f(x)$ obviously has an ordinary derivative at x, which is equal to the common value of the right-hand and left-hand derivatives at x, i.e., the curve $y = f(x)$ has a tangent at the point with abcissa x. In the case where $f'_+(x)$ and $f'_-(x)$ both exist but are unequal, the curve $f(x)$ has a "corner" but we can still speak of right-hand and left-hand tangents (as indicated by the arrows in Fig. 31).

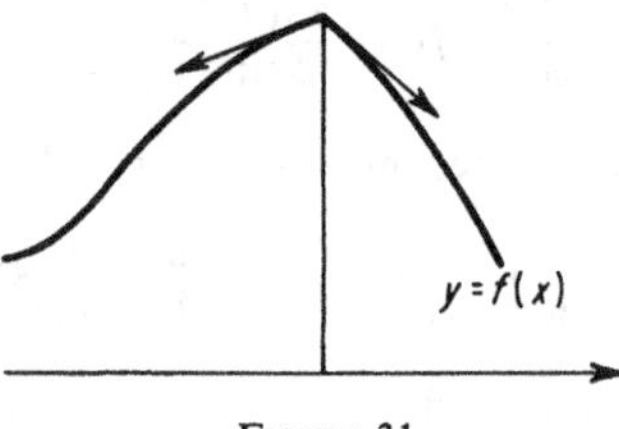

FIGURE 31

Now let x be a point where $f(x)$ has a jump discontinuity. Then, if instead of (5.1), the limit

$$\lim_{\substack{u\to 0\\ u>0}} \frac{f(x+u) - f(x+0)}{u} = f'_+(x) \tag{5.3}$$

exists and is finite, we again say that $f(x)$ has a right-hand derivative at x. Similarly, if instead of (5.2), the limit

$$\lim_{\substack{u\to 0\\ u<0}} \frac{f(x+u) - f(x-0)}{u} = f'_-(x) \tag{5.4}$$

exists and is finite, we say that $f(x)$ has a left-hand derivative at x. The existence of a right-hand derivative at a point of discontinuity $x = x_0$ is equivalent to the existence of a tangent at $x = x_0$ to the curve $y = f_+(x)$, equal to $f(x)$ for $x > x_0$ and equal to $f(x_0 + 0)$ for $x = x_0$. (Thus, the function $f_+(x)$ is defined only for $x \geqslant x_0$.) In just the same way, the existence of a left-hand derivative at $x = x_0$ is equivalent to the existence of a tangent at $x = x_0$ to the curve $y = f_-(x)$, equal to $f(x)$ for $x < x_0$ and equal to $f(x_0 - 0)$ for $x = x_0$. (The function $f_-(x)$ is defined only for $x \leqslant x_0$.)

As an example, consider the function

$$f(x) = \begin{cases} -x^3 & \text{for} \quad x < 1 \\ 0 & \text{for} \quad x = 1 \\ \sqrt{x} & \text{for} \quad x > 1, \end{cases}$$

shown in Fig. 32. This function has a jump discontinuity at $x = 1$, and obviously

$$f_+(x) = \sqrt{x} \quad \text{for} \quad x \geqslant 1,$$
$$f_-(x) = -x^3 \quad \text{for} \quad x \leqslant 1.$$

Therefore, we have

$$f'_+(1) = \left(\frac{1}{2\sqrt{x}}\right)_{x=1} = \frac{1}{2},$$

$$f'_-(1) = (-3x^2)_{x=1} = -3.$$

The corresponding tangents are indicated by arrows in the figure.

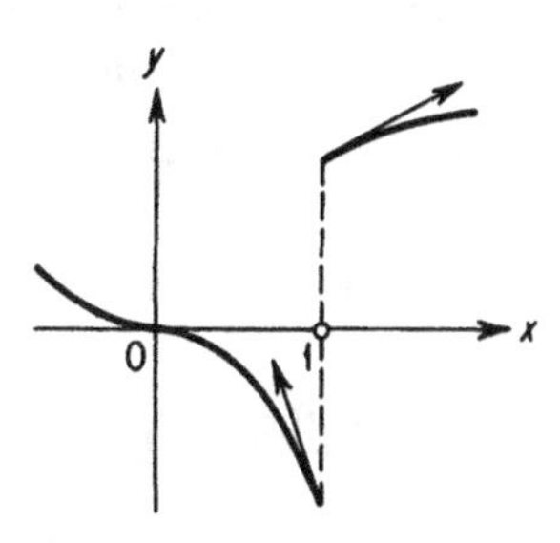

FIGURE 32

6. A Sufficient Condition for Convergence of a Fourier Series at a Continuity Point

We now prove the following important

THEOREM. *Let $f(x)$ be an absolutely integrable function of period 2π. Then, at every continuity point where the right-hand and left-hand derivatives exist, the Fourier series of $f(x)$ converges to the value $f(x)$. In particular, this is the case at every point where $f(x)$ has a derivative.*

Proof. Let x be a continuity point of $f(x)$, where both the right-hand and left-hand derivatives exist. We have to prove that

$$\lim_{n\to\infty} s_n(x) = f(x).$$

By (4.1), this is equivalent to proving that

$$\lim_{n\to\infty} \frac{1}{\pi}\int_{-\pi}^{\pi} f(x+u)\frac{\sin{(n+\frac{1}{2})u}}{2\sin{(u/2)}}\,du = f(x). \tag{6.1}$$

Moreover, since it follows from (3.2) that

$$f(x) = \frac{1}{\pi}\int_{-\pi}^{\pi} f(x)\frac{\sin{(n+\frac{1}{2})u}}{2\sin{(u/2)}}\,du,$$

we can write

$$\lim_{n\to\infty} \frac{1}{\pi}\int_{-\pi}^{\pi} [f(x+u)-f(x)]\frac{\sin{(n+\frac{1}{2})u}}{2\sin{(u/2)}}\,du = 0, \tag{6.2}$$

instead of (6.1). Thus, the problem has been reduced to proving (6.2).

We begin by proving that the function

$$\varphi(u) = \frac{f(x+u) - f(x)}{2 \sin (u/2)} = \frac{f(x+u) - f(x)}{u} \frac{u}{2 \sin (u/2)} \tag{6.3}$$

(x fixed) is absolutely integrable. Since $f(x)$ has a right-hand and a left-hand derivative at the point x, the ratio

$$\frac{f(x+u) - f(x)}{u} \tag{6.4}$$

remains bounded as $u \to 0$. In other words, there exists a number $\delta > 0$ such that

$$\left| \frac{f(x+u) - f(x)}{u} \right| \leqslant M = \text{const}$$

for $-\delta \leqslant u \leqslant \delta$. For $u \neq 0$, this ratio has discontinuities only where $f(x + u)$ is discontinuous. But $f(x + u)$ is absolutely integrable [since $f(x)$ is absolutely integrable], and hence can have only a finite number of discontinuities. Therefore, the function (6.4) is absolutely integrable on $[-\delta, \delta]$.

Outside $[-\delta, \delta]$, the ratio (6.4) is absolutely integrable, being the product of the absolutely integrable function $f(x + u) - f(x)$ and the bounded function $1/u$ ($|u| \geqslant \delta$ implies that $|1/u| \leqslant 1/\delta$). Thus, finally, (6.4) is absolutely integrable both on $[-\delta, \delta]$ and outside $[-\delta, \delta]$, and hence is absolutely integrable on $[-\pi, \pi]$.

On the other hand, the function

$$\frac{u}{2 \sin (u/2)} \tag{6.5}$$

is continuous for $u \neq 0$ and approaches 1 as $u \to 0$,[3] and therefore is a bounded continuous function, which is undefined only for $u = 0$. Thus, the function $\varphi(u)$ defined in (6.3) is absolutely integrable, being the product of the absolutely integrable function (6.4) and the bounded function (6.5).

Now that we have proved that $\varphi(u)$ is absolutely integrable, we note that

$$\int_{-\pi}^{\pi} [f(x+u) - f(x)] \frac{\sin (n + \frac{1}{2})u}{2 \sin (u/2)} du = \int_{-\pi}^{\pi} \varphi(u) \sin (n + \tfrac{1}{2})u \, du.$$

Then, the desired relation (6.2) follows by using (2.9).

---

[3] Use the familiar formula

$$\lim_{\alpha \to 0} \frac{\sin \alpha}{\alpha} = 1.$$

7. A Sufficient Condition for Convergence of a Fourier Series at a Point of Discontinuity

Next we prove the following

THEOREM. *Let $f(x)$ be an absolutely integrable function of period 2π. Then, at every point of discontinuity where $f(x)$ has a right-hand and a left-hand derivative, the Fourier series of $f(x)$ converges to the value*

$$\frac{f(x + 0) + f(x - 0)}{2}.$$

Proof. According to (4.1), we have to prove that

$$\lim_{n\to\infty} \frac{1}{\pi}\int_{-\pi}^{\pi} f(x + u)\frac{\sin (n + \frac{1}{2})u}{2 \sin (u/2)}\,du = \frac{f(x + 0) + f(x - 0)}{2}.$$

It is sufficient to show that

$$\lim_{n\to\infty} \frac{1}{\pi}\int_{0}^{\pi} f(x + u)\frac{\sin (n + \frac{1}{2})u}{2 \sin (u/2)}\,du = \frac{f(x + 0)}{2}, \tag{7.1}$$

$$\lim_{n\to\infty} \frac{1}{\pi}\int_{-\pi}^{0} f(x + u)\frac{\sin (n + \frac{1}{2})u}{2 \sin (u/2)}\,du = \frac{f(x - 0)}{2}. \tag{7.2}$$

We shall only give the proof of (7.1), since the argument leading to (7.2) is essentially the same.

According to (3.3)

$$\frac{f(x + 0)}{2} = \frac{1}{\pi}\int_{0}^{\pi} f(x + 0)\frac{\sin (n + \frac{1}{2})u}{2 \sin (u/2)}\,du.$$

Therefore, we can prove the relation

$$\lim_{n\to\infty} \frac{1}{\pi}\int_{0}^{\pi} [f(x + u) - f(x + 0)]\frac{\sin (n + \frac{1}{2})u}{2 \sin (u/2)}\,du = 0, \tag{7.3}$$

instead of (7.1). First we prove the absolute integrability on the interval $[0, \pi]$ of the following function of the variable u:

$$\varphi(u) = \frac{f(x + u) - f(x + 0)}{2 \sin (u/2)} = \frac{f(x + u) - f(x + 0)}{u}\,\frac{u}{2 \sin (u/2)}.$$

Since $f(x)$ has a right-hand derivative at x, the ratio

$$\frac{f(x + u) - f(x + 0)}{u} \qquad (u > 0) \tag{7.4}$$

remains bounded as $u \to 0$.[4] From this we conclude [just as in the case

[4] To prove (7.2), we have to consider

$$\frac{f(x + u) - f(x - 0)}{u} \qquad (u < 0)$$

instead of (7.4).

of the ratio (6.4) of Sec. 6] that (7.4) is absolutely integrable on $[0, \pi]$. Then, since the function

$$\frac{u}{2 \sin (u/2)}$$

is bounded, the function $\varphi(u)$ is absolutely integrable on $[0, \pi]$. But

$$\int_0^\pi [f(x + u) - f(x + 0)] \frac{\sin (n + \frac{1}{2})u}{2 \sin (u/2)} du = \int_0^\pi \varphi(u) \sin (n + \tfrac{1}{2})u \, du,$$

and therefore the desired relation (7.3) follows by using (2.9).

8. Generalization of the Sufficient Conditions Proved in Secs. 6 and 7

An analysis of the proofs given in Secs. 6 and 7 leads to the conclusion that the existence of right-hand and left-hand derivatives at the point x was needed only to be able to prove the absolute integrability of the ratio

$$\frac{f(x + u) - f(x)}{u} \tag{8.1}$$

in Sec. 6 [see (6.4) *et seq.*] and the absolute integrability of the ratios

$$\begin{aligned} &\frac{f(x + u) - f(x + 0)}{u} \qquad (u > 0), \\ &\frac{f(x + u) - f(x - 0)}{u} \qquad (u < 0) \end{aligned} \tag{8.2}$$

in Sec. 7 [see (7.4) *et seq.*], where x is fixed and the ratios are regarded as functions of u. Therefore, if we *assume* this absolute integrability (relinquishing the requirement that the right-hand and left-hand derivatives exist), we obtain the following more general convergence criterion:

THEOREM. *The Fourier series of an absolutely integrable function $f(x)$ of period 2π converges to $f(x)$ at every point of continuity for which the ratio* (8.1) *is an absolutely integrable function of the variable u, and converges to the value*

$$\frac{f(x + 0) + f(x - 0)}{2}$$

at every point of discontinuity for which both ratios (8.2) *are absolutely integrable.*

9. Convergence of the Fourier Series of a Piecewise Smooth Function (Continuous or Discontinuous)

The following theorem is a consequence of the results of Secs. 6 and 7:

THEOREM. *If $f(x)$ is an absolutely integrable function of period 2π which is piecewise smooth on the interval $[a, b]$, then for all x in $a < x < b$, the Fourier series of $f(x)$ converges to $f(x)$ at points of continuity and to the value*

$$\frac{f(x+0)+f(x-0)}{2}$$

at points of discontinuity. (The convergence may fail at $x = a$ and $x = b$.)

Proof. The theorem is a simple consequence of the fact that a piecewise smooth function on $[a, b]$ (see Ch. 1, Sec. 9) must have a right-hand and a left-hand derivative for every x in $a < x < b$, so that we need only apply the theorems of Secs. 6 and 7. This is obvious at the points where $f(x)$ has a derivative. At the corners of $f(x)$, we use L'Hospital's rule and obtain

$$\lim_{\substack{u\to 0\\ u>0}} \frac{f(x+u)-f(x)}{u} = \lim_{\substack{u\to 0\\ u>0}} f'(\xi) = f'(x+0),$$

since $x < \xi < x + u$, so that $\xi \to x$, $\xi > x$. Similarly, at a point of discontinuity of $f(x)$, we have

$$\lim_{\substack{u\to 0\\ u>0}} \frac{f(x+u)-f(x+0)}{u} = \lim_{\substack{u\to 0\\ u>0}} f'(\xi) = f'(x+0).$$

In other words, $f(x)$ has a right-hand derivative both at a corner and at a point of discontinuity. The existence of the left-hand derivative is proved in the same way.

As for the end points of the interval $[a, b]$, the conditions of the theorem imply only that the *right-hand* derivative exists at $x = a$ and that the *left-hand* derivative exists at $x = b$. Therefore, the criteria of Secs. 6 and 7 cannot be applied at these points. However, if the interval $[a, b]$ is of length 2π, then it is easily seen that $f(x)$ is piecewise smooth on the whole x-axis, since $f(x)$ is periodic. In this case, the Fourier series converges *everywhere*.

Thus, finally, we have proved the first part of the criterion formulated in Ch. 1, Sec. 10. The second part of the criterion, pertaining to the absolute and uniform convergence of the Fourier series in the case where $f(x)$ is continuous, will be proved in the next section.

10. Absolute and Uniform Convergence of the Fourier Series of a Continuous, Piecewise Smooth Function of Period 2π

Let $f(x)$ be a continuous, piecewise smooth function of period 2π. Then, the derivative $f'(x)$ exists everywhere except at the corners of $f(x)$ and is a bounded function (see Ch. 1, Sec. 9). Therefore, applying the formula for integration by parts (which is permissible because of Ch. 1, Sec. 4), we obtain

$$\begin{aligned} a_n &= \frac{1}{\pi}\int_{-\pi}^{\pi} f(x)\cos nx\,dx \\ &= \frac{1}{\pi n}\Big[f(x)\sin nx\Big]_{x=-\pi}^{x=\pi} - \frac{1}{\pi n}\int_{-\pi}^{\pi} f'(x)\sin nx\,dx, \\ b_n &= \frac{1}{\pi}\int_{-\pi}^{\pi} f(x)\sin nx\,dx \\ &= -\frac{1}{\pi n}\Big[f(x)\cos nx\Big]_{x=-\pi}^{x=\pi} + \frac{1}{\pi n}\int_{-\pi}^{\pi} f'(x)\cos nx\,dx. \end{aligned}$$

The terms in brackets in the right-hand side of both formulas vanish. Thus, denoting the Fourier coefficients of the function $f'(x)$ by a_n' and b_n', we find that

$$a_n = -\frac{b_n'}{n}, \qquad b_n = \frac{a_n'}{n} \qquad (n = 1, 2, \ldots). \tag{10.1}$$

Since $f'(x)$ is bounded and hence square integrable, it follows from the theorem of Sec. 1 that the series

$$\sum_{n=1}^{\infty} (a_n'^2 + b_n'^2) \tag{10.2}$$

converges.

Next, consider the obvious inequalities

$$\left(|a_n'| - \frac{1}{n}\right)^2 = a_n'^2 - \frac{2|a_n'|}{n} + \frac{1}{n^2} \geqslant 0,$$

$$\left(|b_n'| - \frac{1}{n}\right)^2 = b_n'^2 - \frac{2|b_n'|}{n} + \frac{1}{n^2} \geqslant 0,$$

which imply

$$\frac{|a_n'|}{n} + \frac{|b_n'|}{n} \leqslant \frac{1}{2}(a_n'^2 + b_n'^2) + \frac{1}{n^2} \qquad (n = 1, 2, \ldots).$$

On the right appears the general term of a convergent series. Therefore the series

$$\sum_{n=1}^{\infty} \left(\frac{|a_n'|}{n} + \frac{|b_n'|}{n}\right)$$

also converges. But then it follows from (10.1) that *for any continuous, piecewise smooth function, the series*

$$\sum_{n=1}^{\infty} (|a_n| + |b_n|) \tag{10.3}$$

converges.

Remark. To prove the convergence of the series (10.3), we used only the convergence of the series (10.2); this series always converges when $f'(x)$ is square integrable [$f'(x)$ may not exist at certain points[5]]. Therefore, (10.3) also converges in this case [when $f(x)$ is a continuous function of period 2π].

We now consider a very simple but very important fact. Suppose we are given a trigonometric series

$$\frac{a_0}{2} + \sum_{n=1}^{\infty} (a_n \cos nx + b_n \sin nx) \tag{10.4}$$

which is not assumed in advance to be the Fourier series of any function. Then the following result holds:

THEOREM 1. *If the series*

$$\sum_{n=1}^{\infty} (|a_n| + |b_n|) \tag{10.5}$$

converges, then the series (10.4) *converges absolutely and uniformly, and therefore has a continuous sum of which it is the Fourier series* (see Theorem 1 of Ch. 1, Sec. 6).

Proof. Since

$$\begin{aligned} |a_n \cos nx + b_n \sin nx| &\leqslant |a_n \cos nx| + |b_n \sin nx| \\ &\leqslant |a_n| + |b_n|, \end{aligned}$$

the terms of the trigonometric series (10.4) do not exceed the terms of the convergent numerical series (10.5). Then, the theorem follows from Weierstrass' M-test (Ch. 1, Sec. 4).

This theorem and the theorem of Sec. 9 imply the following

THEOREM 2. *The Fourier series of a continuous, piecewise smooth function $f(x)$ of period 2π converges to $f(x)$ absolutely and uniformly.*

This proves the second part of the convergence criterion of Ch. 1, Sec. 10. It follows from this theorem that a continuous, piecewise smooth function

[5] I.e., $f'(x)$ may not exist at a finite number of points in each period.

$f(x)$ of period 2π can be well approximated for large n by the partial sums $s_n(x)$ of its Fourier series; in fact, this is just what uniform convergence means! (See Ch. 1, Sec. 4.)

As an illustration, consider the periodic, continuous, piecewise smooth function $f(x)$ equal to $|x|$ for $-\pi \leqslant x \leqslant \pi$. In Example 2 of Ch. 1, Sec. 13, we proved that

$$f(x) = \frac{\pi}{2} - \frac{4}{\pi}\left(\cos x + \frac{\cos 3x}{3^2} + \frac{\cos 5x}{5^2} + \cdots\right).$$

Figure 33 shows the graph of $f(x)$ and of the partial sum

$$s_5(x) = \frac{\pi}{2} - \frac{4}{\pi}\left(\cos x + \frac{\cos 3x}{3^2} + \frac{\cos 5x}{5^2}\right).$$

of its Fourier series. We see that the two curves are already quite close to each other for $n = 5$.

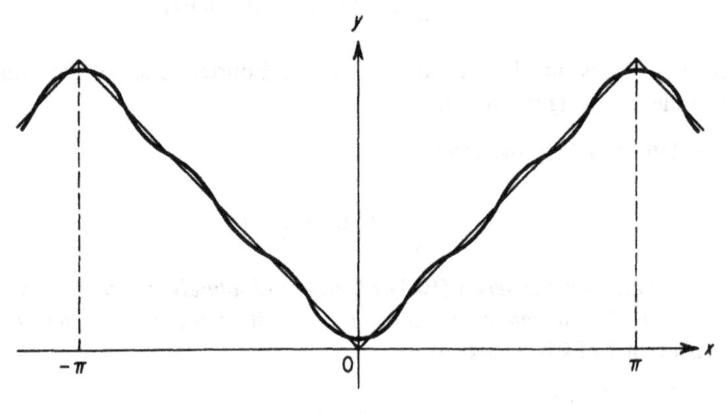

FIGURE 33

The remark preceding Theorem 1 allows us to state the following generalization of Theorem 2:

THEOREM 3. *The Fourier series of a continuous function $f(x)$ of period 2π, whose derivative (which may not exist at certain points) is square integrable, converges absolutely and uniformly to $f(x)$.*

11. Uniform Convergence of the Fourier Series of a Continuous Function of Period 2π with an Absolutely Integrable Derivative

LEMMA 1. *Let $f(x)$ be a continuous function of period 2π, which has an absolutely integrable derivative (that may not exist at certain points),* *and let* $\omega(u)$ $(\alpha \leqslant u \leqslant \beta)$ *be a function with a continuous derivative. Then, for any* $\varepsilon > 0$ *whatsoever, the inequality*

$$\left|\int_\alpha^\beta f(x+u)\omega(u)\sin mu\,du\right| \leqslant \varepsilon \tag{11.1}$$

holds for all values of x, *provided that* m *(which is not necessarily an integer) is sufficiently large.*

Proof. Integration by parts gives

$$\begin{aligned}\int_\alpha^\beta f(x+u)\omega(u)\sin mu\,du &= \frac{1}{m}\left[-f(x+u)\omega(u)\cos mu\right]_{u=\alpha}^{u=\beta} \\ &\quad + \frac{1}{m}\int_\alpha^\beta [f(x+u)\,\omega(u)]'\cos mu\,du.\end{aligned} \tag{11.2}$$

The term in brackets is obviously bounded. Since

$$[f(x+u)\omega(u)]' = f'(x+u)\omega(u) + f(x+u)\omega'(u), \tag{11.3}$$

it is easily seen that the integral on the right is also bounded. In fact, $\omega(u)$ and $f(x+u)\omega'(u)$ are bounded, i.e., have absolute values which do not exceed some constant M. But then, by (11.3)

$$\begin{aligned}\left|\int_\alpha^\beta [f(x+u)\omega(u)]'\cos mu\,du\right| &\leqslant M\int_\alpha^\beta |f'(x+u)|\,du + M(\beta-\alpha) \\ &\leqslant M\int_{-\pi}^{\pi} |f'(u)|\,du + M(\beta-\alpha) \\ &= \text{const},\end{aligned}$$

where we have used the fact that $f(x)$, and hence $|f'(x)|$, is periodic. We have also assumed that $\beta - \alpha \leqslant 2\pi$, which is not essential but is sufficient for our purposes.

Now that we have shown that the term in brackets and the integral in (11.2) are bounded, the validity of (11.1) is obvious.

LEMMA 2. *The integral*

$$I = \int_0^u \frac{\sin mt}{2\sin(t/2)}\,dt \tag{11.4}$$

is bounded for $-\pi \leqslant u \leqslant \pi$ *and any* m.

Proof. We have

$$I = \int_0^u \frac{\sin mt}{t}\,dt + \int_0^u \omega(t)\sin mt\,dt, \tag{11.5}$$

where

$$\omega(t) = \frac{1}{2\sin(t/2)} - \frac{1}{t}.$$

Applying L'Hospital's rule, we see that $\omega(t)$ and $\omega'(t)$ are continuous [if we set $\omega(0) = 0$ and $\omega'(0) = \frac{1}{24}$]. The second integral in (11.5) is obviously bounded. On the other hand, setting $mt = x$, we obtain

$$\int_0^u \frac{\sin mt}{t}\,dt = \int_0^{mu} \frac{\sin x}{x}\,dx,$$

and it is easily seen that the last integral does not exceed the area of the first "hump" of the curve $y = \sin x/x$ (see Fig. 34). Therefore, both integrals on the right in (11.5) are bounded, i.e., the integral I is bounded.

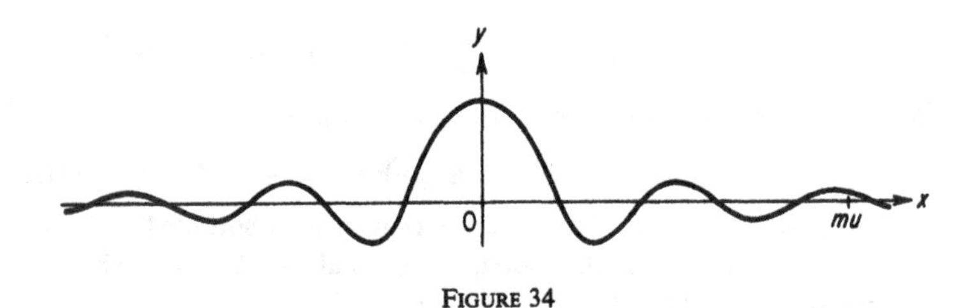

FIGURE 34

THEOREM. *The Fourier series of a continuous function $f(x)$ of period 2π with an absolutely integrable derivative (which may not exist at certain points) converges uniformly to $f(x)$ for all x.*

Proof. Consider the difference

$$s_n(x) - f(x) = \frac{1}{\pi}\int_{-\pi}^{\pi} [f(x + u) - f(x)]\frac{\sin mu}{2\sin(u/2)}\,du, \quad (11.6)$$

already calculated in Sec. 6, where we have set $m = n + \frac{1}{2}$. Choose any $\varepsilon > 0$ whatsoever, and let δ be a number between 0 and π. Then divide the integral in (11.6) into three integrals I_1, I_2, I_3 over the intervals $[-\delta, \delta]$, $[\delta, \pi]$, $[-\pi, -\delta]$, respectively. Integration by parts gives

$$I_1 = \left[(f(x + u) - f(x))\int_0^u \frac{\sin mt}{2\sin(t/2)}\,dt\right]_{u=-\delta}^{u=\delta} - \int_{-\delta}^{\delta}\left[f'(x + u)\int_0^u \frac{\sin mt}{2\sin(t/2)}\,dt\right]du.$$

For the value of the first term on the right we obtain [using the evenness of the function $(\sin mt)/2\sin(t/2)$]

$$[(f(x + \delta) - f(x)) + (f(x - \delta) - f(x))]\int_0^{\delta}\frac{\sin mt}{2\sin(t/2)}\,dt,$$

which shows that this term does not exceed $\varepsilon/2$ in absolute value for all sufficiently small δ [since $f(x)$ is continuous and the integral is bounded, by Lemma 2]. On the other hand, by Lemma 2

$$\left|\int_{-\delta}^{\delta}\left[f'(x+u)\int_0^u \frac{\sin mt}{2\sin(t/2)}\,dt\right]du\right|$$
$$\leqslant M\int_{-\delta}^{\delta}|f'(x+u)|\,du = M\int_{x-\delta}^{x+\delta} f'(t)\,dt \leqslant \frac{\varepsilon}{2}$$

where M is a constant, for all sufficiently small δ, since the integral

$$\int_{x-\delta}^{x+\delta} |f'(t)|\,dt$$

is the increment of the continuous function

$$\int_{x_0}^{x} |f'(t)|\,dt$$

(x_0 fixed) and is therefore small when δ is small.[6] Thus, finally

$$|I_1| \leqslant \varepsilon,$$

for any x whatsoever, provided that the number δ is chosen to be sufficiently small.

Furthermore

$$|I_2| \leqslant \left|\int_\delta^\pi f(x+u)\frac{\sin mu}{2\sin(u/2)}\,du\right| + \left|\int_\delta^\pi f(x)\frac{\sin mu}{2\sin(u/2)}\,du\right| \leqslant \varepsilon$$

for all x, provided that n is sufficiently large. This follows from Lemma 1, if we set

$$\omega(u) = \frac{1}{2\sin(u/2)}$$

and $\alpha = \delta$, $\beta = \pi$. A similar inequality is obtained for the integral I_3. Combining results, we have

$$|s_n(x) - f(x)| = \frac{1}{\pi}|I_1 + I_2 + I_3| \leqslant \frac{3\varepsilon}{\pi} < \varepsilon$$

for all x, provided that n is sufficiently large.

12. Generalization of the Results of Sec. 11

What can we say about the nature of the convergence of a Fourier series if $f(x)$ is continuous and has an absolutely integrable derivative not

[6] Without loss of generality, we can assume that $-\pi \leqslant x \leqslant \pi$.

everywhere but only in some interval? We shall now concern ourselves with this question. First we strengthen Lemma 1 of Sec. 11.

LEMMA. *Let $f(x)$ be an absolutely integrable function of period 2π, and let $\omega(u)$ ($\alpha \leqslant u \leqslant \beta$) be a function with a continuous derivative. Then, for any $\varepsilon > 0$ whatsoever, the inequality*

$$\left| \int_\alpha^\beta f(x+u)\omega(u) \sin mu \, du \right| \leqslant \varepsilon, \tag{12.1}$$

holds for all x, provided that m (which is not necessarily an integer) is sufficiently large.

Proof. Let $|\omega(u)| \leqslant M$, $M = \text{const}$, and choose a continuous, piecewise smooth function $g(x)$ of period 2π which satisfies the inequality

$$\int_{-\pi}^{\pi} |f(x) - g(x)| \, dx \leqslant \frac{\varepsilon}{2M}$$

(see Lemma 2 of Sec. 2 and the remark). Then we have

$$\begin{aligned} \left| \int_\alpha^\beta f(x+u)\omega(u) \sin mu \, du \right| &= \left| \int_\alpha^\beta [f(x+u) - g(x+u)]\omega(u) \sin mu \, du \right. \\ &\quad + \left. \int_\alpha^\beta g(x+u)\omega(u) \sin mu \, du \right| \leqslant \int_\alpha^\beta |[f(x+u) - g(x+u)]\omega(u)| \, du \\ &\quad + \left| \int_\alpha^\beta g(x+u)\omega(u) \sin mu \, du \right|. \end{aligned} \tag{12.2}$$

If m is large enough, then by Lemma 1 of Sec. 11, the last integral does not exceed $\varepsilon/2$. On the other hand

$$\begin{aligned} &\int_\alpha^\beta |[f(x+u) - g(x+u)]\omega(u)| \, du \\ &\qquad \leqslant M \int_\alpha^\beta |f(x+u) - g(x+u)| \, du \leqslant M \int_{-\pi}^{\pi} |f(x) - g(x)| \, dx \leqslant \frac{\varepsilon}{2}, \end{aligned}$$

where we have used the periodicity of the difference $f(x) - g(x)$ and have assumed that $\beta - \alpha \leqslant 2\pi$. Thus, (12.2) implies (12.1).

THEOREM. *Let $f(x)$ be an absolutely integrable function of period 2π, which is continuous and has an absolutely integrable derivative on some interval $[a, b]$. (The derivative may not exist at certain points.) Then, the Fourier series of $f(x)$ converges uniformly to $f(x)$ on every interval $[a + \delta, b - \delta]$ ($\delta > 0$).*

Proof. If the length of the interval $[a, b]$ is not less than 2π, it is clear that $f(x)$ is continuous for all x and has an absolutely integrable derivative, so that by the theorem of Sec. 11, the convergence of its

Fourier series is uniform on the whole x-axis. Thus, we assume that the length of $[a, b]$ is less than 2π.

We introduce an auxiliary continuous function $F(x)$ of period 2π, which equals $f(x)$ for $a \leqslant x \leqslant b$, equals $f(a)$ for $x = a + 2\pi$, and is "linear" on the interval $[b, a + 2\pi]$ (see Fig. 35). Outside $[a, a + 2\pi]$, the values of $F(x)$ are obtained by periodic extension. It is easy to see that $F(x)$ has an absolutely integrable derivative.

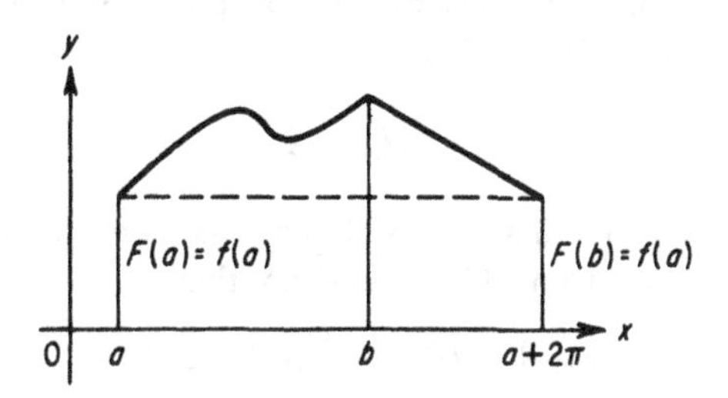

FIGURE 35

Next, we set $\Phi(x) = f(x) - F(x)$. This function is absolutely integrable and $\Phi(x) = 0$ for $a \leqslant x \leqslant b$. Obviously

$$f(x) = F(x) + \Phi(x)$$

and

$$\begin{aligned} s_n(x) - f(x) &= \frac{1}{\pi}\int_{-\pi}^{\pi} [F(x+u) - F(x)] \frac{\sin mu}{2\sin(u/2)}\, du \\ &\quad + \frac{1}{\pi}\int_{-\pi}^{\pi} [\Phi(x+u) - \Phi(x)] \frac{\sin mu}{2\sin(u/2)}\, du \qquad (12.3) \\ &= I_1 + I_2, \end{aligned}$$

where we have set $m = n + \frac{1}{2}$. Let $\varepsilon > 0$ be arbitrary. By the theorem of Sec. 11, the Fourier series of $F(x)$ converges uniformly to $F(x)$, so that

$$|I_1| \leqslant \frac{\varepsilon}{2} \qquad (12.4)$$

for all x, provided that n is sufficiently large.

Now let $a + \delta \leqslant x \leqslant b - \delta$. Then $\Phi(x) = 0$ and therefore

$$I_2 = \frac{1}{\pi}\int_{-\pi}^{\pi} \Phi(x+u) \frac{\sin mu}{2\sin(u/2)}\, du.$$

If $-\delta \leqslant u \leqslant \delta$, we have

$$a \leqslant x + u \leqslant b$$

for the values of x under consideration, and hence

$$\Phi(x + u) = 0.$$

Therefore

$$I_2 = \frac{1}{\pi}\int_{-\pi}^{-\delta} \Phi(x + u)\frac{\sin mu}{2\sin(u/2)}\,du + \frac{1}{\pi}\int_{\delta}^{\pi} \Phi(x + u)\frac{\sin mu}{2\sin(u/2)}\,du,$$

and it remains only to apply the lemma just proved to each of these integrals. As a result, we obtain

$$|I_2| \leqslant \frac{\varepsilon}{2} \tag{12.5}$$

for $a + \delta \leqslant x \leqslant b - \delta$ and all sufficiently large n. Finally, according to (12.3), it follows from (12.4) and (12.5) that

$$|s_n(x) - f(x)| \leqslant |I_1| + |I_2| \leqslant \varepsilon,$$

for all x in the interval $[a + \delta, b - \delta]$, provided that n is sufficiently large. This proves the theorem.

Remark. In particular, the theorem is valid for an absolutely integrable function $f(x)$ of period 2π which is continuous and piecewise smooth on the interval $[a, b]$.

To illustrate this theorem consider the piecewise smooth, even, periodic function $f(x)$ which equals $\pi/4$ for $0 < x < \pi$ and $-\pi/4$ for $-\pi < x < 0$. In Example 5 of Ch. 1, Sec. 13, it was shown that

$$f(x) = \sin x + \frac{\sin 3x}{3} + \frac{\sin 5x}{5} + \frac{\sin 7x}{7} + \cdots$$

for $x \neq k\pi$, while $f(k\pi) = 0$ $(k = 0, \pm 1, \pm 2, \ldots)$.

In Fig. 36 we show the graph of $f(x)$ and of the following partial sums of its Fourier series:

$$s_1(x) = \sin x,$$

$$s_3(x) = \sin x + \frac{\sin 3x}{3},$$

$$s_5(x) = \sin x + \frac{\sin 3x}{3} + \frac{\sin 5x}{5},$$

$$s_7(x) = \sin x + \frac{\sin 3x}{3} + \frac{\sin 5x}{5} + \frac{\sin 7x}{7}.$$

The figure clearly shows the uniform character of the approximation of the partial sums to $f(x)$ on the intervals $[-\pi + \delta, -\delta]$ and $[\delta, \pi - \delta]$ $(\delta > 0)$, on which $f(x)$ is a smooth function. Note that the number δ can be chosen

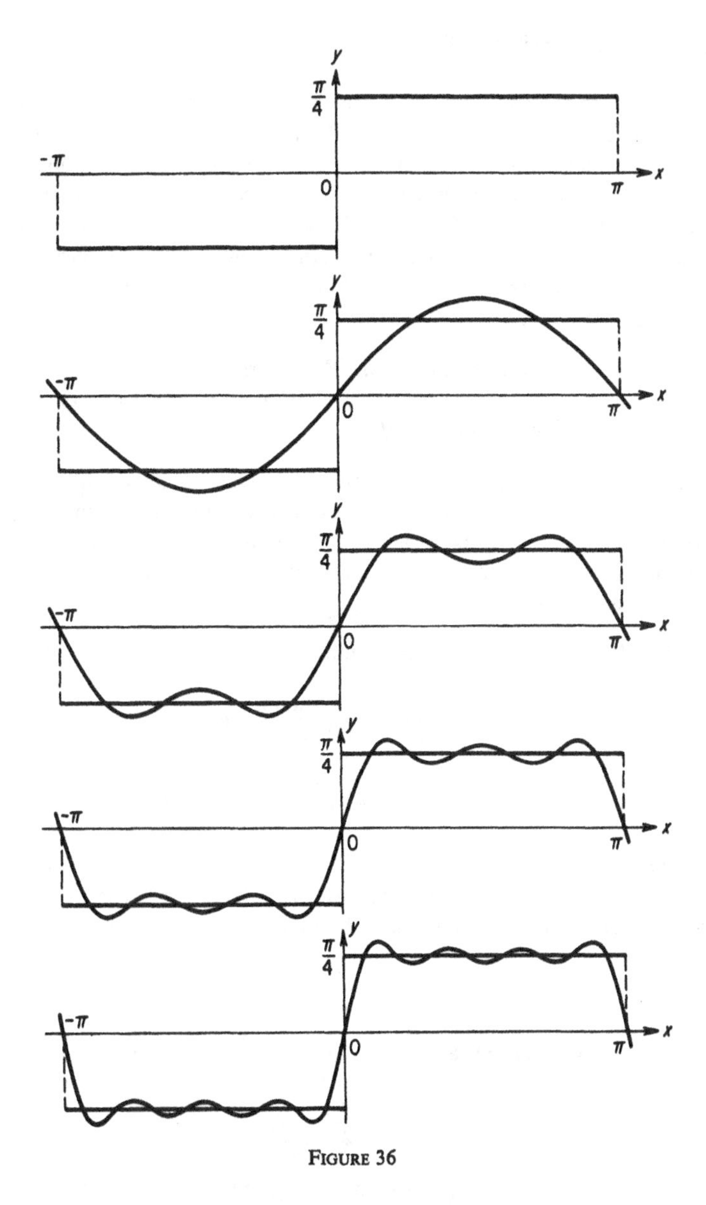

FIGURE 36

to be arbitrarily small (but not equal to zero). It is easily seen that as δ is made smaller, we need to take partial sums with higher indices in order to have a good approximation to $f(x)$ on the intervals just mentioned (more precisely, to approximate $f(x)$ with a specified degree of accuracy).

13. The Localization Principle

A change in the values of a function even on a small interval can change its Fourier coefficients considerably. However, if an absolutely integrable function $f(x)$ has a right-hand and a left-hand derivative at the point x, or is continuous and has an absolutely integrable derivative in a neighborhood of x, then, according to Secs. 6, 7, and 12, its Fourier series remains convergent no matter how the values of $f(x)$ are changed outside a neighborhood of x. This fact is a special case of the following proposition, known as the *localization principle*:

THEOREM. *The behavior of the Fourier series of an absolutely integrable function $f(x)$ at the point x depends only on the values of x in an arbitrarily small neighborhood of x.*

This proves that if the Fourier series of $f(x)$ is originally convergent at x, then no matter how we change the values of the function (while leaving it absolutely integrable) outside a neighborhood (however small) of x, the series remains convergent, while if the series was originally divergent at x, it remains divergent.

Proof. We use the integral formula for the partial sums (see Sec. 4):

$$s_n(x) = \frac{1}{\pi}\int_{-\pi}^{\pi} f(x+u)\frac{\sin mu}{2\sin(u/2)}\,du$$

$$= \frac{1}{\pi}\int_{-\delta}^{\delta} f(x+u)\frac{\sin mu}{2\sin(u/2)}\,du + I_1 + I_2,$$

where we have set $m = n + \frac{1}{2}$. Here δ is an arbitrarily small positive number, and I_1, I_2 are the integrals over the intervals $[\delta, \pi]$ and $[-\pi, -\delta]$ respectively. On these intervals, the function

$$\frac{1}{2\sin(u/2)}$$

is continuous (since $|u| \geqslant \delta$), and therefore, the function

$$\varphi(u) = \frac{f(x+u)}{2\sin(u/2)}$$

is absolutely integrable. But then according to (2.9), the integral

$$I_1 = \frac{1}{\pi}\int_{\delta}^{\pi} \varphi(u)\sin mu\,du$$

approaches zero as $m \to \infty$. The same is true of I_2. Thus, whether or not the partial sums of the Fourier series have a limit at the point x depends on the behavior as $m \to \infty$ of the integral

$$\frac{1}{\pi}\int_{-\delta}^{\delta} f(x+u)\frac{\sin mu}{2\sin(u/2)}\,du,$$

which involves only the values of the function $f(x)$ in the neighborhood $[x-\delta, x+\delta]$ of the point x. This proves the localization principle.

14. Examples of Fourier Series Expansions of Unbounded Functions

Example 1. *Let* $f(x) = -\ln|2\sin(x/2)|$.[7] This function is even and becomes infinite at $x = 2k\pi$ $(k = 0, \pm 1, \pm 2, \ldots)$. Moreover, $f(x)$ is periodic, since

$$\left|2\sin\frac{x+2\pi}{2}\right| = \left|2\sin\left(\frac{x}{2}+\pi\right)\right| = \left|-2\sin\frac{x}{2}\right| = \left|2\sin\frac{x}{2}\right|,$$

so that

$$\ln\left|2\sin\frac{x+2\pi}{2}\right| = \ln\left|2\sin\frac{x}{2}\right|.$$

The graph of $f(x)$ is shown in Fig. 37.

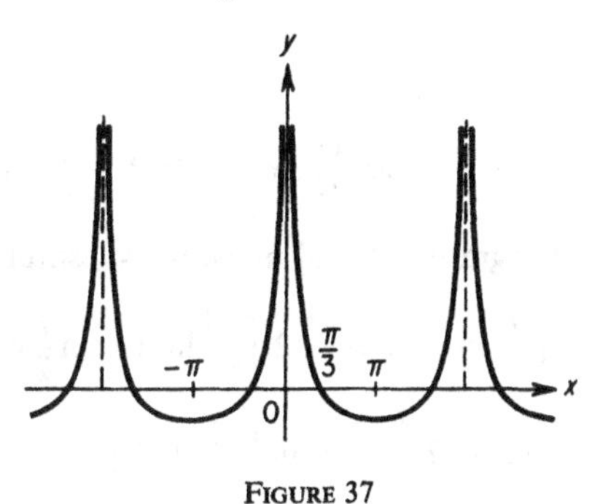

FIGURE 37

To prove that $f(x)$ is integrable, it is sufficient to prove that it is integrable on the interval $[0, \pi/3]$ (see the graph of $f(x)$). Clearly, we have

$$-\int_{\varepsilon}^{\pi/3}\ln\left|2\sin\frac{x}{2}\right|dx = -\int_{\varepsilon}^{\pi/3}\ln\left(2\sin\frac{x}{2}\right)dx$$

$$= -\left[x\ln\left(2\sin\frac{x}{2}\right)\right]_{x=\varepsilon}^{x=\pi/3} + \int_{\varepsilon}^{\pi/3}\frac{x\cos(x/2)}{2\sin(x/2)}\,dx$$

$$= \varepsilon\ln\left(2\sin\frac{\varepsilon}{2}\right) + \int_{\varepsilon}^{\pi/3}\frac{x\cos(x/2)}{2\sin(x/2)}\,dx,$$

[7] $\ln x$ denotes the *natural* logarithm of x. (*Translator*)

where we have dropped the absolute value sign, since $2\sin(x/2) > 1$ for $0 < x < \pi/3$. As $\varepsilon \to 0$, the quantity $\varepsilon \ln(2\sin(\varepsilon/2))$ approaches zero, as can easily be verified by using L'Hospital's rule, while the last integral converges to the integral

$$\int_0^{\pi/3} \frac{x\cos(x/2)}{2\sin(x/2)}\,dx,$$

which obviously has meaning, since the integrand is bounded.[8] Thus

$$\lim_{\varepsilon\to 0} \int_\varepsilon^{\pi/3} \ln\left|2\sin\frac{x}{2}\right| dx$$

exists, i.e., $f(x)$ is integrable on the interval $[0, \pi/3]$. Moreover, $f(x)$ is absolutely integrable on the interval $[0, \pi/3]$, since it does not change sign there (see Fig. 37).

Since $f(x)$ is even, we have

$$b_n = 0 \qquad (n = 1, 2, \ldots),$$

$$a_n = -\frac{2}{\pi}\int_0^\pi \ln\left(2\sin\frac{x}{2}\right)\cos nx\,dx \qquad (n = 0, 1, 2, \ldots).$$

First of all, we calculate the integral

$$I = \int_0^\pi \ln\left(2\sin\frac{x}{2}\right)dx$$

$$= \int_0^\pi \left(\ln 2 + \ln\sin\frac{x}{2}\right)dx = \pi\ln 2 + \int_0^\pi \ln\sin\frac{x}{2}\,dx.$$

We denote the last integral by Y and make the substitution $x = 2t$:

$$Y = 2\int_0^{\pi/2} \ln\sin t\,dt = 2\int_0^{\pi/2} \ln\left(2\sin\frac{t}{2}\cos\frac{t}{2}\right)dt$$

$$= \pi\ln 2 + 2\int_0^{\pi/2} \ln\sin\frac{t}{2}\,dt + 2\int_0^{\pi/2} \ln\cos\frac{t}{2}\,dt.$$

The substitution $t = \pi - u$ gives

$$2\int_0^{\pi/2} \ln\cos\frac{t}{2}\,dt = 2\int_{\pi/2}^{\pi} \ln\sin\frac{u}{2}\,du = 2\int_{\pi/2}^{\pi} \ln\sin\frac{t}{2}\,dt.$$

Therefore $Y = \pi\ln 2 + 2Y$, so that

$$Y = -\pi\ln 2.$$

This implies that $I = 0$, so that $a_0 = 0$.

---

[8] Recall that

$$\lim_{x\to 0} \frac{x}{2\sin(x/2)} = 1.$$

Furthermore, integration by parts gives

$$a_n = -\frac{2}{\pi}\left\{\left[\frac{\ln(2\sin(x/2))\sin nx}{n}\right]_{x=0}^{x=\pi} - \frac{1}{n}\int_0^{\pi}\frac{\sin nx\cos(x/2)}{2\sin(x/2)}\,dx\right\}$$
$$= \frac{1}{n\pi}\int_0^{\pi}\frac{\sin nx\cos(x/2)}{\sin(x/2)}\,dx.$$

(The first term in braces vanishes, since the indeterminacy for $x \to 0$ is easily "removed" by using L'Hospital's rule.) But

$$\sin nx \cos\frac{x}{2} = \frac{1}{2}[\sin(n + \tfrac{1}{2})x + \sin(n - \tfrac{1}{2})x],$$

and therefore

$$a_n = \frac{1}{n\pi}\int_0^{\pi}\frac{\sin(n+\frac{1}{2})x}{2\sin(x/2)}\,dx + \frac{1}{n\pi}\int_0^{\pi}\frac{\sin(n-\frac{1}{2})x}{2\sin(x/2)}\,dx,$$

which by formula (3.3) of Sec. 3 gives

$$a_n = \frac{1}{n} \qquad (n = 1, 2, \ldots).$$

Since the function $f(x)$ is obviously differentiable for $x \neq 2k\pi$ $(k = 0, \pm 1, \pm 2, \ldots)$, it follows from the theorem of Sec. 6 that

$$-\ln\left|2\sin\frac{x}{2}\right| = \cos x + \frac{\cos 2x}{2} + \frac{\cos 3x}{3} + \cdots \tag{14.1}$$

for $x \neq 2k\pi$ $(k = 0, \pm 1, \pm 2, \ldots)$. It should be noted that for $x = 2k\pi$, both sides of (14.1) become infinite. Thus, in this sense, the equation (14.1) can be regarded as valid for all x.

Setting $x = \pi$ in (14.1), we obtain the familiar formula

$$\ln 2 = 1 - \tfrac{1}{2} + \tfrac{1}{3} - \tfrac{1}{4} + \cdots.$$

Example 2. *Let* $f(x) = \ln|2\cos(x/2)|$. This function is even and goes to $-\infty$ for $x = (2k+1)\pi$ $[k = 0, \pm 1, \pm 2, \ldots]$. Setting $x = t - \pi$, we see that

$$\ln\left|2\cos\frac{2}{x}\right| = \ln\left|2\cos\left(\frac{t}{2} - \frac{\pi}{2}\right)\right| = \ln\left|2\sin\frac{t}{2}\right|,$$

i.e., the graph of the function $\ln|2\cos(x/2)|$ is obtained from that of the function $\ln|2\sin(x/2)|$ by shifting it by an amount π. To obtain the Fourier series of $f(x)$, it is sufficient to make the substitution $t = x + \pi$ in the expansion

$$\ln\left|2\sin\frac{t}{2}\right| = -\cos t - \frac{\cos 2t}{2} - \frac{\cos 3t}{3} - \cdots$$

[see (14.1)]. The result is

$$\ln\left|2\cos\frac{x}{2}\right| = \cos x - \frac{\cos 2x}{2} + \frac{\cos 3x}{3} - \cdots \tag{14.2}$$

for $x \neq (2k + 1)\pi$ $[k = 0, \pm 1, \pm 2, \ldots]$. Moreover, since both sides of (14.2) go to $-\infty$ for $x = (2k + 1)\pi$, this equation can be regarded as valid for all x.

15. A Remark Concerning Functions of Period $2l$

In discussing questions of theory in this and later chapters, we do not talk about Fourier expansions of functions of period $T = 2l$. However, the reader who is familiar with the contents of Chs. 1 and 2 will have no trouble in making the transition from the case of the "standard" period 2π to the case of any period.

PROBLEMS

1. Derive formula (3.1) from the formula

$$1 + e^{ix} + \cdots + e^{inx} = \frac{1 - e^{i(n+1)x}}{1 - e^{ix}}.$$

2. For each of the following functions, find the right-hand derivative at zerc $f'_+(0)$, if it exists, and find $\lim_{\substack{x \to 0 \\ x > 0}} f'(x) = f'(0+)$, if this limit exists:

a) $f(x) = x \sin \dfrac{1}{x}$;

b) $f(x) = x^2 \sin \dfrac{1}{x}$;

c) $f(x) = x^3 \sin \dfrac{1}{x}$.

[In each case, $f(0) = 0$.]

3. Show that the theorem of Sec. 6 can be generalized in the following way: If $f(x)$ is an absolutely integrable function of period 2π, if $f(x)$ is continuous at the point x_0, and if there are numbers $c > 0$, $\alpha > 0$ such that

$$|f(x) - f(x_0)| \leqslant c|x - x_0|^\alpha$$

for all x in some neighborhood of x_0 (cf. Ch. 1, Prob. 7), then the Fourier series of $f(x)$ converges to the value $f(x_0)$ at the point x_0.

4. Let $f(x)$, and $g(x)$ be absolutely integrable functions with period 2π, whose Fourier series are

$$f(x) \sim \sum_{n=1}^{\infty} a_n e^{inx}, \quad g(x) \sim \sum_{n=1}^{\infty} b_n e^{inx},$$

and let

$$h(x) = \frac{1}{2\pi}\int_0^{2\pi} f(x-t)g(t)\,dt,$$

$$h(x) \sim \sum_{n=1}^{\infty} c_n e^{inx}.$$

Show that

$$\frac{1}{2\pi}\int_0^{2\pi} |h(x)|\,dx \leqslant \left(\frac{1}{2\pi}\int_0^{2\pi} |f(x)|\,dx\right)\left(\frac{1}{2\pi}\int_0^{2\pi} |g(x)|\,dx\right)$$

and that $c_n = a_n b_n$. In particular, if $f(x)$ and $g(x)$ are square integrable, show that

$$\sum_{n=1}^{\infty} |c_n| < \infty,$$

i.e., that the Fourier series of $h(x)$ is absolutely convergent.

5. Show that the Fourier series

$$\tfrac{1}{2}a_0 + \sum_{n=1}^{\infty} (a_n \cos nx + b_n \sin nx)$$

can be written in the form

$$\tfrac{1}{2}\rho_0 + \sum_{n=1}^{\infty} \rho_n \cos(nx + \theta_n),$$

where $\rho_n = \sqrt{a_n^2 + b_n^2}$. Express θ_n in terms of a_n and b_n.

6. Suppose that $\rho_n \geqslant 0$ and that

$$\sum_{n=1}^{\infty} \rho_n |\cos(nx + \theta_n)| \leqslant M \tag{I}$$

for $a \leqslant x \leqslant b$. Show that

$$\sum_{n=1}^{\infty} \rho_n < \infty.$$

Hint. Integrate equation (I) from a to b, use the fact that

$$|\cos(nx + \theta_n)| \geqslant \cos^2(nx + \theta_n) = \tfrac{1}{2} + \tfrac{1}{2}\cos 2nx \cos 2\theta_n - \tfrac{1}{2}\sin 2nx \sin 2\theta_n,$$

and then use equation (2.9).

7. According to equation (13.7) of Ch. 1, the function

$$f(x) = \tfrac{1}{2}(\pi - x) \qquad (0 < x < 2\pi)$$

has the Fourier expansion

$$f(x) = \sum_{k=1}^{\infty} \frac{\sin kx}{k}.$$

Let $s_n(x)$ be the nth partial sum of the series, i.e.,

$$s_n(x) = \sum_{k=1}^{n} \frac{\sin kx}{k},$$

and let

$$D_n(x) = \frac{\sin (n + \frac{1}{2})x}{2 \sin (x/2)}.$$

Show that

a) $\displaystyle \frac{x}{2} + s_n(x) = \int_0^x D_n(t)\, dt;$

b) $\displaystyle \int_0^x D_n(t)\, dt = \int_0^x \frac{\sin nt}{t}\, dt + \omega_n(x) = \int_0^{nx} \frac{\sin t}{t}\, dt + \omega_n(x),$

where $\omega_n(x) \to 0$ as $n \to \infty$;

c) $\displaystyle \int_0^\infty \frac{\sin t}{t}\, dt = \frac{\pi}{2}.$

8. Using the notation of the preceding problem, show that

$$\lim_{n\to\infty} s_n\left(\frac{\pi}{n}\right) = \int_0^\pi \frac{\sin t}{t}\, dt > \int_0^\infty \frac{\sin t}{t}\, dt = \frac{\pi}{2}.$$

Comment. Thus, near $x = 0$ (a point of discontinuity), the partial sums of the Fourier series of the function $f(x)$ of the preceding problem exceed the value of the function at $x = 0$ by the amount

$$\int_0^\pi \frac{\sin t}{t}\, dt - \frac{\pi}{2} \approx 0.28.$$

This illustrates the so-called *Gibbs phenomenon*, according to which the Fourier series of a discontinuous function "overshoots" its limiting value at a point of jump discontinuity.

4

TRIGONOMETRIC SERIES WITH DECREASING COEFFICIENTS

1. Abel's Lemma

The following result will be needed below:

ABEL'S LEMMA. *Let*

$$u_0 + u_1 + u_2 + \cdots + u_n + \cdots$$

be a numerical series (with real or complex terms), whose partial sums σ_n satisfy the condition

$$|\sigma_n| \leqslant M,$$

where M is a constant. Then, if the positive numbers $\alpha_0, \alpha_1, \alpha_2, \ldots, \alpha_n, \ldots$ approach zero monotonically, the series

$$\alpha_0 u_0 + \alpha_1 u_1 + \alpha_2 u_2 + \cdots + \alpha_n u_n + \cdots \tag{1.1}$$

converges, and its sum s satisfies the inequality

$$|s| \leqslant M\alpha_0 \tag{1.2}$$

Proof. Let

$$s_n = \alpha_0 u_0 + \alpha_1 u_1 + \cdots + \alpha_n u_n;$$

then, since $u_0 = \sigma_0$ and $u_n = \sigma_n - \sigma_{n-1}$ $(n = 1, 2, \ldots)$, we have

$$s_n = \alpha_0\sigma_0 + \alpha_1(\sigma_1 - \sigma_0) + \alpha_2(\sigma_2 - \sigma_1) + \cdots$$
$$+ \alpha_n(\sigma_n - \sigma_{n-1})$$

or

$$s_n = \sigma_0(\alpha_0 - \alpha_1) + \sigma_1(\alpha_1 - \alpha_2) + \cdots + \sigma_{n-1}(\alpha_{n-1} - \alpha_n) + \sigma_n\alpha_n.$$

It follows that

$$s_n - \sigma_n\alpha_n = \sigma_0(\alpha_0 - \alpha_1) + \sigma_1(\alpha_1 - \alpha_2) + \cdots + \sigma_{n-1}(\alpha_{n-1} - \alpha_n). \tag{1.3}$$

Now consider the series

$$\sigma_0(\alpha_0 - \alpha_1) + \sigma_1(\alpha_1 - \alpha_2) + \cdots + \sigma_{n-1}(\alpha_{n-1} - \alpha_n) + \cdots \tag{1.4}$$

This series converges, since the absolute values of its terms do not exceed the corresponding terms of the following convergent series with nonnegative terms:

$$M(\alpha_0 - \alpha_1) + M(\alpha_1 - \alpha_2) + \cdots + M(\alpha_{n-1} - \alpha_n) + \cdots$$
$$= M(\alpha_0 - \alpha_1 + \alpha_1 - \alpha_2 + \cdots + \alpha_{n-1} - \alpha_n + \cdots) = M\alpha_0.$$

The right-hand side of (1.3) is the nth partial sum of the series (1.4), and therefore, as $n \to \infty$, it approaches a definite limit, whose absolute value does not exceed the number $M\alpha_0$. But then the left-hand side of (1.3) also approaches a limit as $n \to \infty$, and

$$\left|\lim_{n\to\infty} (s_n - \sigma_n\alpha_n)\right| \leqslant M\alpha_0.$$

Since

$$|\sigma_n\alpha_n| \leqslant M\alpha_n,$$

we have

$$\lim_{n\to\infty} \sigma_n\alpha_n = 0.$$

Therefore

$$\lim_{n\to\infty} s_n = s$$

exists, i.e., the series (1.1) converges, and s satisfies the inequality (1.2).

2. Formula for the Sum of Sines. Auxiliary Inequalities

We now prove the formula

$$\sin x + \sin 2x + \cdots + \sin nx = \frac{\cos (x/2) - \cos (n + \frac{1}{2})x}{2 \sin (x/2)}. \tag{2.1}$$

To do so, we denote the sum on the left by S. Then, we obviously have

$$2S\sin\frac{x}{2} = 2\sin x\sin\frac{x}{2} + 2\sin 2x\sin\frac{x}{2} + \cdots + 2\sin nx\sin\frac{x}{2}.$$

Using the formula

$$2\sin\alpha\sin\beta = \cos(\alpha-\beta) - \cos(\alpha+\beta),$$

we obtain

$$\begin{aligned}2S\sin\frac{x}{2} &= \left(\cos\frac{x}{2} - \cos\frac{3}{2}x\right) + \left(\cos\frac{3}{2}x - \cos\frac{5}{2}x\right) + \cdots \\ &\quad + \left(\cos\left(n-\frac{1}{2}\right)x - \cos\left(n+\frac{1}{2}\right)x\right) \\ &= \cos\frac{x}{2} - \cos\left(n+\frac{1}{2}\right)x.\end{aligned}$$

It follows that

$$S = \frac{\cos(x/2) - \cos(n+\frac{1}{2})x}{2\sin(x/2)},$$

which proves (2.1).

Since obviously

$$\left|\frac{\cos(x/2) - \cos(n+\frac{1}{2})x}{2\sin(x/2)}\right| \leqslant \frac{|\cos(x/2)| + |\cos(n+\frac{1}{2})x|}{|2\sin(x/2)|} < \frac{1}{|\sin(x/2)|}$$

for $x \neq 2k\pi$ $(k = 0, \pm 1, \pm 2, \ldots)$, we obtain the inequality

$$\left|\sum_{k=1}^{n}\sin kx\right| \leqslant \frac{1}{|\sin(x/2)|} \tag{2.2}$$

for $x \neq 2k\pi$. This shows that the sum of sines is bounded for every fixed x. (For $x = 2k\pi$, the sum vanishes and hence is also bounded.)

We now recall the formula

$$\tfrac{1}{2} + \cos x + \cos 2x + \cdots + \cos nx = \frac{\sin(n+\frac{1}{2})x}{2\sin(x/2)},$$

(see Ch. 3, Sec. 3), from which it follows at once that

$$\left|\tfrac{1}{2} + \sum_{k=1}^{n}\cos kx\right| \leqslant \frac{1}{|2\sin(x/2)|} \tag{2.3}$$

for $x \neq 2k\pi$ $(k = 0, \pm 1, \pm 2, \ldots)$. (For $x = 2k\pi$, the sum is obviously $n + \frac{1}{2}$, and hence is not bounded as n increases.)

3. Convergence of Trigonometric Series with Monotonically Decreasing Coefficients

Consider the two trigonometric series

$$\frac{a_0}{2} + \sum_{n=1}^{\infty} a_n \cos nx, \tag{3.1}$$

$$\sum_{n=1}^{\infty} b_n \sin nx, \tag{3.2}$$

which we do not even assume to be the Fourier series of any functions.

THEOREM 1. *If the coefficients a_n and b_n are positive and decrease monotonically to zero as $n \to \infty$,*[1] *then the series* (3.1) *and* (3.2) *converge for any x, except possibly the values $x = 2k\pi$ ($k = 0, \pm 1, \pm 2, \ldots$) in the case of the series* (3.1).

Proof. If the sums of the coefficients a_n and b_n converge, then the theorem follows from Theorem 1 of Ch. 3, Sec. 10. In the general case, consider the series

$$\tfrac{1}{2} + \cos x + \cos 2x + \cdots + \cos nx + \cdots, \tag{3.3}$$

whose partial sums $\sigma_n(x)$ are bounded for any $x \neq 2k\pi$. Then, to prove the theorem for the series (3.1), we apply Abel's lemma. By (2.2), the same argument applies to the series (3.2).

Remark. Of course, Theorem 1 and the theorems which follow remain valid if the requirement that the coefficients be nonincreasing is satisfied not for all n, but only starting from a certain value of n. In particular, the theorem is true in the case where some of the early coefficients vanish. For example, the series

$$\sum_{n=2}^{\infty} \frac{\cos nx}{\ln n}$$

converges for $x \neq 2k\pi$. For $x = 2k\pi$, this series becomes

$$\sum_{n=2}^{\infty} \frac{1}{\ln n}$$

and hence diverges.

We now make Theorem 1 more precise.

---

[1] I.e., $a_1 \geqslant a_2 \geqslant \cdots$, $b_1 \geqslant b_2 \geqslant \cdots$ and $\lim_{n\to\infty} a_n = \lim_{n\to\infty} b_n = 0$. (*Translator*)

THEOREM 2. *If the coefficients a_n and b_n are positive and decrease monotonically to zero as $n \to \infty$, then the series* (3.1) *and* (3.2) *converge uniformly on any interval* $[a, b]$ *which does not contain points of the form* $x = 2k\pi$ $(k = 0, \pm 1, \pm 2, \ldots)$.

Proof. If the sums of the coefficients a_n and b_n converge, then (3.1) and (3.2) converge uniformly on the whole x-axis by Theorem 1 of Ch. 3, Sec. 10. In the general case, since the sums of the series (3.1) and (3.2) are periodic functions, it is sufficient to prove the theorem for every interval $[a, b]$ contained in the interval $[0, 2\pi]$. The proof is the same for both series, and therefore we confine ourselves to the case of the series (3.1).

Let $\varepsilon > 0$ be arbitrary. For $a \leqslant x \leqslant b$, we consider the *remainder*

$$s(x) - s_n(x) = a_{n+1} \cos (n + 1)x + a_{n+2} \cos (n + 2)x + \cdots \quad (3.4)$$

of the series (3.1), and apply Abel's lemma. To do this, we set

$$\tau_m(x) = \sigma_{n+m}(x) - \sigma_n(x),$$

where $\sigma_n(x)$ and $\sigma_{n+m}(x)$ are partial sums of the series (3.3). Then by (2.3),

$$|\tau_m(x)| \leqslant |\sigma_{n+m}(x)| + |\sigma_n(x)| \leqslant \frac{1}{\sin (x/2)}.$$

Since $0 < a \leqslant x \leqslant b < 2\pi$, we have

$$\sin \frac{x}{2} \geqslant \mu > 0,$$

where μ is the smaller of the numbers $\sin (a/2)$ and $\sin (b/2)$ (see Fig. 38).

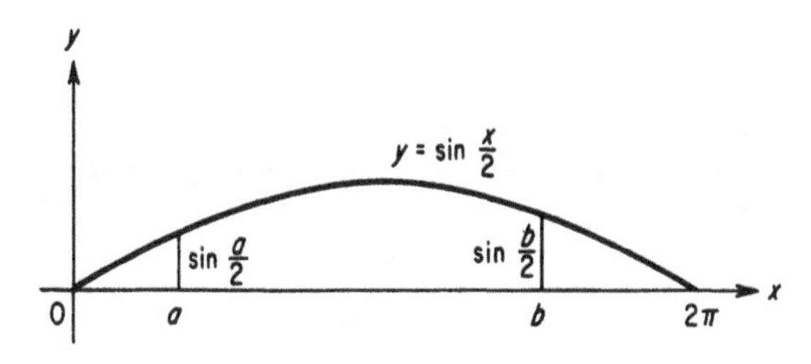

FIGURE 38

Therefore, setting $M = 1/\mu$, we find

$$|\tau_m(x)| \leqslant M = \text{const}$$

for all x in the interval $[a, b]$. Thus, Abel's lemma is applicable to the numbers $a_{n+1}, a_{n+2}, \ldots, a_{n+m}, \ldots$ and to the series (3.4), and gives

$$|s(x) - s_n(x)| \leqslant Ma_{n+1}$$

for any x in the interval $[a, b]$. Since $a_n \to 0$ as $n \to \infty$, we have

$$Ma_{n+1} \leqslant \varepsilon$$

for all sufficiently large n. In other words, for all sufficiently large n and for any x in $[a, b]$, the inequality

$$|s(x) - s_n(x)| \leqslant \varepsilon$$

holds, which proves the uniform convergence of the series (3.1).

Theorems 1 and 2 imply

THEOREM 3. *If the coefficients a_n and b_n are positive and decrease monotonically to zero as $n \to \infty$, then the functions*

$$f(x) = \frac{a_0}{2} + \sum_{n=1}^{\infty} a_n \cos nx, \quad g(x) = \sum_{n=1}^{\infty} b_n \sin nx$$

of period 2π are continuous for all x, except possibly for the values $x = 2k\pi$ $(k = 0, \pm 1, \pm 2, \ldots)$.

Proof. If the sums of the coefficients a_n and b_n converge, then the theorem follows from Theorem 1 of Ch. 3, Sec. 10. In the general case, we can include any point $x_0 \neq 2k\pi$ in an interval $[a, b]$ which does not contain points of the form $x = 2k\pi$. In this interval, the two series converge uniformly (by Theorem 2) and hence their sums are continuous (see Ch. 1, Sec. 4). In particular, they are continuous at $x = x_0$. Since x_0 is any point different from a point of the form $x = 2k\pi$, the theorem is proved.

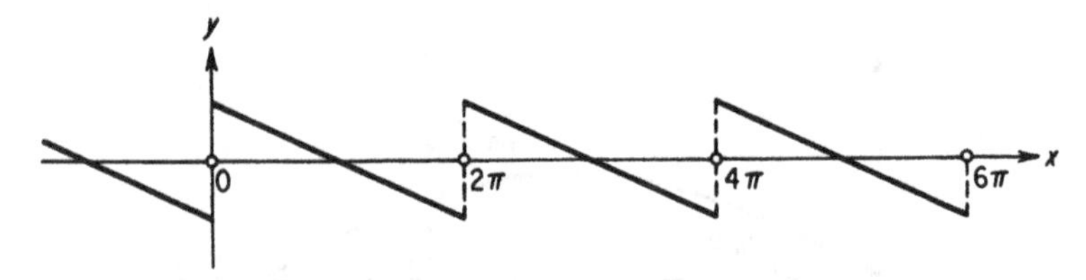

FIGURE 39

As an illustration, consider the expressions

$$f(x) = -\ln\left|2 \sin \frac{x}{2}\right| = \sum_{n=1}^{\infty} \frac{\cos nx}{n},$$

$$g(x) = \frac{\pi - x}{2} = \sum_{n=1}^{\infty} \frac{\sin nx}{n} \qquad (0 < x < 2\pi)$$

with which we are already familiar [see Example 1 of Ch. 3, Sec. 14 and formula (13.7) of Ch. 1]. The graph of $f(x)$ is shown in Fig. 37, and the graph of $g(x)$, together with its periodic extension, is shown in Fig. 39.

*4. Some Consequences of the Theorems of Sec. 3

We now present some interesting consequences of the theorems proved in Sec. 3.

THEOREM 1. *If the coefficients a_n and b_n are positive and decrease monotonically to zero as $n \to \infty$, then the series*

$$\frac{a_0}{2} + \sum_{n=1}^{\infty} (-1)^n a_n \cos nx, \tag{4.1}$$

$$\sum_{n=1}^{\infty} (-1)^n b_n \sin nx \tag{4.2}$$

have the following properties:

1) *They converge for all values of x except possibly at the points $x = (2k + 1)\pi$ $[k = 0, \pm 1, \pm 2, \ldots]$ in the case of the series (4.1);*

2) *The convergence is uniform on every interval $[a, b]$ not containing these points;*

3) *The sums of the series are continuous, except possibly at these points.*

Proof. We substitute $x = t - \pi$ in (3.1) and (3.2), thereby obtaining the series

$$\frac{a_0}{2} + \sum_{n=1}^{\infty} a_n \cos n(t - \pi) = \frac{a_0}{2} + \sum_{n=1}^{\infty} a_n[\cos n\pi \cos nt + \sin n\pi \sin nt]$$

$$= \frac{a_0}{2} - a_1 \cos t + a_2 \cos 2t - a_3 \cos 3t + \cdots,$$

$$\sum_{n=1}^{\infty} b_n \sin n(t - \pi) = \sum_{n=1}^{\infty} b_n[\cos n\pi \sin nt - \sin n\pi \cos nt]$$

$$= - b_1 \sin t + b_2 \sin 2t - b_3 \sin 3t + \cdots.$$

These are alternating series to which Theorems 1, 2, and 3 of Sec. 3 apply, provided the points $t = \pi + 2k\pi = (2k + 1)\pi$ now play the role of the points $x = 2k\pi$.

Theorem 1 obviously remains valid if in (4.1) and (4.2), we write $(-1)^{n+1}$ instead of $(-1)^n$. To illustrate Theorem 1, consider the series

$$\ln\left|2\cos\frac{x}{2}\right| = \cos x - \frac{\cos 2x}{2} + \frac{\cos 3x}{3} - \cdots$$

and

$$\frac{x}{2} = \sin x - \frac{\sin 2x}{2} + \frac{\sin 3x}{3} - \cdots \qquad (-\pi < x < \pi)$$

with which we are already familiar [see Example 2 of Ch. 3, Sec. 14 and formula (13.9) of Ch. 1].

The theorem just proved can be generalized even further. In fact, consider the series of the form

$$\begin{aligned} &a_1 \cos px + a_2 \cos(p+m)x + a_3 \cos(p+2m)x + \cdots \\ &\qquad + a_{n+1}\cos(p+nm)x + \cdots, \\ &b_1 \sin px + b_2 \sin(p+m)x + b_3 \sin(p+2m)x + \cdots \\ &\qquad + b_{n+1}\sin(p+nm)x + \cdots, \end{aligned} \tag{4.3}$$

where p and m are any numbers, and the coefficients a_n and b_n are positive and converge monotonically to zero. In both of these series, the coefficients of x form an arithmetic progression with difference m. The following are examples of series of this type:

$$\begin{aligned} &\cos x + \frac{\cos 5x}{2} + \frac{\cos 9x}{3} + \frac{\cos 13x}{4} + \cdots \qquad (p = 1, m = 4), \\ &\frac{\sin 2x}{\ln 2} + \frac{\sin 5x}{\ln 3} + \frac{\sin 8x}{\ln 4} + \frac{\sin 11x}{\ln 5} + \cdots \qquad (p = 2, m = 3). \end{aligned} \tag{4.4}$$

Observing that

$$\begin{aligned} \cos(p+nm)x &= \cos px \cos nmx - \sin px \sin nmx, \\ \sin(p+nm)x &= \sin px \cos nmx + \cos px \sin nmx, \end{aligned}$$

we can rewrite the series (4.3) as

$$\cos px \sum_{n=0}^{\infty} a_{n+1}\cos nmx - \sin px \sum_{n=0}^{\infty} a_{n+1}\sin nmx,$$

$$\sin px \sum_{n=0}^{\infty} b_{n+1}\cos nmx + \cos px \sum_{n=0}^{\infty} b_{n+1}\sin nmx.$$

If we now set $mx = t$ or $x = t/m$, we obtain

$$\begin{aligned} &\cos\frac{pt}{m} \sum_{n=0}^{\infty} a_{n+1}\cos nt - \sin\frac{pt}{m} \sum_{n=0}^{\infty} a_{n+1}\sin nt, \\ &\sin\frac{pt}{m} \sum_{n=0}^{\infty} b_{n+1}\cos nt - \cos\frac{pt}{m} \sum_{n=0}^{\infty} b_{n+1}\sin nt. \end{aligned} \tag{4.5}$$

Theorems 1, 2, and 3 of Sec. 3 are applicable to all four series appearing in (4.5). It follows that for any p, the series (4.5) converge and have continuous sums for all t, except possibly values of the form $t = 2k\pi$, and that the convergence of the series (4.5) is uniform on any interval which does not contain these values. Thus, returning to the variable x, we have

THEOREM 2. *If the coefficients a_n and b_n are positive and decrease monotonically to zero as $n \to \infty$, then the series* (4.3) *converge and have continuous sums for all values of x, except possibly the values $x = 2k\pi/m$ $(k = 0, \pm 1, \pm 2, \ldots)$, and the convergence of the series is uniform on any interval not containing these points.*

For example, for the first of the series (4.4), the "exceptional" points are $x = 2k\pi/4 = k\pi/2$ and for the second series, they are $x = 2k\pi/3$.

In the same way as we obtained Theorem 2, we can prove

THEOREM 3. *If the coefficients a_n and b_n are positive and decrease monotonically to zero as $n \to \infty$, then the series of the form*

$$a_1 \cos px - a_2 \cos (p + m)x + a_3 \cos (p + 2m)x - \cdots,$$
$$b_1 \sin px - b_2 \sin (p + m)x + b_3 \sin (p + 2m)x - \cdots$$

converge and have continuous sums for all x, except possibly

$$x = \frac{2(k + 1)\pi}{m} \qquad (k = 0, \pm 1, \pm 2, \ldots),$$

and the convergence is uniform on any interval not containing these points.

In the applications, the following theorem, which we cite without proof, is often useful:

THEOREM 4. *With the hypotheses of the preceding theorems, if the series*

$$\sum_{n=1}^{\infty} \frac{a_n}{n}, \quad \sum_{n=1}^{\infty} \frac{b_n}{n}$$

converge, then the corresponding trigonometric series define absolutely integrable functions (and are therefore the Fourier series of these functions). (See Theorem 2 of Ch. 1, Sec. 6.)

5. Applications of Functions of a Complex Variable to the Evaluation of Certain Trigonometric Series

Let $F(z)$ be an analytic (differentiable) function of the complex variable $z = x + iy$, without singularities[2] for $|z| \leqslant 1$. Then, it is well known that

---

[2] A point z is said to be a *singularity* of the function $F(z)$ if in the complex plane there exists no circle with center at z (no matter how small its radius) within which $F(z)$ is analytic.

$F(z)$ can be expanded in a power series

$$F(z) = c_0 + c_1 z + c_2 z^2 + \cdots + c_n z^n + \cdots \tag{5.1}$$

for $|z| \leqslant 1$, i.e., within the circle of radius 1 with center at the origin of the complex plane. Suppose that the coefficients of this series are real numbers, and set $z = e^{ix}$. Then (5.1) still holds, since

$$|e^{ix}| = |\cos x + i \sin x| = \sqrt{\cos^2 x + \sin^2 x} = 1.$$

Thus, for any x, we have

$$\begin{aligned} F(e^{ix}) &= c_0 + c_1 e^{ix} + c_2 e^{i2x} + \cdots + c_n e^{inx} + \cdots \\ &= c_0 + c_1 (\cos x + i \sin x) + c_2 (\cos 2x + i \sin 2x) + \cdots \\ &\qquad + c_n (\cos nx + i \sin nx) + \cdots \\ &= (c_0 + c_1 \cos x + c_2 \cos 2x + \cdots + c_n \cos nx + \cdots) \\ &\qquad + i(c_1 \sin x + c_2 \sin 2x + \cdots + c_n \sin nx + \cdots). \end{aligned} \tag{5.2}$$

We now separate the real and imaginary parts of $F(e^{ix})$, i.e., we write $F(e^{ix})$ in the form

$$F(e^{ix}) = f(x) + ig(x),$$

where $f(x)$ and $g(x)$ are real functions. Then, it obviously follows from (5.2) that

$$\begin{aligned} f(x) &= c_0 + c_1 \cos x + c_2 \cos 2x + \cdots + c_n \cos nx + \cdots, \\ g(x) &= c_1 \sin x + c_2 \sin 2x + \cdots + c_n \sin nx + \cdots. \end{aligned}$$

This fact can be used to find the sums of certain trigonometric series.

Example 1. It is well known that

$$e^z = 1 + z + \frac{z^2}{2!} + \cdots + \frac{z^n}{n!} + \cdots$$

for all z. Then, by (5.2) we have

$$\begin{aligned} e^{e^{ix}} &= \left(1 + \cos x + \frac{\cos 2x}{2!} + \cdots + \frac{\cos nx}{n!} + \cdots\right) \\ &\qquad + i\left(\sin x + \frac{\sin 2x}{2!} + \cdots + \frac{\sin nx}{n!} + \cdots\right). \end{aligned}$$

But

$$\begin{aligned} e^{e^{ix}} &= e^{\cos x + i \sin x} = e^{\cos x} e^{i \sin x} \\ &= e^{\cos x} [\cos (\sin x) + i \sin (\sin x)], \end{aligned}$$

and therefore

$$e^{\cos x} \cos (\sin x) = 1 + \cos x + \frac{\cos 2x}{2!} + \cdots + \frac{\cos nx}{n!} + \cdots,$$

$$e^{\cos x} \sin (\sin x) = \sin x + \frac{\sin 2x}{2!} + \cdots + \frac{\sin nx}{n!} + \cdots.$$

In this example, we started from a given function of a complex variable and found that its real and imaginary parts can be expanded in trigonometric series. In other words, we have found a new approach (as compared with Chs. 1 and 3) to the problem of expanding a function in trigonometric series. Moreover, in many cases, the argument given at the beginning of this section also allows us to solve the inverse problem, i.e., to find the sum of a given trigonometric series. In fact, suppose we are given two convergent series

$$\begin{gathered} c_0 + c_1 \cos x + c_2 \cos 2x + \cdots + c_n \cos nx + \cdots, \\ c_1 \sin x + c_2 \sin 2x + \cdots + c_n \sin nx + \cdots. \end{gathered}$$

Using these series, we form the series

$$\begin{aligned} (c_0 + c_1 \cos x + c_2 \cos 2x + \cdots) &+ i(c_1 \sin x + c_2 \sin 2x + \cdots) \\ &= c_0 + c_1(\cos x + i \sin x) + c_2(\cos 2x + i \sin 2x) + \cdots \\ &= c_0 + c_1 e^{ix} + c_2 e^{i2x} + \cdots, \end{aligned}$$

with complex terms. Denoting e^{ix} by z, we obtain the power series

$$c_0 + c_1 z + c_2 z^2 + \cdots.$$

If we know the sum $F(z)$ of this series for the values of z of interest, then, setting

$$F(e^{ix}) = f(x) + ig(x),$$

we obviously have

$$\begin{aligned} f(x) &= c_0 + c_1 \cos x + c_2 \cos 2x + \cdots, \\ g(x) &= c_1 \sin x + c_2 \sin 2x + \cdots. \end{aligned}$$

Example 2. Find the sums of the series

$$\begin{gathered} 1 + \frac{\cos x}{p} + \frac{\cos 2x}{p^2} + \cdots + \frac{\cos nx}{p^n} + \cdots, \\ \frac{\sin x}{p} + \frac{\sin 2x}{p^2} + \cdots + \frac{\sin nx}{p^n} + \cdots, \end{gathered} \tag{5.3}$$

where p is a real constant with absolute value greater than 1. The two series (5.3) converge for all x. Consider the series

$$\left(1 + \frac{\cos x}{p} + \frac{\cos 2x}{p^2} + \cdots\right) + i\left(\frac{\sin x}{p} + \frac{\sin 2x}{p^2} + \cdots\right) = 1 + \frac{e^{ix}}{p} + \frac{e^{2ix}}{p^2} + \cdots.$$

We have

$$1 + \frac{z}{p} + \frac{z^2}{p^2} + \cdots = \frac{1}{1 - (z/p)} = \frac{p}{p - z} = F(z),$$

since the series on the left is a geometric series, which converges for $|z/p| < 1$. Therefore

$$F(e^{ix}) = \frac{p}{p - e^{ix}} = \frac{p}{(p - \cos x) - i \sin x}$$

$$= p\frac{(p - \cos x) + i \sin x}{(p - \cos x)^2 + \sin^2 x} = p\frac{(p - \cos x) + i \sin x}{p^2 - 2p \cos x + 1},$$

and we find that

$$\frac{p(p - \cos x)}{p^2 - 2p \cos x + 1} = 1 + \frac{\cos x}{p} + \frac{\cos 2x}{p^2} + \cdots + \frac{\cos nx}{p^n} + \cdots,$$

$$\frac{p \sin x}{p^2 - 2p \cos x + 1} = \frac{\sin x}{p} + \frac{\sin 2x}{p^2} + \cdots + \frac{\sin nx}{p^n} + \cdots$$

for all x.

6. A Stronger Form of the Results of Sec. 5

At the beginning of Sec. 5, we assumed that the function $F(z)$ had no singularities for $|z| \leqslant 1$. Suppose now that $F(z)$ has no singularities only for $|z| < 1$, while $F(z)$ has both *regular points*[3] and singularities for $|z| = 1$ (in geometrical language, on the unit circle C with center at the origin). Then, for $|z| < 1$, the expansion (5.1) is still valid. *Moreover,* (5.1) *is still valid at every regular point z on the circle C at which the series* (5.1) *converges.* We now prove this result, for the benefit of the reader who may not already know it. First we need the following

LEMMA. *Let*

$$u_0 + u_1 + u_2 + \cdots + u_n + \cdots \tag{6.1}$$

be a convergent series (with real or complex terms). Then the series

$$u_0 + u_1 r + u_2 r^2 + \cdots + u_n r^n + \cdots \tag{6.2}$$

converges for $0 \leqslant r \leqslant 1$, and its sum $\sigma(r)$ is continuous on the interval [0, 1].

Proof. If

$$\sigma = u_0 + u_1 + u_2 + \cdots + u_n + \cdots,$$

then for every $\varepsilon > 0$, there exists an integer N such that if $n \geqslant N$

$$|\sigma - \sigma_n| \leqslant \frac{\varepsilon}{2}, \tag{6.3}$$

[3] I.e., points which are not singularities. (*Translator*)

where σ_n is the partial sum of the series (6.1). Consider the remainder of the series (6.1)

$$R_n = \sigma - \sigma_n = u_{n+1} + u_{n+2} + \cdots, \tag{6.4}$$

and its partial sums

$$u_{n+1} + u_{n+2} + \cdots + u_{n+m} \qquad (m = 1, 2, \ldots).$$

By (6.3) we obviously have

$$|u_{n+1} + u_{n+2} + \cdots + u_{n+m}| = |R_n - R_{n+m}| \leqslant |R_n| + |R_{n+m}| \leqslant \varepsilon.$$

Thus, every partial sum of the series (6.4), just like its entire sum, is $\leqslant \varepsilon$ in absolute value. Since the numbers

$$r^{n+1}, r^{n+2}, \ldots$$

decrease monotonically to zero as $n \to \infty$ for any r $(0 \leqslant r < 1)$, Abel's lemma is applicable to the series

$$R_n(r) = u_{n+1}r^{n+1} + u_{n+2}r^{n+2} + \cdots.$$

It follows that this series, and hence the series (6.2) converges, and moreover that

$$|R_n(r)| \leqslant \varepsilon r^{n+1} \leqslant \varepsilon \qquad (0 \leqslant r < 1).$$

Now, letting $\sigma_n(r)$ denote the partial sum of the series (6.2), we have

$$|\sigma(r) - \sigma_n(r)| = |R_n(r)| \leqslant \varepsilon \tag{6.5}$$

for $0 \leqslant r < 1$. Noting that $\sigma(1) = \sigma$ and $\sigma_n(1) = \sigma_n$, and using (6.3), we can assert that the inequality (6.5) holds everywhere on the interval $0 \leqslant r \leqslant 1$. This proves that the series (6.2) converges uniformly on $0 \leqslant r \leqslant 1$ and hence implies that the function $\sigma(r)$ is continuous on $0 \leqslant r \leqslant 1$. This proves the lemma.

To prove the proposition formulated at the beginning of this section, we assume that the series

$$c_0 + c_1 z + c_2 z^2 + \cdots + c_n z^n + \cdots$$

converges at a point z lying on the circle C. By the lemma just proved, the function

$$c_0 + c_1 r z^2 + c_2 r^2 z^2 + \cdots + c_n r^n z^n + \cdots$$

is a continuous function of r for $0 \leqslant r \leqslant 1$, and therefore

$$\begin{aligned} \lim_{\substack{r\to 1\\ r<1}} F(rz) &= \lim_{\substack{r\to 1\\ r<1}} (c_0 + c_1 r z + c_2 r^2 z^2 + \cdots) \\ &= c_0 + c_1 z + c_2 z^2 + \cdots. \end{aligned} \tag{6.6}$$

On the other hand, since the point z is a regular point, $F(z)$ is continuous at z, and therefore

$$\lim_{\substack{r \to 1 \\ r < 1}} F(rz) = F(z), \tag{6.7}$$

since $rz \to z$ as $r \to 1$. Comparing (6.6) and (6.7), we obtain

$$F(z) = c_0 + c_1 z + c_2 z^2 + \cdots + c_n z^n + \cdots,$$

as was to be shown.

This result makes the argument of Sec. 5 legitimate for any regular point $z = e^{ix}$ (such a point lies on C, since $|e^{ix}| = 1$) for which the series in (5.1) converges. To illustrate the situation, we now consider two examples.

Example 1. It is well known that

$$\ln(1 + z) = z - \frac{z^2}{2} + \frac{z^3}{3} - \frac{z^4}{4} + \cdots$$

for $|z| < 1$, and that the function $\ln(1 + z)$ is analytic at all points of the unit circle C except the point $z = -1$, or equivalently, the point $z = e^{i(2k+1)\pi}$. By (5.2), if $z = e^{ix}$ where $x \neq (2k + 1)\pi$, we have

$$\begin{aligned} \ln(1 + e^{ix}) = &\left(\cos x - \frac{\cos 2x}{2} + \frac{\cos 3x}{3} - \cdots\right) \\ &+ i\left(\sin x - \frac{\sin 2x}{2} + \frac{\sin 3x}{3} - \cdots\right), \end{aligned} \tag{6.8}$$

where we are justified in writing the equation, since the two series on the right actually converge for $x \neq (2k + 1)\pi$. (See Theorem 1 of Sec. 4.) On the other hand, for $-\pi < x < \pi$, we obviously have

$$\begin{aligned} 1 + e^{ix} &= (1 + \cos x) + i \sin x \\ &= 2\cos^2\frac{x}{2} + 2i \sin\frac{x}{2}\cos\frac{x}{2} = 2\cos\frac{x}{2}\left(\cos\frac{x}{2} + i \sin\frac{x}{2}\right). \end{aligned}$$

Therefore[4]

$$\ln(1 + e^{ix}) = \ln\left(2\cos\frac{x}{2}\right) + i\frac{x}{2}$$

for $-\pi < x < \pi$. Then, by (6.8) we find that

$$\begin{aligned} \ln\left(2\cos\frac{x}{2}\right) &= \cos x - \frac{\cos 2x}{2} + \frac{\cos 3x}{3} - \cdots, \\ \frac{x}{2} &= \sin x - \frac{\sin 2x}{2} + \frac{\sin 3x}{3} - \cdots \end{aligned} \tag{6.9}$$

[4] We have used the following familiar property of the logarithm: If $z = \rho e^{i\theta}$, $-\pi < \theta < \pi$, then $\ln z = \ln \rho + i\theta$. In our case $\rho = 2\cos(x/2)$, $\theta = x/2$.

for $-\pi < x < \pi$, and we have obtained two expansions with which we are already familiar [see Example 2 of Ch. 3, Sec. 14 and formula (13.9) of Ch. 1, Sec. 13].

By a similar method, we can find the two expansions

$$\left.\begin{aligned} -\ln\left(2\sin\frac{x}{2}\right) &= \cos x + \frac{\cos 2x}{2} + \frac{\cos 3x}{3} + \cdots, \\ \frac{\pi - x}{2} &= \sin x + \frac{\sin 2x}{2} + \frac{\sin 3x}{3} + \cdots, \end{aligned}\right\} \quad (0 < x < 2\pi) \quad (6.10)$$

which we have also encountered previously. In this case, we have to start from the function

$$\ln\frac{1}{1-z} = -\ln(1-z) = z + \frac{z^2}{2} + \frac{z^3}{3} + \cdots \qquad (z \neq 1),$$

for which

$$f(x) = -\ln\left(2\sin\frac{x}{2}\right), \quad g(x) = \frac{\pi - x}{2} \qquad (0 < x < 2\pi).$$

However, a much simpler way of obtaining the expansions (6.10) is to make the substitution $x = t - \pi$ in (6.9).

Example 2. Find the sums of the series

$$\frac{\cos 2x}{1\cdot 2} + \frac{\cos 3x}{2\cdot 3} + \cdots + \frac{\cos(n+1)x}{n(n+1)} + \cdots,$$

$$\frac{\sin 2x}{1\cdot 2} + \frac{\sin 3x}{2\cdot 3} + \cdots + \frac{\sin(n+1)x}{n(n+1)} + \cdots.$$

These series converge for all x. Consider the new series

$$\left(\frac{\cos 2x}{1\cdot 2} + \frac{\cos 3x}{2\cdot 3} + \cdots\right) + i\left(\frac{\sin 2x}{1\cdot 2} + \frac{\sin 3x}{2\cdot 3} + \cdots\right) = \frac{e^{2ix}}{1\cdot 2} + \frac{e^{3ix}}{2\cdot 3} + \cdots + \frac{e^{(n+1)ix}}{n(n+1)} + \cdots.$$

It follows from the identity

$$\frac{1}{n(n+1)} = \frac{1}{n} - \frac{1}{n+1}$$

that

$$\begin{aligned} \frac{z^2}{1\cdot 2} + \frac{z^3}{2\cdot 3} + \cdots + \frac{z^{n+1}}{n(n+1)} + \cdots &= \left(z^2 + \frac{z^3}{2} + \cdots + \frac{z^{n+1}}{n} + \cdots\right) \\ &\quad - \left(\frac{z^2}{2} + \frac{z^3}{3} + \cdots + \frac{z^{n+1}}{n+1} + \cdots\right) \\ &= -z\ln(1-z) + \ln(1-z) + z \\ &= (1-z)\ln(1-z) + z = F(z) \end{aligned}$$

(see Example 1) for all z satisfying the conditions $|z| \leqslant 1, z \neq -1$. Therefore,

$$\begin{aligned}
F(e^{ix}) &= (1 - e^{ix}) \ln (1 - e^{ix}) + e^{ix} \\
&= [(1 - \cos x) - i \sin x] \left[\ln \left(2 \sin \frac{x}{2}\right) - i \frac{\pi - x}{2}\right] \\
&\quad + (\cos x + i \sin x) \\
&= \left[(1 - \cos x) \ln \left(2 \sin \frac{x}{2}\right) - \frac{\pi - x}{2} \sin x + \cos x\right] \\
&\quad + i \left[\frac{\pi - x}{2} (\cos x - 1) - \sin x \ln \left(2 \sin \frac{x}{2}\right) + \sin x\right]
\end{aligned}$$

for $0 < x < 2\pi$ (see Example 1), and consequently

$$(1 - \cos x) \ln \left(2 \sin \frac{x}{2}\right) - \frac{\pi - x}{2} \sin x + \cos x = \frac{\cos 2x}{1 \cdot 2} + \frac{\cos 3x}{2 \cdot 3} + \cdots,$$

$$\frac{\pi - x}{2} (\cos x - 1) - \sin x \ln \left(2 \sin \frac{x}{2}\right) + \sin x = \frac{\sin 2x}{1 \cdot 2} + \frac{\sin 3x}{2 \cdot 3} + \cdots.$$

PROBLEMS

1. For which values of x do the following series converge:

$$\text{a)} \quad \sum_{n=1}^{\infty} \frac{\cos nx}{\sqrt{n}}; \quad \text{b)} \quad \sum_{n=1}^{\infty} \frac{\sin nx}{\sqrt{n}}; \quad \text{c)} \quad \sum_{n=1}^{\infty} \frac{\cos nx + \sin nx}{\sqrt{n}}?$$

2. For which values of x are the sums of the series in Prob. 1 continuous? Are these series the Fourier series of square integrable functions?

3. For which values of x do the following series converge:

$$\text{a)} \quad \sum_{n=1}^{\infty} (-1)^n \frac{\cos nx}{n + \sqrt{n}}; \quad \text{b)} \quad \sum_{n=2}^{\infty} \frac{\cos nx + (-1)^n \sin nx}{\ln n};$$

$$\text{c)} \quad \sum_{n=1}^{\infty} \frac{\sin 3nx}{n}; \quad \text{d)} \quad \sum_{n=0}^{\infty} (-1)^{n+1} \frac{\cos (2n + 3)x}{n + 2}?$$

4. For which values of x are the sums of the series in Prob. 3 continuous? Construct the graph of the sum of the series c).

5. For the complex variable $z = x + iy$, the trigonometric and hyperbolic functions are defined by the series

$$\sin z = z - \frac{z^3}{3!} + \frac{z^5}{5!} - \cdots,$$

$$\cos z = 1 - \frac{z^2}{2!} + \frac{z^4}{4!} - \cdots,$$

$$\sinh z = z + \frac{z^3}{3!} + \frac{z^5}{5!} + \cdots \quad \text{(hyperbolic sine)},$$

$$\cosh z = 1 + \frac{z^2}{2!} + \frac{z^4}{4!} + \cdots \quad \text{(hyperbolic cosine)}.$$

Using the formulas

$$\sin(\alpha + \beta) = \sin\alpha\cos\beta + \cos\alpha\sin\beta,$$
$$\cos(\alpha + \beta) = \cos\alpha\cos\beta - \sin\alpha\sin\beta,$$

which are also valid for complex α and β, prove that

$$\begin{aligned}\sin(\alpha + i\beta) &= \sin\alpha\cosh\beta + i\cos\alpha\sinh\beta,\\ \cos(\alpha + i\beta) &= \cos\alpha\cosh\beta - i\sin\alpha\sinh\beta.\end{aligned} \tag{I}$$

6. Find the sums of the series

a) $$\cos x - \frac{\cos 3x}{3!} + \frac{\cos 5x}{5!} - \cdots,$$

b) $$\sin x - \frac{\sin 3x}{3!} + \frac{\sin 5x}{5!} - \cdots.$$

Hint. $F(z) = \sin z$ (see Sec. 5). Use the first of the formulas (I) of Prob. 5.

7. Find the sums of the series

a) $$1 - \frac{\cos 2x}{2!} + \frac{\cos 4x}{4!} - \cdots,$$

b) $$\frac{\sin 2x}{2!} - \frac{\sin 4x}{4!} + \frac{\sin 6x}{6!} - \cdots.$$

Hint. $F(z) = \cos z$. Use the second of the formulas (I) of Prob. 5.

8. Find the sums of the series

a) $$1 + \frac{\cos x}{1 \cdot 2} - \frac{\cos 2x}{2 \cdot 3} + \frac{\cos 3x}{3 \cdot 4} - \cdots,$$

b) $$\frac{\sin x}{1 \cdot 2} - \frac{\sin 2x}{2 \cdot 3} + \frac{\sin 3x}{3 \cdot 4} - \cdots.$$

Hint. Use the method of Example 2 of Sec. 6 and the result of Example 1 of the same section.

9. Find the sums of the series

a) $$\frac{\cos 2x}{3} - \frac{\cos 3x}{8} + \cdots + (-1)^n \frac{\cos nx}{n^2 - 1} + \cdots,$$

b) $$\frac{\sin 2x}{3} - \frac{\sin 3x}{8} + \cdots + (-1)^n \frac{\sin nx}{n^2 - 1} + \cdots.$$

Hint. Use the identity

$$\frac{1}{n^2 - 1} = \frac{1}{2}\left(\frac{1}{n - 1} - \frac{1}{n + 1}\right)$$

and the results of Example 1 of Sec. 6.

10. Find the sums of the series

a) $$\frac{2\cos 2x}{3} - \frac{3\cos 3x}{8} + \cdots + (-1)^n \frac{n\cos nx}{n^2 - 1} + \cdots,$$

b) $$\frac{2\sin 2x}{3} - \frac{3\sin 3x}{8} + \cdots + (-1)^n \frac{n\sin nx}{n^2 - 1} + \cdots$$

Hint. Use the identity

$$\frac{n}{n^2 - 1} = \frac{1}{2}\left(\frac{1}{n-1} + \frac{1}{n+1}\right)$$

and the results of Example 1 of Sec. 6.

11. Find the sums of the series

a) $$\frac{\cos x}{1+p} + \frac{\cos 2x}{2+p} + \cdots + \frac{\cos nx}{n+p} + \cdots,$$

b) $$\frac{\sin x}{1+p} + \frac{\sin 2x}{2+p} + \cdots + \frac{\sin nx}{n+p} + \cdots,$$

where p is a positive integer.

Hint.

$$\begin{aligned} F(z) &= \frac{z}{1+p} + \frac{z^2}{2+p} + \cdots + \frac{z^n}{n+p} + \cdots \\ &= \frac{1}{z^p}\left(\frac{z^{1+p}}{1+p} + \frac{z^{2+p}}{2+p} + \cdots + \frac{z^{n+p}}{n+p} + \cdots\right) \\ &= \frac{1}{z^p}\left(-\ln(1-z) - \sum_{n=1}^{p} \frac{z^n}{n}\right). \end{aligned}$$

Also use the results of Example 1 of Sec. 6.

12. Show that if $\sum_{n=1}^{\infty} a_n$ converges, then so does $\sum_{n=1}^{\infty} \frac{a_n}{n}$.

13. Using the notation of Sec. 1, show that if $|\sigma_n| \leqslant M$, if $\alpha_n \to 0$ as $n \to \infty$, and if

$$\sum_{n=0}^{\infty} |\alpha_n - \alpha_{n+1}| < \infty,$$

then $\sum_{n=0}^{\infty} \alpha_n u_n$ converges.

14. Show that if $|a_1| \geqslant |a_2| \geqslant \cdots \to 0$ and if

$$\sum_{n=1}^{\infty} a_n \cos nx$$

converges absolutely at even a single point x_0, then

$$\sum_{n=1}^{\infty} |a_n| < \infty.$$

Show that the same is true in the case of the series

$$\sum_{n=1}^{\infty} a_n \sin nx,$$

provided that x_0 is not an integral multiple of π.

5

OPERATIONS ON FOURIER SERIES

1. Approximation of Functions by Trigonometric Polynomials

In Ch. 3, we established certain conditions under which a continuous (and sometimes even a discontinuous) function of period 2π can be represented as a sum of a trigonometric series. This raises the following question: Was it only because of the inadequacy of the argument used that we were not able to prove that the Fourier series of any continuous function $f(x)$ always converges to $f(x)$? It turns out that the answer to this question is negative, and in fact there exist examples of continuous functions with discontinuous Fourier series.

We now adopt a different approach, i.e., we consider the *approximate* representation of functions by trigonometric series. Here we immediately obtain the following remarkable result.

THEOREM. *Let $f(x)$ be a continuous function of period 2π. Then, given any $\varepsilon > 0$, there exists a trigonometric polynomial*

$$\sigma_n(x) = \alpha_0 + \sum_{k=1}^{n} (\alpha_k \cos kx + \beta_k \sin kx) \tag{1.1}$$

for which

$$|f(x) - \sigma_n(x)| \leqslant \varepsilon$$

for any x.

Proof. Consider the graph of the function $y = f(x)$ on the interval $[-\pi, \pi]$. Divide $[-\pi, \pi]$ into subintervals by the points

$$x_0 = -\pi < x_1 < x_2 < \cdots < x_{m-1} < x_m = \pi,$$

and construct the continuous function $g(x)$ for which $g(x_k) = f(x_k)$ $(k = 0, 1, 2, \ldots, m)$ and which is linear on each subinterval $[x_{k-1}, x_k]$. The graph of the function $y = g(x)$ is a broken line with vertices on

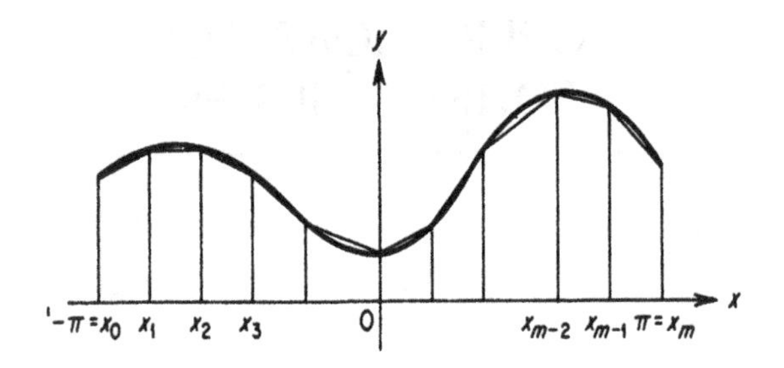

FIGURE 40

the curve $y = f(x)$ (see Fig. 40). The subintervals into which we divided $[-\pi, \pi]$ can be made so small that the inequality

$$|f(x) - g(x)| \leqslant \frac{\varepsilon}{2} \tag{1.2}$$

is satisfied for any x in the interval $[-\pi, \pi]$. Moreover, the function $g(x)$ can be periodically extended onto the whole x-axis. Then, obviously, the condition (1.2) is also valid for the periodic extension of $g(x)$, which is continuous and piecewise smooth on the whole x-axis.

We now form the Fourier series of $g(x)$. According to Theorem 2 of Ch. 3, Sec. 10, this series converges uniformly to $g(x)$. But then for all sufficiently large n we have

$$|g(x) - \sigma_n(x)| \leqslant \frac{\varepsilon}{2} \tag{1.3}$$

for any x, where $\sigma_n(x)$ denotes the nth partial sum of the Fourier series of the function $g(x)$. Let n be an index for which (1.3) is valid. Then it follows from (1.2) and (1.3) that

$$\begin{aligned} |f(x) - \sigma_n(x)| &= |(f(x) - g(x)) + (g(x) - \sigma_n(x))| \\ &\leqslant |f(x) - g(x)| + |g(x) - \sigma_n(x)| \leqslant \varepsilon \end{aligned}$$

for any x, which proves the theorem.

We now note the following consequences of this theorem:

COROLLARY 1. *If $f(x)$ is continuous on the interval $[a, a + 2\pi]$ and $f(a) = f(a + 2\pi)$, then for any $\varepsilon > 0$, there exists a trigonometric polynomial of the form* (1.1) *for which*

$$|f(x) - \sigma_n(x)| \leqslant \varepsilon \tag{1.4}$$

for any x in the interval $[a, a + 2\pi]$.

To see this, it suffices to make the periodic extension of $f(x)$ onto the whole x-axis (because of the condition $f(a) = f(a + 2\pi)$, this does not destroy the continuity) and then apply the theorem to the extended function.

COROLLARY 2. *If $f(x)$ is continuous on an interval $[a, b]$ of length less than 2π, then for any $\varepsilon > 0$, there exists a trigonometric polynomial of the form* (1.1) for which

$$|f(x) - \sigma_n(x)| \leqslant \varepsilon$$

for any x in $[a, b]$.

Proof. Extend $f(x)$ from $[a, b]$ onto $[a, a + 2\pi]$ in such a way that continuity is preserved and the extended function satisfies $f(a) = f(a + 2\pi)$. For example, this can be done by *setting* $f(a + 2\pi) = f(a)$ and then taking the function to be linear on the interval $[b, a + 2\pi]$ (see Fig. 41). Then Corollary 1 is applicable to the extension of $f(x)$

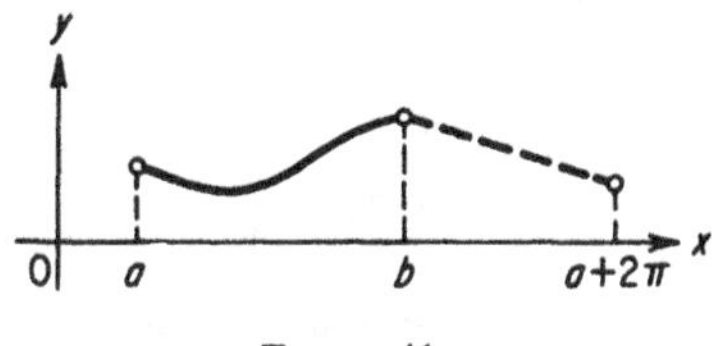

FIGURE 41

obtained in this way, and therefore the inequality (1.4) is valid for all x in $[a, a + 2\pi]$, in particular, to all x in $[a, b]$.

What has just been proved can be formulated succinctly as follows: *Under the conditions of the theorem (or its corollaries), the function $f(x)$ can be uniformly approximated by a trigonometric polynomial of the form* (1.1) *with any degree of accuracy specified in advance.*

2. Completeness of the Trigonometric System

In view of Ch. 2, Sec. 9 (the completeness criterion for an orthogonal system), to prove that the trigonometric system is complete it is sufficient to prove that *given any function* $f(x)$ *which is continuous on* $[-\pi, \pi]$ *and any* $\varepsilon > 0$, *there exists a trigonometric polynomial* $\sigma_n(x)$ *for which*

$$\int_{-\pi}^{\pi} [f(x) - \sigma_n(x)]^2 \, dx \leqslant \varepsilon. \tag{2.1}$$

We now prove that this is actually the case.

If $f(-\pi) = f(\pi)$, then by Corollary 1 of the theorem of Sec. 1, there exists a trigonometric polynomial $\sigma_n(x)$ for which

$$|f(x) - \sigma_n(x)| \leqslant \sqrt{\varepsilon/2\pi}$$

for any x. Therefore, we have

$$\int_{-\pi}^{\pi} [f(x) - \sigma_n(x)]^2 \, dx \leqslant \int_{-\pi}^{\pi} \frac{\varepsilon}{2\pi} \, dx = \varepsilon,$$

as required.

Next, suppose that $f(-\pi) \neq f(\pi)$. Let M be the maximum of the absolute value of $f(x)$ on the interval $[-\pi, \pi]$, and choose the number $h > 0$ so small that the condition

$$4M^2h \leqslant \frac{\varepsilon}{4}$$

holds. Now, let $g(x)$ be the continuous function which coincides with $f(x)$ on the interval $[-\pi, \pi - h]$, is equal to $f(-\pi)$ at $x = \pi$, and is linear on the interval $[\pi - h, \pi]$. (This construction is like the one shown in Fig. 41.) Obviously, $|g(x)| \leqslant M$, and hence

$$\int_{-\pi}^{\pi} [f(x) - g(x)]^2 \, dx = \int_{\pi - h}^{\pi} [f(x) - g(x)]^2 \, dx \leqslant \int_{\pi - h}^{\pi} 4M^2 \, dx$$

$$= 4M^2h \leqslant \frac{\varepsilon}{4}. \tag{2.2}$$

On the other hand, $g(x)$ is continuous and takes equal values at the end points of the interval $[-\pi, \pi]$. Therefore, there exists a trigonometric polynomial $\sigma_n(x)$ for which

$$\int_{-\pi}^{\pi} [g(x) - \sigma_n(x)]^2 \, dx \leqslant \frac{\varepsilon}{4}. \tag{2.3}$$

But then it follows from (2.2), (2.3) and the elementary inequality

$$(a + b)^2 \leqslant 2(a^2 + b^2)$$

that

$$\int_{-\pi}^{\pi} [f(x) - \sigma_n(x)]^2 \, dx = \int_{-\pi}^{\pi} [(f(x) - g(x)) + (g(x) - \sigma_n(x))]^2 \, dx$$

$$\leqslant 2 \int_{-\pi}^{\pi} [f(x) - g(x)]^2 \, dx + 2 \int_{-\pi}^{\pi} [g(x) - \sigma_n(x)]^2 \, dx \leqslant \varepsilon,$$

as was to be proved.

3. Parseval's Theorem. The Most Important Consequences of the Completeness of the Trigonometric System

Since the trigonometric system is complete, Bessel's inequality [see (1.2) of Ch. 3] becomes the equality

$$\frac{1}{\pi}\int_{-\pi}^{\pi} f^2(x)\,dx = \frac{a_0^2}{2} + \sum_{n=1}^{\infty} (a_n^2 + b_n^2), \tag{3.1}$$

where $f(x)$ is an arbitrary square integrable function, and a_n, b_n are its Fourier coefficients. Formula (3.1) is known as *Parseval's theorem.*[1] Recalling Theorem 1 of Ch. 2, Sec. 7 and the fact that

$$\|1\| = \sqrt{2\pi}, \quad \|\cos nx\| = \|\sin nx\| = \sqrt{\pi} \qquad (n = 1, 2, \ldots),$$

we obtain

THEOREM 1. *If $f(x)$ and $F(x)$ are square integrable functions defined on* $[-\pi, \pi]$, for which

$$f(x) \sim \frac{a_0}{2} + \sum_{n=1}^{\infty} (a_n \cos nx + b_n \sin nx),$$

$$F(x) \sim \frac{A_0}{2} + \sum_{n=1}^{\infty} (A_n \cos nx + B_n \sin nx),$$

then

$$\frac{1}{\pi}\int_{-\pi}^{\pi} f(x)F(x)\,dx = \frac{a_0A_0}{2} + \sum_{n=1}^{\infty} (a_nA_n + b_nB_n).$$

The following result is an immediate consequence of Theorem 2 of Ch. 2, Sec. 7:

THEOREM 2. *The trigonometric Fourier series of any square integrable function $f(x)$ converges to $f(x)$ in the mean*, i.e.,

$$\lim_{n\to\infty}\int_{-\pi}^{\pi}\left[f(x) - \left(\frac{a_0}{2} + \sum_{k=1}^{n} (a_k \cos kx + b_k \sin kx)\right)\right]^2 dx = 0.$$

This theorem is all the more remarkable, in view of the fact already noted in Sec. 1 that a trigonometric Fourier series does not always converge in the *ordinary* sense even for a continuous function.

In Ch. 2, Sec. 7, we proved that a Fourier series can converge in the mean

[1] Attributed by the author to the Russian mathematician A. M. Lyapunov "who first gave a rigorous proof of the formula (for the case of bounded functions)." Theorem 1 below, of which (3.1) is a special case, is the result usually called *Parseval's theorem* in the English-language literature. (*Translator*)

to only one function $f(x)$ [to within a change of $f(x)$ at a finite number of points[2]]. This implies

THEOREM 3. *Any square integrable function $f(x)$ is completely defined (except for its values at a finite number of points) by its trigonometric Fourier series, whether or not the series converges.*

In Theorem 3 we say "defined" but this does not mean that we yet know how to actually determine a function from a knowledge of its Fourier series. The problem of actually "reconstructing" a function from its Fourier series will be solved completely in the next chapter; however, in some special cases, the problem can be solved by using the method given in Ch. 4, Secs. 5 and 6.

Finally, we note two more consequences of the completeness of the trigonometric system (see Theorems 1 and 2 of Ch. 2, Sec. 8):

THEOREM 4. *Any continuous function $f(x)$ which is orthogonal to all the functions of the trigonometric system must be identically zero.*

In other words, except for the trivial function which is identically zero, we *cannot add* an extra continuous function to the trigonometric system and still have an orthogonal "enlarged" system. Theorem 4 can also be paraphrased as follows: *If all the Fourier coefficients of a continuous function are zero, then the function vanishes identically.*

THEOREM 5. *If the trigonometric Fourier series of a continuous function $f(x)$ is uniformly convergent, then the sum of the series equals $f(x)$.*

*4. Approximation of Functions by Polynomials

As a rather simple consequence of the results of Sec. 1, we can obtain the following result which is often very useful:

THEOREM.[3] *Let $f(x)$ be a function which is continuous on the interval $[a, b]$. Then, for any $\varepsilon > 0$, there exists a polynomial*

$$p_m(x) = c_0 + c_1x + \cdots c_mx^m,$$

for which

$$|f(x) - p_m(x)| \leqslant \varepsilon \tag{4.1}$$

everywhere on the interval $[a, b]$.

---

[2] See the footnote to Ch. 2, Sec. 7. (*Translator*)

[3] This proposition is usually known as the *Weierstrass approximation theorem.* (*Translator*)

Proof. By making the transformation

$$t = \pi \frac{x - a}{b - a}, \tag{4.2}$$

or equivalently

$$x = \frac{b - a}{\pi} t + a,$$

we take the interval $[a, b]$ of the x-axis into the interval $[0, \pi]$ of the t-axis. (If the length of $[a, b]$ is less than 2π, we do not have to make this transformation.) We now set

$$f\left(\frac{b - a}{\pi} t + a\right) = F(t).$$

By Corollary 2 of the theorem of Sec. 1, there exists a trigonometric polynomial

$$\sigma_n(t) = \alpha_0 + \sum_{k=1}^{n} (\alpha_k \cos kt + \beta_k \sin kt),$$

for which

$$|F(t) - \sigma_n(t)| \leqslant \frac{\varepsilon}{2} \tag{4.3}$$

everywhere on $[0, \pi]$. Holding n fixed, we choose a positive number $\omega > 0$ which is so small that the condition

$$\omega \sum_{k=1}^{n} (|\alpha_k| + |\beta_k|) \leqslant \frac{\varepsilon}{2}. \tag{4.4}$$

Now, it is known that for all z

$$\cos z = 1 - \frac{z^2}{2!} + \frac{z^4}{4!} - \cdots,$$

$$\sin z = z - \frac{z^3}{3!} + \frac{z^5}{5!} - \cdots,$$

where these series converge uniformly on every interval of finite length, in particular, on $0 \leqslant z \leqslant n\pi$. But then, for fixed n and all sufficiently larg[illegible] we have

$$\begin{aligned} \left|\cos kt - \left(1 - \frac{k^2t^2}{2!} + \frac{k^4t^4}{4!} - \cdots + (-1)^l \frac{k^{2l}t^{2l}}{(2l)!}\right)\right| &\leqslant \omega, \\ \left|\sin kt - \left(kt - \frac{k^3t^3}{3!} + \frac{k^5t^5}{5!} - \cdots + (-1)^l \frac{k^{2l+1}t^{2l+1}}{(2l+1)!}\right)\right| &\leqslant \omega, \end{aligned} \tag{4.5}$$

where $k = 1, 2, \ldots, n$, for any t in the interval $[0, \pi]$. (Since there are only a finite number of the functions $\cos kt$, $\sin kt$ for $k = 1, 2, \ldots, n$, i.e., just $2n$ of them, the index l can be chosen large enough to make *all* the inequalities (4.5) hold *simultaneously*.)

We denote the polynomials of degree $2l$ and $2l + 1$ appearing in parentheses in the inequalities (4.5) by $r_k(t)$ and $s_k(t)$, respectively. Then

$$\begin{aligned} |\cos kt - r_k(t)| &\leqslant \omega, \\ |\sin kt - s_k(t)| &\leqslant \omega \end{aligned} \tag{4.6}$$

$(k = 1, 2, \ldots, n)$ for any t in $[0, \pi]$. Consider the sum

$$P_m(t) = \alpha_0 + \sum_{k=1}^{n} [\alpha_k r_k(t) + \beta_k s_k(t)],$$

which is a polynomial of degree $m = 2l + 1$. By (4.4) and (4.6), this polynomial satisfies the inequality

$$\begin{aligned} |\sigma_n(t) - P_m(t)| &= \left| \sum_{k=1}^{n} \alpha_k(\cos kt - r_k(t)) + \beta_k(\sin kt - s_k(t)) \right| \\ &\leqslant \omega \sum_{k=1}^{n} (|\alpha_k| + |\beta_k|) \leqslant \frac{\varepsilon}{2} \end{aligned}$$

everywhere on $[0, \pi]$. Therefore, it follows from (4.3) that

$$\begin{aligned} |F(t) - P_m(t)| &= |(F(t) - \sigma_n(t)) + (\sigma_n(t) - P_m(t))| \\ &\leqslant |F(t) - \sigma_n(t)| + |\sigma_n(t) - P_m(t)| \leqslant \varepsilon \end{aligned}$$

everywhere on $[0, \pi]$. Using the formula (4.2) to return to the variable x, we obtain

$$\left| f(x) - P_m\left(\pi \frac{x - a}{b - a}\right) \right| \leqslant \varepsilon \tag{4.7}$$

everywhere on $[a, b]$. It is easy to see that the function

$$p_m(x) = P_m\left(\pi \frac{x - a}{b - a}\right)$$

is also a polynomial of degree m. Consequently, (4.7) is just the inequality (4.1) needed for the proof.

5. Addition and Subtraction of Fourier Series. Multiplication of a Fourier Series by a Number

To obtain the Fourier series of the sum or difference of two functions whose Fourier series are known, it is sufficient to add or subtract the two known series. In fact, suppose that

$$f(x) \sim \frac{a_0}{2} + \sum_{n=1}^{\infty} (a_n \cos nx + b_n \sin nx),$$
$$F(x) \sim \frac{A_0}{2} + \sum_{n=1}^{\infty} (A_n \cos nx + B_n \sin nx). \tag{5.1}$$

Then, the Fourier coefficients α_n and β_n of the function $f(x) \pm F(x)$ are given by

$$\alpha_n = \frac{1}{\pi}\int_{-\pi}^{\pi} [f(x) \pm F(x)] \cos nx\, dx$$
$$= \frac{1}{\pi}\int_{-\pi}^{\pi} f(x) \cos nx\, dx \pm \frac{1}{\pi}\int_{-\pi}^{\pi} F(x) \sin nx\, dx = a_n \pm A_n$$

and similarly

$$\beta_n = b_n \pm B_n,$$

which proves our assertion.

It can be shown in just the same way that the Fourier series of the function $kf(x)$, where k is a constant, is obtained from the Fourier series of $f(x)$ by multiplying all its terms by k.

Despite the simplicity of these considerations, they testify to a very important fact, namely, that although convergence is not assumed, it is possible to operate on Fourier series as if they were convergent, i.e., as if the sign $=$ appeared instead of $\sim$. The same sort of phenomenon will also be encountered in subsequent sections of this chapter.

*6. Products of Fourier Series

How do we form the Fourier series of the product $f(x)F(x)$, if we know the Fourier series of the factors? To answer this question, we argue as follows: We begin by assuming that $f(x)$ and $F(x)$ are square integrable functions, in which case we know that the product $f(x)F(x)$ is an integrable function (see Ch. 2, Sec. 4). [Note that if we fail to impose this requirement on $f(x)$ and $F(x)$, the product $f(x)F(x)$ may turn out to be nonintegrable, and then the problem of finding the Fourier series of $f(x)F(x)$ becomes meaningless.]

Now, suppose that $f(x)$ and $F(x)$ obey the relations (5.1) and that

$$f(x)F(x) \sim \frac{\alpha_0}{2} + \sum_{n=1}^{\infty} (\alpha_n \cos nx + \beta_n \sin nx).$$

Our problem is to express the coefficients α_n and β_n in terms of a_n, b_n, A_n and B_n. Using Theorem 1 of Sec. 3, we find that

$$\alpha_0 = \frac{1}{\pi}\int_{-\pi}^{\pi} f(x)F(x)\,dx = \frac{a_0A_0}{2} + \sum_{n=1}^{\infty}(a_nA_n + b_nB_n). \tag{6.1}$$

To calculate

$$\alpha_n = \frac{1}{\pi}\int_{-\pi}^{\pi} f(x)F(x)\cos nx\,dx, \tag{6.2}$$

it is sufficient to know the Fourier coefficients of the function $F(x)\cos nx$, since then we can use Theorem 1 of Sec. 3 again, this time applied to the product $f(x)F(x)\cos nx$. The Fourier coefficients of $F(x)\cos nx$ are

$$\frac{1}{\pi}\int_{-\pi}^{\pi} F(x)\cos nx\cos mx\,dx$$
$$= \frac{1}{2}\left[\frac{1}{\pi}\int_{-\pi}^{\pi} F(x)\cos(m+n)x\,dx + \frac{1}{\pi}\int_{-\pi}^{\pi} F(x)\cos(m-n)x\,dx\right],$$

which becomes

$$\tfrac{1}{2}(A_{m+n} + A_{m-n}) \quad \text{for} \quad m \geqslant n,$$
$$\tfrac{1}{2}(A_{n+m} + A_{n-m}) \quad \text{for} \quad m < n.$$

If we agree to write

$$A_{-k} = A_k,$$

then we have

$$\frac{1}{\pi}\int_{-\pi}^{\pi} F(x)\cos nx\cos mx\,dx = \tfrac{1}{2}(A_{m+n} + A_{m-n}).$$

Similarly, we find

$$\frac{1}{\pi}\int_{-\pi}^{\pi} F(x)\cos nx\sin mx\,dx = \frac{1}{2}\left[\frac{1}{\pi}\int_{-\pi}^{\pi} F(x)\sin(m+n)x\,dx\right.$$
$$\left. + \frac{1}{\pi}\int_{-\pi}^{\pi} F(x)\sin(m-n)x\,dx\right]$$
$$= \tfrac{1}{2}(B_{m+n} + B_{m-n}),$$

where we have set

$$B_{-k} = -B_k.$$

Thus, we now have the Fourier coefficients of the function $F(x)\cos nx$. Therefore, applying Theorem 1 of Sec. 3 to the integral (6.2), we obtain

$$\alpha_n = \frac{a_0A_n}{2} + \frac{1}{2}\sum_{m=1}^{\infty}[a_m(A_{m+n} + A_{m-n}) + b_m(B_{m+n} + B_{m-n})]. \tag{6.3}$$

In just the same way, we find that

$$\beta_n = \frac{a_0 B_n}{2} + \frac{1}{2}\sum_{m=1}^{\infty} [a_m(B_{m+n} - B_{m-n}) - b_m(A_{m+n} - A_{m-n})]. \tag{6.4}$$

The formulas (6.1), (6.3), and (6.4) give the solution of our problem. We note that these formulas can also be obtained by *formally* multiplying together the two series (5.1) (i.e., by acting as if these series are convergent and multiplication is justified[4]), then replacing products of sines and cosines by their sums and differences, and finally collecting similar terms.

7. Integration of Fourier Series

In the applications, one encounters cases where only the Fourier series of a function is known, but not the function itself. In this connection, the following problems arise:

1) Given the Fourier series of the function $f(x)$ of period 2π, calculate

$$\int_a^b f(x)\,dx,$$

where $[a, b]$ is an arbitrary interval;

2) Given the Fourier series of the function $f(x)$, find the Fourier series of the function

$$F(x) = \int_0^x f(x)\,dx.$$

The solution to the first problem is given by

THEOREM 1. *If the absolutely integrable function $f(x)$ of period 2π is specified by giving its Fourier series*

$$f(x) \sim \frac{a_0}{2} + \sum_{n=1}^{\infty} (a_n \cos nx + b_n \sin nx), \tag{7.1}$$

then

$$\int_a^b f(x)\,dx$$

[4] It is well known that the product of two convergent series $s = u_1 + u_2 + \cdots + u_n + \cdots$ and $\sigma = v_1 + v_2 + \cdots + v_n + \cdots$ is given by the formula

$$s\sigma = u_1v_1 + (u_1v_2 + u_2v_1) + \cdots + (u_1v_n + u_2v_{n-1} + \cdots + u_{n-1}v_2 + u_nv_1) + \cdots$$
$$= \sum_{n=1}^{\infty} (u_1v_n + u_2v_{n-1} + \cdots + u_{n-1}v_2 + u_nv_1),$$

which is valid whenever the last series converges. In particular, the formula *always* holds if the two series converge absolutely.

can be found by term by term integration of the series (7.1), *whether or not the series converges, i.e.,*

$$\int_a^b f(x)\,dx = \frac{a_0}{2}(b-a) + \sum_{n=1}^{\infty} \frac{a_n(\sin nb - \sin na) - b_n(\cos nb - \cos na)}{n}. \tag{7.2}$$

Proof. If $f(x)$ is square integrable, then this theorem is a special case of Theorem 3 of Ch. 2, Sec. 8. However, in the general case, we argue as follows: First we set

$$F(x) = \int_0^x \left[f(x) - \frac{a_0}{2}\right] dx. \tag{7.3}$$

This function is continuous and has an absolutely integrable derivative (which possibly does not exist at a finite number of points). Moreover

$$\begin{aligned} F(x+2\pi) &= \int_0^x \left[f(x) - \frac{a_0}{2}\right] dx + \int_x^{x+2\pi} \left[f(x) - \frac{a_0}{2}\right] dx \\ &= F(x) + \int_{-\pi}^{\pi} \left[f(x) - \frac{a_0}{2}\right] dx \\ &= F(x) + \int_{-\pi}^{\pi} f(x)\,dx - \pi a_0 = F(x), \end{aligned}$$

so that $F(x)$ has period 2π. Therefore, $F(x)$ can be expanded in a Fourier series (see Ch. 3, Sec. 11):

$$F(x) = \frac{A_0}{2} + \sum_{n=1}^{\infty} (A_n \cos nx + B_n \sin nx).$$

Integrating by parts for $n \geqslant 1$ gives

$$\begin{aligned} A_n &= \frac{1}{\pi}\int_{-\pi}^{\pi} F(x)\cos nx\,dx \\ &= \frac{1}{\pi}\left[F(x)\frac{\sin nx}{n}\right]_{x=-\pi}^{x=\pi} - \frac{1}{\pi n}\int_{-\pi}^{\pi}\left[f(x) - \frac{a_0}{2}\right]\sin nx\,dx = -\frac{b_n}{n}, \end{aligned}$$

and similarly

$$B_n = \frac{a_n}{n}.$$

Therefore

$$F(x) = \frac{A_0}{2} + \sum_{n=1}^{\infty} \frac{a_n \sin nx - b_n \cos nx}{n}$$

or by (7.3)

$$\int_0^x f(x)\,dx = \frac{a_0 x}{2} + \frac{A_0}{2} + \sum_{n=1}^{\infty} \frac{a_n \sin nx - b_n \cos nx}{n}. \tag{7.4}$$

To obtain (7.2), it remains only to set $x = b$, then $x = a$ and subtract the resulting formulas.

The solution to the second problem is given by

THEOREM 2. *Let the absolutely integrable function $f(x)$ be specified by giving its Fourier series*

$$f(x) \sim \frac{a_0}{2} + \sum_{n=1}^{\infty} (a_n \cos nx + b_n \sin nx),$$

whether convergent or not. Then the integral of $f(x)$ has the Fourier expansion

$$\int_0^x f(x)\,dx = \sum_{n=1}^{\infty} \frac{b_n}{n} + \sum_{n=1}^{\infty} \frac{-b_n \cos nx + [a_n + (-1)^{n+1} a_0] \sin nx}{n}$$

$$(-\pi < x < \pi). \tag{7.5}$$

Proof. We set $x = 0$ in (7.4), thereby obtaining

$$\frac{A_0}{2} = \sum_{n=1}^{\infty} \frac{b_n}{n}. \tag{7.6}$$

On the other hand, by formula (13.9) of Ch. 1, we have

$$\frac{x}{2} = \sum_{n=1}^{\infty} (-1)^{n+1} \frac{\sin nx}{n} \tag{7.7}$$

for $-\pi < x < \pi$. Substituting (7.6) and (7.7) in (7.4) gives (7.5).

Remark. In passing, we have proved that the series

$$\sum_{n=1}^{\infty} \frac{b_n}{n}$$

converges for any absolutely integrable function. This fact is sometimes useful, since in some cases it allows us to distinguish the Fourier series of absolutely integrable functions from other trigonometric series. For example, the series

$$\sum_{n=2}^{\infty} \frac{\sin nx}{\ln n}$$

which converges everywhere (see Sec. 3 of Ch. 4) cannot be the Fourier series of an absolutely integrable function, since the series

$$\sum_{n=2}^{\infty} \frac{1}{n \ln n}$$

diverges.

Next, we note an important special case of Theorem 2.

THEOREM 3. *If $a_0 = 0$ and the other conditions are the same as in Theorem 2, then*

$$\int_0^x f(x)\,dx = \sum_{n=1}^{\infty} \frac{b_n}{n} + \sum_{n=1}^{\infty} \frac{-b_n \cos nx + a_n \sin nx}{n}, \tag{7.8}$$

for all x,[5] *i.e., the Fourier series of the integral can be obtained by term by term integration of the series for $f(x)$.*

Proof. The formula (7.8) is obtained from (7.5) by simply setting $a_0 = 0$. Its validity for *all* x [and not just for $-\pi < x < \pi$, as in (7.5)] follows from the periodicity of the integral on the left in (7.8), which itself follows from

$$\begin{aligned}\int_0^{x+2\pi} f(x)\,dx &= \int_0^x f(x)\,dx + \int_x^{x+2\pi} f(x)\,dx \\ &= \int_0^x f(x)\,dx + \pi a_0 = \int_0^x f(x)\,dx.\end{aligned}$$

We can use (7.8) to obtain many new Fourier series, by starting from a known series. For example, we know that

$$\frac{x}{2} = \sin x - \frac{\sin 2x}{2} + \frac{\sin 3x}{3} - \cdots \qquad (-\pi < x < \pi).$$

Integrating, we obtain

$$\begin{aligned}\frac{x^2}{4} &= \left(1 - \frac{1}{2^2} + \frac{1}{3^2} - \cdots\right) \\ &\quad - \left(\cos x - \frac{\cos 2x}{2^2} + \frac{\cos 3x}{3^2} - \cdots\right) \\ &= C - \sum_{n=1}^{\infty} (-1)^{n+1} \frac{\cos nx}{n^2}, \quad C = \text{const.}\end{aligned}$$

To find C, we integrate the last equation from $-\pi$ to π. Since the series on

[5] If we are interested only in the values $-\pi < x < \pi$, then as in Theorem 2, we do not have to require that $f(x)$ be periodic. However, if we are interested in all values of x, then for (7.8) to be valid, we have to assume that $f(x)$ is periodic.

the right converges uniformly, the term by term integration is justified, and we obtain

$$\int_{-\pi}^{\pi} \frac{x^2}{4}\, dx = 2\pi C - \sum_{n=1}^{\infty} (-1)^{n+1} \frac{1}{n^2} \int_{-\pi}^{\pi} \cos nx\, dx = 2\pi C.$$

Therefore, we have

$$C = \frac{1}{8\pi} \int_{-\pi}^{\pi} x^2\, dx = \frac{\pi^2}{12},$$

so that

$$\frac{\pi^2}{12} - \frac{x^2}{4} = \sum_{n=1}^{\infty} (-1)^{n-1} \frac{\cos nx}{n^2},$$

an expression that we have already obtained in Ch. 1, Sec. 13. Integrating this formula again gives

$$\int_0^x \left(\frac{\pi^2}{12} - \frac{x^2}{4}\right) dx = \sum_{n=1}^{\infty} (-1)^{n+1} \frac{\sin nx}{n^3}$$

or

$$\frac{\pi^2 x}{12} - \frac{x^3}{12} = \sum_{n=1}^{\infty} (-1)^{n+1} \frac{\sin nx}{n^3}.$$

8. Differentiation of Fourier Series. The Case of a Continuous Function of Period 2π

THEOREM 1. *Let $f(x)$ be a continuous function of period 2π, with an absolutely integrable derivative (which may not exist at certain points[6]). Then the Fourier series of $f'(x)$ can be obtained from the Fourier series of the function $f(x)$ by term by term differentiation.*

Proof. Let

$$f(x) = \frac{a_0}{2} + \sum_{n=1}^{\infty} (a_n \cos nx + b_n \sin nx), \tag{8.1}$$

where we can write the equality because of the theorem of Ch. 3, Sec. 11. Let a_n' and b_n' denote the Fourier coefficients of $f'(x)$. First of all, we have

$$a_0' = \frac{1}{\pi} \int_{-\pi}^{\pi} f'(x)\, dx = f(\pi) - f(-\pi) = 0.$$

[6] In other words, $f'(x)$ may not exist at a finite number of points (in each period).

Then, integrating by parts, we obtain

$$\begin{aligned} a_n' &= \frac{1}{\pi}\int_{-\pi}^{\pi} f'(x)\cos nx\,dx \\ &= \frac{1}{\pi}\Big[f(x)\cos nx\Big]_{x=-\pi}^{x=\pi} + \frac{n}{\pi}\int_{-\pi}^{\pi} f(x)\sin nx\,dx = nb_n, \\ b_n' &= \frac{1}{\pi}\int_{-\pi}^{\pi} f'(x)\sin nx\,dx \\ &= \frac{1}{\pi}\Big[f(x)\sin nx\Big]_{x=-\pi}^{x=\pi} - \frac{n}{\pi}\int_{-\pi}^{\pi} f(x)\cos nx\,dx = -na_n. \end{aligned} \tag{8.2}$$

Therefore

$$f'(x) \sim \sum_{n=1}^{\infty} n(b_n \cos nx - a_n \sin nx),$$

and this is the series obtained from (8.1) by term by term differentiation.

Remark. Under the conditions of Theorem 1

$$a_n = -\frac{b_n'}{n}, \qquad b_n = \frac{a_n'}{n}, \tag{8.3}$$

which follows at once from (8.2). Moreover, since the Fourier coefficients of an absolutely integrable function converge to zero as $n \to \infty$ (see Ch. 3, Sec. 2), we can write

$$\lim_{n\to\infty} na_n = \lim_{n\to\infty} nb_n = 0,$$

i.e., a_n and b_n go to zero faster than $1/n$ as $n \to \infty$.

THEOREM 2. *Let $f(x)$ be a continuous function of period 2π, which has m derivatives, where $m - 1$ derivatives are continuous and the m'th derivative is absolutely integrable (the m'th derivative may not exist at certain points). Then, the Fourier series of all m derivatives can be obtained by term by term differentiation of the Fourier series of $f(x)$, where all the series, except possibly the last, converge to the corresponding derivatives. Moreover, the Fourier coefficients of the function $f(x)$ satisfy the relations*

$$\lim_{n\to\infty} n^m a_n = \lim_{n\to\infty} n^m b_n = 0. \tag{8.4}$$

Proof. To prove the first assertion, it is sufficient to make m applications of Theorem 1. The convergence of all the series obtained by term by term differentiation, except possibly the last, to the corresponding derivatives follows from the differentiability of these derivatives [up to the $(m - 1)$th derivative].

The relations (8.4) are obtained as a result of successively applying the relations (8.3) m times. Thus

$$a_n = -\frac{b'_n}{n} = -\frac{a''_n}{n^2} = \frac{b'''_n}{n^3} = \cdots = \frac{\alpha_n}{n^m},$$
$$b_n = \frac{a'_n}{n} = -\frac{b''_n}{n^2} = -\frac{a'''_n}{n^3} = \cdots = \frac{\beta_n}{n^m}, \tag{8.5}$$

where $a'_n, a''_n, \ldots, b'_n, b''_n, \ldots$ are the Fourier coefficients of the functions $f'(x), f''(x), \ldots,$ while α_n and β_n denote the appropriate Fourier coefficients of the function $f^{(m)}(x)$, taken with the proper sign. Since $f^{(m)}(x)$ is absolutely integrable, $\alpha_n \to 0$ and $\beta_n \to 0$ as $n \to \infty$, which implies (8.4).

Remark. Under the conditions of Theorem 2, the series for $f(x)$ and all the series obtained from $f(x)$ by term by term differentiation, except possibly for the last, converge uniformly (this follows from Ch. 3, Sec. 11).

The next proposition is in a certain sense the converse of Theorem 2:

THEOREM 3. *Given the trigonometric series*

$$\frac{a_0}{2} + \sum_{n=1}^{\infty} (a_n \cos nx + b_n \sin nx), \tag{8.6}$$

if the relations

$$|n^m a_n| \leqslant M, \quad |n^m b_n| \leqslant M \qquad (m \geqslant 2;\ M = \text{const}) \tag{8.7}$$

are satisfied by the coefficients a_n and b_n, then the sum of the series (8.6) *is a continuous function of period 2π, with $m - 2$ continuous derivatives, which can be obtained by term by term differentiation of the series* (8.6).

Proof. Denote the sum of the series (8.6) by $f(x)$. By (8.7), we can write

$$f(x) = \frac{a_0}{2} + \sum \left(\frac{\alpha_n}{n^m} \cos nx + \frac{\beta_n}{n^m} \sin nx\right),$$

where

$$|\alpha_n| \leqslant M, \quad |\beta_n| \leqslant M, \quad M = \text{const}.$$

If we formally differentiate this series, then after k differentiations the coefficients have absolute values which do not exceed the numbers

$$\frac{M}{n^{m-k}}.$$

It follows that the sum of the absolute values of the coefficients converges for $k = 1, 2, \ldots, m - 2$. Therefore, by Theorem 1 of Ch. 3, Sec. 10, the series obtained by term by term differentiation of the series (8.6) are uniformly convergent for $k = 1, 2, \ldots, m - 2$. But then, it follows from Theorem 2 of Ch. 1, Sec. 4 that the function $f(x)$ is differentiable $m - 2$ times (and hence continuous), that the derivatives are continuous, and that the term by term differentiation is legitimate.

*9. Differentiation of Fourier Series. The Case of a Function Defined on the Interval $[-\pi, \pi]$

THEOREM 1. *Let $f(x)$ be a continuous function defined on the interval $[-\pi, \pi]$ with an absolutely integrable derivative (which may not exist at certain points). Then*

$$f'(x) \sim \frac{c}{2} + \sum_{n=1}^{\infty} [(nb_n + (-1)^n c) \cos nx - na_n \sin nx], \tag{9.1}$$

where a_n and b_n are the Fourier coefficients of the function $f(x)$ and the constant c is given by the formula

$$c = \frac{1}{\pi} [f(\pi) - f(-\pi)]. \tag{9.2}$$

Proof. Let

$$f'(x) \sim \frac{a_0'}{2} + \sum_{n=1}^{\infty} (a_n' \cos nx + b_n' \sin nx),$$

so that

$$a_0' = \frac{1}{\pi} \int_{-\pi}^{\pi} f'(x)\, dx = \frac{1}{\pi} [f(\pi) - f(-\pi)].$$

Obviously we have[7]

$$f'(x) - \frac{a_0'}{2} \sim \sum_{n=1}^{\infty} (a_n' \cos nx + b_n' \sin nx). \tag{9.3}$$

The Fourier series of the function

$$\int_0^x \left(f'(x) - \frac{a_0'}{2}\right) dx = f(x) - \frac{a_0' x}{2} - f(0) \tag{9.4}$$

[7] Although (9.3) is not an equality, it is still possible to transpose terms from the right-hand side to the left-hand side, as can be verified by calculating the Fourier coefficients of the function appearing in the left-hand side. In our case, the constant term of the function $f'(x) - \frac{1}{2}a_0'$ vanishes, and all the other coefficients remain the same as for $f'(x)$.

can be obtained from the series (9.3) by term by term integration (see Theorem 3 of Sec. 7). Therefore, conversely, the series (9.3) can be obtained by term by term differentiation of the series for the function (9.4). But we have

$$f(x) = \frac{a_0}{2} + \sum_{n=1}^{\infty} (a_n \cos nx + b_n \sin nx),$$

and by (7.7)

$$f(x) - \frac{a_0' x}{2} - f(0) = \frac{a_0}{2} - f(0) + \sum_{n=1}^{\infty} \left[a_n \cos nx + \left(b_n + \frac{(-1)^n a_0'}{n}\right) \sin nx\right].$$

Therefore

$$f'(x) - \frac{a_0'}{2} \sim \sum_{n=1}^{\infty} [-na_n \sin nx + (nb_n + (-1)^n a_0') \cos nx],$$

which implies (9.1), if we set $c = a_0'$.

COROLLARY. *If* $c = 0$, *i.e., if* $f(\pi) = f(-\pi)$, *then* (9.1) *gives*

$$f'(x) \sim \sum_{n=1}^{\infty} n(b_n \cos nx - a_n \sin nx).$$

In other words, if $f(\pi) = f(-\pi)$, the Fourier series of $f(x)$ can be differentiated term by term. But this is immediately clear, since if $f(\pi) = f(-\pi)$, the periodic extension of $f(x)$ onto the whole x-axis is continuous and hence Theorem 1 of Sec. 8 can be applied.

Remark. Theorem 1 is especially important in the case where the Fourier series of $f(x)$ is given instead of $f(x)$ itself, for it shows that a knowledge of the Fourier series of $f(x)$ is all that is needed to form the Fourier series of $f'(x)$. However, in this case, the calculation of the constant c can turn out to be difficult if we use the formula (9.2). We can avoid the use of (9.2) by recalling that the Fourier coefficients of an absolutely integrable function converge to zero as $n \to \infty$ (see Ch. 3, Sec. 2). Therefore, it follows from (9.1) that

$$\lim_{n\to\infty} [nb_n + (-1)^n c] = 0,$$

whence

$$c = \lim_{n\to\infty} [(-1)^{n+1} nb_n].$$

It is usually an easy matter to calculate this limit. Moreover, it is not hard to see that the existence of this limit is equivalent to the quantity nb_n having limits for even and odd n separately, which are equal in absolute value and opposite in sign.

Theorem 1 presupposes that we know that the function $f(x)$ is continuous on $[-\pi, \pi]$ and that $f(x)$ has an absolutely integrable derivative. In the applications, one often encounters situations where we only know the Fourier series of $f(x)$. In this case, the problem becomes more complicated, i.e., from a knowledge of the Fourier series we have to ascertain whether $f(x)$ is differentiable and whether its derivative is absolutely integrable, and if the answer is affirmative, we have to form the Fourier series of the derivative. The following theorem often helps to solve this problem:

THEOREM 2. *Suppose that we are given the series*

$$\frac{a_0}{2} + \sum_{n=1}^{\infty} (a_n \cos nx + b_n \sin nx). \tag{9.5}$$

If the series

$$\frac{c}{2} + \sum_{n=1}^{\infty} [(nb_n + (-1)^n c) \cos nx - na_n \sin nx], \tag{9.6}$$

where

$$c = \lim_{n\to\infty} [(-1)^{n+1} nb_n], \tag{9.7}$$

is the Fourier series of an absolutely integrable function $\varphi(x)$,[8] *then the series* (9.5) *is the Fourier series of the function*

$$f(x) = \int_0^x \varphi(x)\,dx + \frac{a_0}{2} + \sum_{n=1}^{\infty} a_n$$

which is continuous for $-\pi < x < \pi$. *Moreover,* (9.5) *converges to* $f(x)$, *and obviously* $f'(x) = \varphi(x)$ *at all the continuity points of* $\varphi(x)$.

Proof. We can apply Theorem 2 of Sec. 7 to the series

$$\varphi(x) \sim \frac{c}{2} + \sum_{n=1}^{\infty} [(nb_n + (-1)^n c) \cos nx - na_n \sin nx]$$

obtaining

$$\begin{aligned}\int_0^x \varphi(x)\,dx &= -\sum_{n=1}^{\infty} a_n + \sum_{n=1}^{\infty} \frac{na_n \cos nx + [nb_n + (-1)^n c + (-1)^{n+1} c] \sin nx}{n} \\ &= -\sum_{n=1}^{\infty} a_n + \sum_{n=1}^{\infty} (a_n \cos nx + b_n \sin nx)\end{aligned}$$

[8] It is not assumed that the series (9.6) converges.

for $-\pi < x < \pi$. Thus, we have

$$\int_0^x \varphi(x)\,dx + \sum_{n=1}^{\infty} a_n = \sum_{n=1}^{\infty} (a_n \cos nx + b_n \sin nx)$$

and hence

$$\int_0^x \varphi(x)\,dx + \frac{a_0}{2} + \sum_{n=1}^{\infty} a_n = \frac{a_0}{2} + \sum_{n=1}^{\infty} (a_n \cos nx + b_n \sin nx).$$

Example 1. The series

$$\sum_{n=2}^{\infty} (-1)^n \frac{n \sin nx}{n^2 - 1}$$

is the Fourier series of a continuous[9] and differentiable function for $-\pi < x < \pi$. To see this, we first use (9.7) to find

$$c = \lim_{n\to\infty} \left(-\frac{n^2}{n^2 - 1}\right) = -1,$$

and then form the series (9.6). The result is

$$-\frac{1}{2} + \cos x + \sum_{n=2}^{\infty} [(-1)^n \frac{n^2}{n^2 - 1} + (-1)^{n+1}] \cos nx$$

or

$$-\frac{1}{2} + \cos x + \sum_{n=2}^{\infty} (-1)^n \frac{\cos nx}{n^2 - 1}.$$

This series converges absolutely and uniformly (since the sum of the absolute values of the coefficients is obviously convergent), and therefore it has a continuous sum $\varphi(x)$ of which it is the Fourier series (see Theorem 1 of Ch. 1, Sec. 6). According to Theorem 2

$$f(x) = \int_0^x \varphi(x)\,dx = \sum_{n=2}^{\infty} (-1)^n \frac{n \sin nx}{n^2 - 1} \tag{9.8}$$

and

$$f'(x) = \varphi(x) = -\frac{1}{2} + \cos x + \sum_{n=2}^{\infty} (-1)^n \frac{\cos nx}{n^2 - 1}. \tag{9.9}$$

Incidentally, we note that it is sometimes possible to find the sum of

[9] The continuity of the sum of the series can be inferred from Theorem 1 of Ch. 4, Sec. 4.

a Fourier series by differentiating it. Thus, in this example, if we apply Theorem 2 to the series (9.9), we obtain

$$f''(x) = -\sin x - \sum_{n=2}^{\infty} (-1)^n \frac{n \sin nx}{n^2 - 1}$$

so that

$$f''(x) = -\sin x - f(x)$$

or

$$f''(x) + f(x) = -\sin x.$$

Solving this differential equation for $f(x)$ gives

$$f(x) = c_1 \cos x + c_2 \sin x + \tfrac{1}{2}x \cos x. \tag{9.10}$$

To find c_1, we set $x = 0$, obtaining $f(0) = c_1$. According to (9.8), $f(0) = 0$ so that $c_1 = 0$. To find c_2, we differentiate (9.10) and compare the result with (9.9):

$$c_2 \cos x + \frac{\cos x}{2} - \frac{x \sin x}{2} = -\frac{1}{2} + \cos x + \sum_{n=2}^{\infty} (-1)^n \frac{\cos nx}{n^2 - 1}.$$

For $x = 0$, this gives

$$\begin{aligned} c_2 &= \sum_{n=2}^{\infty} (-1)^n \frac{1}{n^2 - 1} = \frac{1}{2} \sum_{n=2}^{\infty} (-1)^n \left(\frac{1}{n-1} - \frac{1}{n+1} \right) \\ &= \frac{1}{2} \left[\left(1 - \frac{1}{3}\right) - \left(\frac{1}{2} - \frac{1}{4}\right) + \left(\frac{1}{3} - \frac{1}{5}\right) - \left(\frac{1}{4} - \frac{1}{6}\right) + \cdots \right] = \frac{1}{4}. \end{aligned}$$

Thus we have

$$f(x) = \frac{\sin x}{4} + \frac{x \cos x}{2}.$$

We now note another useful criterion for the differentiability of a function defined by a trigonometric series:

THEOREM 3. *Consider the series*

$$\frac{a_0}{2} + \sum_{n=1}^{\infty} (-1)^n (a_n \cos nx + b_n \sin nx), \tag{9.11}$$

where a_n and b_n are positive. If the quantities na_n, nb_n do not increase (after a certain n) and converge to zero as $n \to \infty$, then the series (9.11) *converges for $-\pi < x < \pi$ and has a differentiable sum $f(x)$ for which*

$$f'(x) = \sum_{n=1}^{\infty} (-1)^n n(b_n \cos nx - a_n \sin nx), \tag{9.12}$$

i.e., the series (9.11) *can be differentiated term by term.*

Proof. By hypothesis, a_n and b_n decrease monotonically to zero as $n \to \infty$. Therefore, by Theorem 1 of Ch. 4, Sec. 4, the series (9.11) and the series in the right-hand side of (9.12) converge uniformly on every interval $[a, b]$ inside the interval $[-\pi, \pi]$. It follows that the series (9.11) can be differentiated term by term for $-\pi < x < \pi$ (see Theorem 2 of Ch. 1, Sec. 4), which proves (9.12).

Example 2. It is an immediate consequence of Theorem 3 that the series

$$\sum_{n=2}^{\infty} (-1)^n \frac{\cos nx}{n \ln n}$$

has a differentiable sum for $-\pi < x < \pi$, and that

$$f'(x) = -\sum_{n=2}^{\infty} (-1)^n \frac{\sin nx}{\ln n}.$$

*10. Differentiation of Fourier Series. The Case of a Function Defined on the Interval $[0, \pi]$

As a simple consequence of Theorem 1 of Sec. 8, we obtain

THEOREM 1. *Let $f(x)$ be continuous and have an absolutely integrable derivative on $[0, \pi]$ (the derivative may not exist at certain points), and let $f(x)$ be expanded in Fourier cosine series or Fourier sine series. Then the cosine series can always be differentiated term by term, while the sine series can be differentiated term by term if $f(0) = f(\pi) = 0$.*

Proof. In both cases, the extension of $f(x)$ onto $[-\pi, 0]$ (which is even for the cosine series and odd for the sine series) leads to a function which is continuous on $[-\pi, \pi]$ and which takes equal values at the end points of $[-\pi, \pi]$. Therefore, in both cases, the subsequent periodic extension of $f(x)$ onto the whole x-axis leads to a continuous function of period 2π with an absolutely integrable derivative. We can now apply Theorem 1 of Sec. 8.

THEOREM 2. *Let $f(x)$ be continuous and have an absolutely integrable derivative on $[0, \pi]$ (the derivative may not exist at certain points), and let $f(x)$ be expanded in the Fourier sine series*

$$f(x) = \sum_{n=1}^{\infty} b_n \sin nx \qquad (0 < x < \pi).$$

Then

$$f'(x) \sim \frac{c}{2} + \sum_{n=1}^{\infty} [nb_n - d + (c + d)(-1)^n] \cos nx, \tag{10.1}$$

where

$$c = \frac{2}{\pi}[f(\pi) - f(0)], \quad d = \frac{2}{\pi}f(0). \tag{10.2}$$

Proof. If

$$f'(x) \sim \frac{a_0'}{2} + \sum_{n=1}^{\infty} a_n' \cos nx,$$

then

$$f'(x) - \frac{a_0'}{2} \sim \sum_{n=1}^{\infty} a_n' \cos nx. \tag{10.3}$$

We know that

$$\sum_{n=1}^{\infty} (-1)^{n+1} \frac{\sin nx}{n} = \frac{x}{2},$$

$$\sum_{k=0}^{\infty} \frac{\sin (2k+1)x}{2k+1} = \frac{\pi}{4} = \frac{1}{2}\sum_{n=1}^{\infty} [1 - (-1)^n] \frac{\sin nx}{n} \tag{10.4}$$

$$(0 < x < \pi).$$

[See (13.9) and (13.11) of Ch. 1, Sec. 13.] Therefore

$$\begin{aligned} \int_0^x \left(f'(x) - \frac{a_0'}{2}\right) dx &= f(x) - \frac{a_0'x}{2} - f(0) \\ &= \sum_{n=1}^{\infty} b_n \sin nx - a_0' \sum_{n=1}^{\infty} (-1)^{n+1} \frac{\sin nx}{n} \\ &\quad - \frac{2}{\pi} f(0) \sum_{n=1}^{\infty} [1 - (-1)^n] \frac{\sin nx}{n} \\ &= \sum_{n=1}^{\infty} \left[nb_n - \frac{2}{\pi}f(0) + \left(a_0' + \frac{2}{\pi}f(0)\right)(-1)^n\right] \frac{\sin nx}{n}, \end{aligned} \tag{10.5}$$

which is the Fourier series of the function

$$\int_0^x \left(f'(x) - \frac{a_0'}{2}\right) dx.$$

But this series can also be obtained by integrating the series (10.3) term by term (see Sec. 7), and hence, conversely, (10.3) can be obtained by differentiating (10.5) term by term. Therefore

$$f'(x) - \frac{a_0'}{2} \sim \sum_{n=1}^{\infty} \left[nb_n - \frac{2}{\pi}f(0) + \left(a_0' + \frac{2}{\pi}f(0)\right)(-1)^n\right] \cos nx,$$

which implies (10.1) and (10.2) if we set

$$c = a_0' = \frac{2}{\pi}\int_0^\pi f'(x)\,dx = \frac{2}{\pi}[f(\pi) - f(0)],$$

$$d = \frac{2}{\pi}f(0).$$

COROLLARY. *If*

$$-d + (c + d)(-1)^n = 0 \qquad (n = 1, 2, \ldots),$$

then, instead of (10.1), *we obtain*

$$f'(x) \sim \sum_{n=1}^{\infty} nb_n \cos nx,$$

i.e., the Fourier series of $f'(x)$ is obtained simply by term by term differentiation of the Fourier series of $f(x)$.

Proof: This case corresponds to the condition

$$f(0) = f(\pi) = 0$$

already considered in Theorem 1. In fact, for even n, we immediately obtain $c = 0$. But then for odd n, we have $-2d = 0$ or $d = 0$, and $f(0) = f(\pi)$ follows from (10.2).

Remark. Instead of (10.2), we can determine the constants c and d by using the formulas

$$\begin{aligned} c &= -\lim_{n\to\infty} nb_n, \quad \text{where } n \text{ is even}, \\ d &= \tfrac{1}{2}(\lim_{n\to\infty} nb_n - c), \quad \text{where } n \text{ is odd}. \end{aligned} \tag{10.6}$$

To see this, we note that the Fourier coefficients of the absolutely integrable function $f'(x)$ converge to zero as $n \to \infty$. Therefore, it follows from (10.1) that

$$\lim_{n\to\infty} (nb_n + c) = 0$$

for even n, which implies the first formula (10.6), while for odd n

$$\lim_{n\to\infty} (nb_n - c - 2d) = 0,$$

which implies the second formula (10.6).

Theorems 1 and 2 have the following converses:

THEOREM 3. *Suppose that we are given the series*

$$\frac{a_0}{2} + \sum_{n=1}^{\infty} a_n \cos nx. \tag{10.7}$$

If the series[10]

$$-\sum_{n=1}^{\infty} na_n \sin nx$$

is the Fourier series of an absolutely integrable function $\varphi(x)$, *then* (10.7) *is the Fourier series of the function*

$$f(x) = \int_0^x \varphi(x)\,dx + \frac{a_0}{2} + \sum_{n=1}^{\infty} a_n,$$

which is continuous on $[0, \pi]$.[11] *Moreover,* (10.7) *converges to this function, and obviously* $f'(x) = \varphi(x)$ *at all the continuity points of* $\varphi(x)$.

This theorem is a simple consequence of Theorem 2 of Sec. 9, obtained by setting $b_n = 0$ $(n = 1, 2, \ldots)$.

THEOREM 4. *Suppose that we are given the series*

$$\sum_{n=1}^{\infty} b_n \sin nx. \tag{10.8}$$

If the limits (10.6) *exist and if the series*[10]

$$\frac{c}{2} + \sum_{n=1}^{\infty} [nb_n - d + (c + d)(-1)^n] \cos nx \tag{10.9}$$

is the Fourier series of an absolutely integrable function $\varphi(x)$, *then* (10.8) *is the Fourier series of the function*

$$f(x) = \int_0^x \varphi(x)\,dx + \frac{\pi d}{2} \qquad (0 < x < \pi).$$

Moreover, (10.8) *converges to this function, and obviously* $f'(x) = \varphi(x)$ *at all the continuity points of* $\varphi(x)$.

Proof. We can apply Theorem 2 of Sec. 7 to the series

$$\varphi(x) \sim \frac{c}{2} + \sum_{n=1}^{\infty} [nb_n - d + (c + d)(-1)^n] \cos nx,$$

obtaining

$$\begin{aligned}\int_0^x \varphi(x)\,dx &= \sum_{n=1}^{\infty} \frac{[nb_n - d + (c + d)(-1)^n + (-1)^{n+1}c] \sin nx}{n} \\ &= \sum_{n=1}^{\infty} b_n \sin nx - \sum_{n=1}^{\infty} [1 - (-1)^n] d \frac{\sin nx}{n} \\ &= \sum_{n=1}^{\infty} b_n \sin nx - \frac{\pi d}{2}\end{aligned}$$

---

[10] It is not assumed that this series converges.

[11] Since the sum of the series (10.7) is an even function, it will also be continuous on $[-\pi, \pi]$ and hence on the whole x-axis.

for $0 < x < \pi$ [see (10.4)]. Therefore

$$\int_0^x \varphi(x)\,dx + \frac{\pi d}{2} = \sum_{n=1}^{\infty} b_n \sin nx$$

for $0 < x < \pi$.

As a special case of this theorem, we have the following result:

THEOREM 5. *Given the series* (10.8), *if the limit*

$$\lim_{n\to\infty} nb_n = h \tag{10.10}$$

exists, and if the series

$$-\frac{h}{2} + \sum_{n=1}^{\infty} (nb_n - h)\cos nx \tag{10.11}$$

is the Fourier series of an absolutely integrable function $\varphi(x)$, *then* (10.8) *is the Fourier series of the function*

$$f(x) = \int_0^x \varphi(x)\,dx + \frac{\pi h}{2} \qquad (0 < x < \pi).$$

Moreover, (10.8) *converges to this function, and obviously* $f'(x) = \varphi(x)$ *at all the continuity points of* $\varphi(x)$.

To prove Theorem 5, we note that when the limit (10.10) exists, the formulas (10.6) give $c = -h$, $d = h$, and then the series (10.9) becomes (10.11).

Example 1. The series

$$\sum_{n=1}^{\infty} \frac{n^3 \sin nx}{n^4 + 1} \qquad (0 < x < \pi)$$

is the Fourier series of a function which can be differentiated any number of times. In fact, in this case (10.10) gives

$$h = \lim_{n\to\infty} \frac{n^4}{n^4 + 1} = 1,$$

and the series (10.11) is

$$-\frac{1}{2} + \sum_{n=1}^{\infty} \left(\frac{n^4}{n^4 + 1} - 1\right) \cos nx$$

or

$$-\frac{1}{2} - \sum_{n=1}^{\infty} \frac{\cos nx}{n^4 + 1}.$$

This series converges absolutely and uniformly, and therefore has a continuous sum $\varphi(x)$. By Theorem 5

$$f(x) = \sum_{n=1}^{\infty} \frac{n^3 \sin nx}{n^4 + 1} = \int_0^x \varphi(x)\,dx + \frac{\pi}{2},$$

$$f'(x) = \varphi(x)$$

for $0 < x < \pi$.

Next, we note that the Fourier coefficients of the function $\varphi(x)$ satisfy the inequality

$$|n^4 a_n| = \frac{n^4}{n^4 + 1} \leqslant 1.$$

Therefore, by Theorem 3 of Sec. 8, the function $\varphi(x)$ has two continuous derivatives, and in fact

$$\varphi'(x) = \sum_{n=1}^{\infty} \frac{n \sin nx}{n^4 + 1},$$

$$\varphi''(x) = \sum_{n=1}^{\infty} \frac{n^2 \cos nx}{n^4 + 1}.$$

We can now apply Theorem 3 to the last series, obtaining

$$\varphi'''(x) = -\sum_{n=1}^{\infty} \frac{n^3 \sin nx}{n^4 + 1} = -f(x).$$

Obviously

$$\varphi'''(x) = f^{(iv)}(x),$$

so that $f(x)$ satisfies the differential equation

$$f^{(iv)}(x) = -f(x) \qquad (0 < x < \pi),$$

which implies that the derivatives of $f(x)$ exist up to any order.

As in Sec. 9, the results of this section can be used to evaluate the sums of trigonometric series:

Example 2. *Find the sum of the series*

$$\sum_{n=1}^{\infty} \frac{\cos nx}{n^2 + 1}. \tag{10.12}$$

This series is uniformly convergent, and therefore has a continuous sum $F(x)$. Differentiating the series term by term, we obtain

$$-\sum_{n=1}^{\infty} \frac{n \sin nx}{n^2 + 1}. \tag{10.13}$$

We now apply Theorem 5 to this last series. Then, formula (10.10) gives

$$h = \lim_{n\to\infty}\left(-\frac{n^2}{n^2+1}\right) = -1,$$

and for the series (10.11), we obtain

$$\frac{1}{2} + \sum_{n=1}^{\infty}\left(-\frac{n^2}{n^2+1} + 1\right)\cos nx$$

or

$$\frac{1}{2} + \sum_{n=1}^{\infty}\frac{\cos nx}{n^2+1} = \frac{1}{2} + F(x).$$

Thus, according to Theorem 5, the sum $f(x)$ of the series (10.13) is

$$f(x) = \frac{x}{2} + \int_0^x F(x)\,dx - \frac{\pi}{2} \qquad (0 < x < \pi).$$

But then Theorem 3 is applicable to the series (10.12), and therefore

$$F'(x) = \frac{x}{2} + \int_0^x F(x)\,dx - \frac{\pi}{2} \qquad (0 < x < \pi) \tag{10.14}$$

or

$$F''(x) - F(x) = \frac{1}{2}.$$

The solution of this differential equation is

$$F(x) = c_1e^x + c_2e^{-x} - \frac{1}{2}, \tag{10.15}$$

and hence

$$F'(x) = c_1e^x - c_2e^{-x}.$$

Setting $x = 0$ in these expressions and using (10.12) and (10.14), we find that

$$\sum_{n=1}^{\infty}\frac{1}{n^2+1} = c_1 + c_2 - \frac{1}{2},$$

$$-\frac{\pi}{2} = c_1 - c_2,$$

from which we obtain

$$c_1 = \frac{1}{2}\left(\sum_{n=1}^{\infty}\frac{1}{n^2+1} + \frac{1}{2} - \frac{\pi}{2}\right),$$

$$c_2 = \frac{1}{2}\left(\sum_{n=1}^{\infty}\frac{1}{n^2+1} + \frac{1}{2} + \frac{\pi}{2}\right).$$

Then, with these values of c_1 and c_2, the function (10.15) gives the sum of the series (10.12).

11. Improving the Convergence of Fourier Series

In the applications, the most convenient trigonometric series are those with *rapidly* decreasing coefficients. In this case, the first few terms of a series suffice to give its sum quite accurately, since the sum of *all* the remaining terms will be small if the coefficients approach zero rapidly enough. Thus, the *faster* the coefficients decrease, the fewer terms of the series are needed to give its sum to any degree of accuracy. It should also be noted that in differentiating trigonometric series, the situation is the simplest when the series have rapidly decreasing coefficients (see Theorem 3 of Sec. 8).

These considerations lead us quite naturally to the following problem: Suppose we are given a trigonometric series

$$f(x) = \frac{a_0}{2} + \sum_{n=1}^{\infty} (a_n \cos nx + b_n \sin nx), \tag{11.1}$$

whose sum we denote by $f(x)$. How can we subtract from this series another trigonometric series whose sum $\varphi(x)$ is known (in finite terms) in such a way that the series which is left, i.e., the series related to $f(x)$ and $\varphi(x)$ by the formula

$$f(x) - \varphi(x) = \sum_{n=1}^{\infty} (\alpha_n \cos nx + \beta_n \sin nx),$$

has quite rapidly decreasing coefficients? Once this problem has been solved, then operations on $f(x)$ can be replaced by operations on the *known* function $\varphi(x)$ and on a series with rapidly decreasing coefficients.

The fact that this problem can be solved in cases of practical interest can be made plausible by the following argument: Suppose that we are given a function $f(x)$, defined on $[-\pi, \pi]$ (or on $[0, \pi]$), and suppose that $f(x)$ has several derivatives. Then, the periodic extension of $f(x)$ over the whole x-axis may lead to a discontinuous function (or to a function with discontinuous derivatives), with the result that the Fourier coefficients fall off slowly. It is not hard to see that by subtracting a suitably chosen *linear* function from $f(x)$, we can convert $f(x)$ into a function with equal values at the end points of the interval $[-\pi, \pi]$, and hence into a function which can be extended *continuously* over the whole x-axis, i.e., into a function with Fourier coefficients which fall off faster than those of the original function. Similarly, by subtracting a suitably chosen *polynomial* from $f(x)$, we can arrange for not only the function but also several of its derivatives to have equal values at the end points of the interval. But then both the function and these derivatives can be extended *continuously* over the whole x-axis, and Theorem 2 of Sec. 8 can be applied. This already guarantees that the coefficients fall off quite rapidly.

Thus, there is hope for solving our problem. However, in the actual problem, we are given a series and not a function. Therefore, the form of the function $\varphi(x)$ has to be inferred from its series expansion; this constitutes a difficulty which, however, can very often be overcome. When the problem has finally been solved, we say that we have *improved* the convergence of the series (11.1).

We now show two ways of improving the convergence of Fourier series. The first method is based on the fact that the difference between two infinitesimals of the same order is an infinitesimal of a higher order than the order of either of the original infinitesimals.[12]

Example 1. *Improve the convergence of the series*

$$f(x) = \sum_{n=2}^{\infty} (-1)^n \frac{n^3}{n^4 - 1} \sin nx.$$

In this case, the quantity $n^3/(n^4 - 1)$ is of order $1/n$ as $n \to \infty$ [since $n^3/(n^4 - 1) \div 1/n = n^4/(n^4 - 1) \to 1$ as $n \to \infty$]. A simple calculation shows that

$$\frac{n^3}{n^4 - 1} - \frac{1}{n} = \frac{1}{n^5 - n}.$$

Therefore we have

$$f(x) = \sum_{n=2}^{\infty} (-1)^n \frac{\sin nx}{n} + \sum_{n=2}^{\infty} (-1)^n \frac{\sin nx}{n^5 - n}.$$

But according to formula (13.9) of Ch. 1, Sec. 13

$$\sum_{n=1}^{\infty} (-1)^{n+1} \frac{\sin nx}{n} = \frac{x}{2} \qquad (-\pi < x < \pi),$$

and hence

$$f(x) = -\frac{x}{2} + \sin x + \sum_{n=2}^{\infty} (-1)^n \frac{\sin nx}{n^5 - n} \qquad (-\pi < x < \pi).$$

In the last series, obviously

$$|b_n n^5| \leqslant M \qquad (M = \text{const}),$$

i.e., the Fourier coefficients are of order $1/n^5$.

Example 2. *Improve the convergence of the series*

$$f(x) = \sum_{n=1}^{\infty} \frac{n^4 - n^2 + 1}{n^2(n^4 + 1)} \cos nx.$$

[12] Two infinitesimals are said to be *of the same order* if their ratio converges to 1.

In this case, the Fourier coefficients are of order $1/n^2$ and

$$\frac{n^4 - n^2 + 1}{n^2(n^4 + 1)} - \frac{1}{n^2} = \frac{1}{n^4 + 1}.$$

Therefore we have

$$f(x) = \sum_{n=1}^{\infty} \frac{\cos nx}{n^2} - \sum_{n=1}^{\infty} \frac{\cos nx}{n^4 + 1}.$$

But according to formula (13.8) of Ch. 1, Sec. 13

$$\sum_{n=1}^{\infty} \frac{\cos nx}{n^2} = \frac{3x^2 - 6\pi x + 2\pi^2}{12} \qquad (0 \leqslant x \leqslant 2\pi),$$

and hence

$$f(x) = \frac{3x^2 - 6\pi x - 2\pi^2}{12} - \sum_{n=1}^{\infty} \frac{\cos nx}{n^4 + 1} \qquad (0 \leqslant x \leqslant 2\pi).$$

In the last series

$$|a_n n^4| \leqslant 1,$$

i.e., the Fourier coefficients are of order $1/n^4$.

The second method of improving convergence[13] is based on representing the Fourier coefficients as sums of the form

$$\frac{A}{n} + \frac{B}{n^2} + \cdots \qquad (A = \text{const}, \quad B = \text{const}, \ldots).$$

Example 3. *Improve the convergence of the series*

$$f(x) = \sum_{n=1}^{\infty} \frac{\sin nx}{n + a} \qquad (a = \text{const}, \quad a > 0).$$

Obviously we have[14]

$$\frac{1}{n + a} = \frac{1}{n} \cdot \frac{1}{1 + \frac{a}{n}} = \frac{1}{n}\left(1 - \frac{a}{n} + \frac{a^2}{n^2} - \cdots\right).$$

We truncate the series in parentheses at the term a^2/n^2, and then add up the remainder, obtaining

$$\frac{1}{n + a} = \frac{1}{n}\left(1 - \frac{a}{n} + \frac{a^2}{n^2} - \frac{a^3}{n^2(n + a)}\right) = \frac{1}{n} - \frac{a}{n^2} + \frac{a^2}{n^3} - \frac{a^3}{n^3(n + a)}.$$

---

[13] Actually, it can be shown that the second method reduces to repeated application of the first method.

[14] Of course, we can only talk about an infinite series if $a/n < 1$, but the result we need is *always* true, as can easily be verified.

Therefore

$$f(x) = \sum_{n=1}^{\infty} \frac{\sin nx}{n} - a \sum_{n=1}^{\infty} \frac{\sin nx}{n^2} + a^2 \sum_{n=1}^{\infty} \frac{\sin nx}{n^3} - a^3 \sum_{n=1}^{\infty} \frac{\sin nx}{n^3(n+a)},$$

and the sums of the first three series, and more generally the sums of series of the form

$$\sum_{n=1}^{\infty} \frac{\sin nx}{n^p}, \qquad \sum_{n=1}^{\infty} \frac{\cos nx}{n^p}$$

(where p is a positive integer), can easily be obtained by using known expansions. In fact, according to formulas (13.7), (13.8) of Ch. 1, Sec. 13, and (14.1) of Ch. 3, Sec. 14

$$\sum_{n=1}^{\infty} \frac{\sin nx}{n} = \frac{\pi - x}{2},$$

$$\sum_{n=1}^{\infty} \frac{\cos nx}{n^2} = \frac{3x^2 - 6\pi x + 2\pi^2}{12},$$

$$\sum_{n=1}^{\infty} \frac{\cos nx}{n} = -\ln\left(2 \sin \frac{x}{2}\right) \qquad (0 < x < 2\pi).$$

Integrating the second and the third series, we obtain

$$\sum_{n=1}^{\infty} \frac{\sin nx}{n^3} = \int_0^x \frac{3x^2 - 6\pi x + 2\pi^2}{12}\, dx = \frac{x^3 - 3\pi x^2 + 2\pi^2 x}{12},$$

$$\sum_{n=1}^{\infty} \frac{\sin nx}{n^2} = -\int_0^x \ln\left(2 \sin \frac{x}{2}\right) dx$$

for $0 < x < 2\pi$. Therefore

$$f(x) = \frac{\pi - x}{2} + a \int_0^x \ln\left(2 \sin \frac{x}{2}\right) dx$$

$$+ \frac{a^2}{12}(x^3 + 3\pi x^2 - 2\pi^2 x) - a^3 \sum_{n=1}^{\infty} \frac{\sin nx}{n^3(n-a)},$$

where the coefficients of the last series are of order $1/n^4$.

12. A List of Trigonometric Expansions

In dealing with Fourier series, it is convenient to have a list of commonly encountered series. Such a list is particularly useful if we are trying to improve the convergence of a series. In the list given below, we have gathered together all the expansions obtained previously, and we have added some new ones.

1) $$\sum_{n=1}^{\infty} \frac{\cos nx}{n} = -\ln\left(2\sin\frac{x}{2}\right) \qquad (0 < x < 2\pi), \qquad \text{[Ch. 3, (14.1)]}$$

2) $$\sum_{n=1}^{\infty} \frac{\sin nx}{n} = \frac{\pi - x}{2} \qquad (0 < x < 2\pi), \qquad \text{[Ch. 1, (13.7)]}$$

3) $$\sum_{n=1}^{\infty} \frac{\cos nx}{n^2} = \frac{3x^2 - 6\pi x + 2\pi^2}{12} \qquad (0 \leqslant x \leqslant 2\pi), \ \text{[Ch. 1, (13.8)]}$$

4) $$\sum_{n=1}^{\infty} \frac{\sin nx}{n^2} = -\int_0^x \ln\left(2\sin\frac{x}{2}\right) dx \quad (0 \leqslant x \leqslant 2\pi), \quad \text{[See Sec. 11]}$$

5) $$\sum_{n=1}^{\infty} \frac{\cos nx}{n^3} = \int_0^x dx \int_0^x \ln\left(2\sin\frac{x}{2}\right) dx + \sum_{n=1}^{\infty} \frac{1}{n^3} \qquad (0 \leqslant x \leqslant 2\pi)$$

$$\left(\sum_{n=1}^{\infty} \frac{1}{n^3} = \frac{\pi^3}{25.79436\ldots} = 1.20205\ldots\right),$$

which is obtained by term by term integration of the preceding series,

6) $$\sum_{n=1}^{\infty} \frac{\sin nx}{n^3} = \frac{x^3 - 3\pi x^2 + 2\pi^2 x}{12} \qquad (0 \leqslant x \leqslant 2\pi), \quad \text{[See Sec. 11]}$$

7) $$\sum_{n=1}^{\infty} (-1)^{n+1} \frac{\cos nx}{n} = \ln\left(2\cos\frac{x}{2}\right) \quad (-\pi < x < \pi), \ \text{[Ch. 3, (14.2)]}$$

8) $$\sum_{n=1}^{\infty} (-1)^{n+1} \frac{\sin nx}{n} = \frac{x}{2} \qquad (-\pi < x < \pi), \qquad \text{[Ch. 1, (13.9)]}$$

9) $$\sum_{n=1}^{\infty} (-1)^{n+1} \frac{\cos nx}{n^2} = \frac{\pi^2 - 3x^2}{12} \quad (-\pi \leqslant x \leqslant \pi), \ \text{[Ch. 1, (13.10)]}$$

10) $$\sum_{n=1}^{\infty} (-1)^{n+1} \frac{\sin nx}{n^2} = \int_0^x \ln\left(2\cos\frac{x}{2}\right) dx \qquad (-\pi \leqslant x \leqslant \pi),$$

obtained by term by term integration of the series 7),

11) $$\sum_{n=1}^{\infty} (-1)^{n+1} \frac{\cos nx}{n^3}$$

$$= \sum_{n=1}^{\infty} (-1)^{n+1} \frac{1}{n^3} - \int_0^x dx \int_0^x \ln\left(2\cos\frac{x}{2}\right) dx \qquad (-\pi \leqslant x \leqslant \pi),$$

obtained by term by term integration of the series 10),

12) $$\sum_{n=1}^{\infty} (-1)^{n+1} \frac{\sin nx}{n^3} = \frac{\pi^2 x - x^3}{12} \qquad (-\pi \leqslant x \leqslant \pi),$$

obtained by term by term integration of the series 9),

13) $$\sum_{n=0}^{\infty} \frac{\cos(2n+1)x}{2n+1} = -\frac{1}{2}\ln\tan\frac{x}{2} \qquad (0 < x < \pi),$$

obtained by addition of the series 1) and 7),

14) $$\sum_{n=0}^{\infty} \frac{\sin(2n+1)x}{2n+1} = \frac{\pi}{4} \qquad (0 < x < \pi), \qquad \text{[Ch. 1, (13.11)]}$$

15) $$\sum_{n=0}^{\infty} \frac{\cos(2n+1)x}{(2n+1)^2} = \frac{\pi^2 - 2\pi x}{8} \qquad (0 \leqslant x \leqslant \pi), \quad \text{[Ch. 1, (13.12)]}$$

16) $$\sum_{n=0}^{\infty} \frac{\sin(2n+1)x}{(2n+1)^2} = -\frac{1}{2}\int_0^x \ln\tan\frac{x}{2}\,dx \qquad (0 \leqslant x \leqslant \pi),$$

obtained by term by term integration of the series 13),

17) $$\sum_{n=0}^{\infty} \frac{\cos(2n+1)x}{(2n+1)^3}$$
$$= \frac{1}{2}\int_0^x dx \int_0^x \ln\tan\frac{x}{2}\,dx + \sum_{n=0}^{\infty}\frac{1}{(2n+1)^3} \qquad (0 \leqslant x \leqslant \pi),$$

obtained by term by term integration of the series 16),

18) $$\sum_{n=0}^{\infty} \frac{\sin(2n+1)x}{(2n+1)^3} = \frac{\pi^2 x - \pi x^2}{8} \qquad (0 \leqslant x \leqslant \pi),$$

obtained by term by term integration of the series 15).

If in the formulas 13) through 18), we replace x by t and then set $t = \frac{1}{2}\pi - x$, we obtain the expansions

19) $$\sum_{n=0}^{\infty} (-1)^n \frac{\cos(2n+1)x}{2n+1} = \frac{\pi}{4} \qquad \left(-\frac{\pi}{2} < x < \frac{\pi}{2}\right),$$

20) $$\sum_{n=0}^{\infty} (-1)^n \frac{\sin(2n+1)x}{2n+1} = -\frac{1}{2}\ln\tan\left(\frac{\pi}{4} - \frac{x}{2}\right) \qquad \left(-\frac{\pi}{2} < x < \frac{\pi}{2}\right),$$

21) $$\sum_{n=0}^{\infty} (-1)^n \frac{\cos(2n+1)x}{(2n+1)^2} = -\frac{1}{2}\int_0^{\frac{1}{2}\pi - x} \ln\tan\frac{t}{2}\,dt \qquad \left(-\frac{\pi}{2} \leqslant x \leqslant \frac{\pi}{2}\right),$$

22) $$\sum_{n=0}^{\infty} (-1)^n \frac{\sin(2n+1)x}{(2n+1)^2} = \frac{\pi x}{4} \qquad \left(-\frac{\pi}{2} \leqslant x \leqslant \frac{\pi}{2}\right),$$

23) $$\sum_{n=0}^{\infty}(-1)^n\frac{\cos(2n+1)x}{(2n+1)^3}=\frac{\pi^3-4\pi x^2}{32}\qquad\left(-\frac{\pi}{2}\leqslant x\leqslant\frac{\pi}{2}\right),$$

24) $$\sum_{n=0}^{\infty}(-1)^n\frac{\sin(2n+1)x}{(2n+1)^3}=\frac{1}{2}\int_0^{\frac{1}{2}\pi-x}dx\int_0^x\ln\tan\frac{x}{2}\,dx$$
$$+\sum_{n=0}^{\infty}\frac{1}{(2n+1)^3}\qquad\left(-\frac{\pi}{2}\leqslant x\leqslant\frac{\pi}{2}\right).$$

13. Approximate Calculation of Fourier Coefficients

In practical problems, functions which have to be expanded in Fourier series are often given as tables or graphs, i.e., *approximately*, rather than analytically. In this case, the Fourier series cannot be obtained by direct application of the usual formulas

$$\begin{aligned} a_n &= \frac{1}{\pi}\int_0^{2\pi} f(x)\cos nx\,dx \qquad (n = 0, 1, 2, \ldots),\\ b_n &= \frac{1}{\pi}\int_0^{2\pi} f(x)\sin nx\,dx \qquad (n = 1, 2, \ldots) \end{aligned} \tag{13.1}$$

and we are faced with the problem of calculating a_n and b_n approximately. (In most cases, it is sufficient for practical purposes to know only tne first few Fourier coefficients.) To solve this problem, we go from the exact formulàs (13.1) to approximate formulas, by using the technique of approximate integration. The usual method is to use the rectangular rule or the trapezoidal rule. In our case, the rectangular rule reduces to the following procedure:

Let the interval $[0, 2\pi]$ be divided into m equal parts by the points

$$0,\ \frac{2\pi}{m},\ 2\frac{2\pi}{m},\ \ldots,\ (m-1)\frac{2\pi}{m},\ 2\pi, \tag{13.2}$$

and suppose that the values of the function $f(x)$ at these points are known to be

$$y_0,\ y_1,\ y_2,\ \ldots,\ y_{m-1},\ y_m.$$

Then we have

$$\begin{aligned} a_n &\approx \frac{2}{m}\sum_{k=0}^{m-1} y_k\cos\frac{2k\pi}{m}n,\\ b_n &\approx \frac{2}{m}\sum_{k=0}^{m-1} y_k\sin\frac{2k\pi}{m}n, \end{aligned} \tag{13.3}$$

where $\approx$ denotes approximate equality. For example, suppose $m = 12$.

Then, the numbers (13.2) are

$$0, \frac{\pi}{6}, \frac{\pi}{3}, \frac{\pi}{2}, \frac{2\pi}{3}, \frac{5\pi}{6}, \pi, \frac{7\pi}{6}, \frac{4\pi}{3}, \frac{3\pi}{2}, \frac{5\pi}{3}, \frac{11\pi}{6}, 2\pi$$

or in degrees

$$0°, 30°, 60°, 90°, 120°, 150°, 180°, 210°, 240°,$$
$$270°, 300°, 330°, 360°.$$

In this case, it is easy to see that all the factors multiplying the ordinates in (13.3) reduce to

$$0, \pm 1, \pm \sin 30° = \pm 0.5, \pm \sin 60° = \pm 0.866,$$

and it is easily verified that

$$\begin{aligned}
6a_0 &\approx y_0 + y_1 + y_2 + y_3 + y_4 + y_5 + y_6 + y_7 + y_8 + y_9 + y_{10} + y_{11},\\
6a_1 &\approx (y_0 - y_6) + 0.866(y_1 + y_{11} - y_5 - y_7) + 0.5(y_2 + y_{10} - y_4 - y_8),\\
6a_2 &\approx (y_0 + y_6 - y_3 - y_9) + 0.5(y_1 + y_5 + y_7 + y_{11} - y_2 - y_4 - y_8 - y_{10}),\\
6a_3 &\approx y_0 + y_4 + y_8 - y_2 - y_6 - y_{10},\\
6b_1 &\approx 0.5(y_1 + y_5 - y_7 - y_{11}) + 0.866(y_2 + y_4 - y_8 - y_{10}) + (y_3 - y_9),\\
6b_2 &\approx 0.866(y_1 + y_2 + y_7 + y_8 - y_4 - y_5 - y_{10} - y_{11}),\\
6b_3 &\approx y_1 + y_5 + y_9 - y_3 - y_7 - y_{11},\\
&\text{etc.}
\end{aligned} \tag{13.4}$$

To simplify the calculations, it is convenient to perform them according to the following scheme: First we write the ordinates $y_0, y_1, y_2, \ldots$ in the order shown below. Then we form the sum and difference of each pair of ordinates such that one ordinate appears below another. (The number zero is understood to appear in positions where there is no entry.)

	y_0	y_1	y_2	y_3	y_4	y_5	y_6
		y_{11}	y_{10}	y_9	y_8	y_7	
Sum	u_0	u_1	u_2	u_3	u_4	u_5	u_6
Difference		v_1	v_2	v_3	v_4	v_5	

Then we write down these sums and differences and form new sums and differences in a similar way:

	u_0	u_1	u_2	u_3
	u_6	u_5	u_4	
Sum	s_0	s_1	s_2	s_3
Difference	t_0	t_1	t_2	

	v_1	v_2	v_3
	v_5	v_4	
Sum	σ_1	σ_2	σ_3
Difference	τ_1	τ_2	

Using these quantities, we can write

$$\begin{aligned}
6a_0 &\approx s_0 + s_1 + s_2 + s_3, \\
6a_1 &\approx t_0 + 0.866t_1 + 0.5t_2, \\
6a_2 &\approx s_0 - s_3 + 0.5(s_1 - s_2), \\
6a_3 &\approx t_0 - t_2, \\
6b_1 &\approx 0.5\sigma_1 + 0.866\sigma_2 + \sigma_3, \\
6b_2 &\approx 0.866(\tau_1 + \tau_2), \\
6b_3 &\approx \sigma_1 - \sigma_3,
\end{aligned}$$

instead of (13.4).

We have given a scheme for carrying out the calculations when twelve ordinates are given. If this scheme is used for smooth functions for which we know the *exact* values of the Fourier coefficients, we find that the resulting *approximate* values of the coefficients $a_0, a_1, b_1, a_2, b_2, a_3, b_3$, are quite close to the exact values. To obtain more exact results, and also in cases where a larger number of Fourier coefficients are needed, we can use a scheme with a larger number of ordinates. The scheme with 24 ordinates is frequently used.

PROBLEMS

1. Calculate the sums of the following series:

a) $\displaystyle\sum_{n=1}^{\infty} \frac{1}{n^4}$, b) $\displaystyle\sum_{n=0}^{\infty} \frac{1}{(2n+1)^4}$, c) $\displaystyle\sum_{n=1}^{\infty} \frac{(-1)^{n+1}}{n^4}$,

d) $\displaystyle\sum_{n=1}^{\infty} \frac{1}{n^6}$, e) $\displaystyle\sum_{n=1}^{\infty} \frac{1}{(2n+1)^6}$, f) $\displaystyle\sum_{n=0}^{\infty} \frac{(-1)^{n+1}}{n^6}$.

Hint. Use the list in Sec. 12 and Parseval's theorem (Sec. 3).

2. Calculate the following integrals:

a) $\displaystyle\int_0^{\pi} \ln^2 \left(2 \sin \frac{x}{2}\right) dx$,

b) $\displaystyle\int_0^{\pi} \ln^2 \left(2 \cos \frac{x}{2}\right) dx$,

c) $\displaystyle\int_0^{\pi} \ln^2 \tan \frac{x}{2}\, dx$.

Hint. Use the list in Sec. 12 and Parseval's theorem.

3. Suppose that the function $f(x)$ is continuous and has three continuous derivatives on the interval $[0, \pi]$, and $f(0) = f(\pi) = 0, f''(0) = f''(\pi) = 0$. Show that the Fourier series of $f(x)$ can be differentiated twice term by term, and show that the series

$$\sum_{n=1}^{\infty} n^2 |b_n|$$

converges.

4. A function is defined by the series

$$f(x) = \sum_{n=1}^{\infty} \left(\frac{\cos nx}{n^3} + (-1)^n \frac{\sin nx}{n+1} \right).$$

Write the Fourier series of its derivative.

Hint. Use Theorem 2 of Sec. 9.

5. Show that the function defined by the series

$$f(x) = \sum_{n=1}^{\infty} (-1)^n \frac{\sin nx}{n(\sqrt{n}+1)}, \qquad (-\pi < x < \pi)$$

is differentiable, and write the Fourier series of its derivative.

Hint. Use Theorem 3 of Sec. 9.

6. Show that each of the following functions is differentiable, and write the Fourier series of its derivative:

a) $$f(x) = \sum_{n=1}^{\infty} \frac{n+1}{n^2+n+1} \sin nx \qquad (0 < x < \pi);$$

b) $$f(x) = \sum_{n=1}^{\infty} \frac{n^2}{5n^3+1} \sin nx \qquad (0 < x < \pi).$$

Hint. Use Theorem 5 of Sec. 10.

7. Find the sum of the series

$$\sum_{n=2}^{\infty} \frac{\cos nx}{n^2-1}.$$

Hint. Proceed as in Example 2 of Sec. 10.

8. Improve the convergence of the following series:

a) $$\sum_{n=1}^{\infty} \frac{n^3+n+1}{n(n^3+1)} \sin nx; \qquad \text{b)} \quad \sum_{n=1}^{\infty} \frac{n^3}{n^4+1} \sin nx;$$

c) $$\sum_{n=1}^{\infty} \frac{\cos nx}{n+a}$$ (In this example and those that follow, $a > 0$);

d) $$\sum_{n=1}^{\infty} (-1)^{n+1} \frac{\sin nx}{n+a}; \qquad \text{e)} \quad \sum_{n=1}^{\infty} (-1)^{n+1} \frac{\cos nx}{n+a};$$

f) $$\sum_{n=1}^{\infty} \frac{\sin (2n+1)x}{n+a}; \qquad \text{g)} \quad \sum_{n=1}^{\infty} \frac{\cos (2n+1)x}{n+a}.$$

Hint. In a) and b), proceed as in Examples 1 and 2 of Sec. 11. In c), d), and e), use

$$\frac{1}{n+a} = \frac{1}{n} \cdot \frac{1}{1+\frac{a}{n}} = \frac{1}{n}\left(1 - \frac{a}{n} + \frac{a^2}{n^2} - \cdots\right)$$

$$= \frac{1}{n}\left(1 - \frac{a}{n} + \frac{a^2}{n(n+a)}\right) = \frac{1}{n} - \frac{a}{n^2} + \frac{a^2}{n^2(n+a)},$$

divide each series into three series, and use the list in Sec. 12. In f) and g), use

$$\frac{1}{2(n+a)} = \frac{1}{2n+1+2a-1} = \frac{1}{2n+1}\cdot\frac{1}{1+\dfrac{2a-1}{2n+1}}$$

$$= \frac{1}{2n+1}\left(1 - \frac{2a-1}{2n+1} + \frac{(2a-1)^2}{2(2n+1)(n+a)}\right),$$

divide each series into three series and use the list in Sec. 12.

9. Show that if $f(x)$ is square integrable and if

$$f(x) \sim \frac{a_0}{2} + \sum_{n=1}^{\infty} (a_n \cos nx + b_n \sin nx),$$

then

$$\frac{1}{2h}\int_{x-h}^{x+h} f(u)\,du = \frac{a_0}{2} + \sum_{n=1}^{\infty} (a_n \cos nx + b_n \sin nx)\frac{\sin nh}{nh},$$

where $|h| \leqslant \pi$.

10. Show that if $f(x)$ has a square integrable derivative $f'(x)$ and if

$$\int_{-\pi}^{\pi} f(x)\,dx = 0,$$

then

$$\int_{-\pi}^{\pi} |f'(x)|^2\,dx \geqslant \int_{-\pi}^{\pi} |f(x)|^2\,dx.$$

11. Let $f(x)$ have period 2π and Fourier coefficients a_n and b_n. Show that

$$\frac{1}{\pi}\int_0^{2\pi} [f(x+h) - f(x-h)]^2\,dx = 4\sum_{n=1}^{\infty} \rho_n^2 \sin^2 nh, \tag{I}$$

where $\rho_n^2 = a_n^2 + b_n^2$. Furthermore, show that if $f(x)$ satisfies a Hölder condition of order α, i.e., if there are positive numbers c and α such that

$$|f(x) - f(y)| \leqslant c|x-y|^\alpha$$

for all x and y, then

$$\sum_{n=1}^{N} \rho_n^2 \sin^2 \frac{\pi n}{2N} \leqslant \frac{c_0}{N^{2\alpha}},$$

where $c_0 > 0$. (Cf. Ch. 1, Prob. 7.)

12. Let $f(x)$ satisfy all the conditions of the preceding problem. Show that

a) $$\sum_{n=1+2^{k-1}}^{2k} \rho_n^2 \leqslant \frac{2c_1}{2^{2\alpha k}};$$

b) $$\sum_{n=1+2^{k-1}}^{2k} \rho_n \leqslant \frac{c_2}{2^{k[\alpha-(1/2)]}};$$

c) $$\sum_{n=2}^{\infty} \rho_n \leqslant c_2 \sum_{k=1}^{\infty} \frac{1}{2^{k[\alpha-(1/2)]}}.$$

Use c) to prove that if $\alpha > \frac{1}{2}$, then the Fourier series of $f(x)$ converges absolutely (*Bernstein's theorem*).

6

SUMMATION OF TRIGONOMETRIC FOURIER SERIES

1. Statement of the Problem

Given a trigonometric Fourier series

$$\frac{a_0}{2} + \sum_{n=1}^{\infty} (a_n \cos nx + b_n \sin nx) \tag{1.1}$$

about which we know only that it is the Fourier series of some function $f(x)$, can we find the function $f(x)$? If we know in advance that the series (1.1) converges to $f(x)$, then we can obtain $f(x)$ as the limit of the partial sums of the series. The situation is different when we have not succeeded in proving that the series converges or when the series turns out to be divergent, for then we either do not know whether or not the partial sums have a limit or else we actually know that the limit does not exist. Thus, we have to find an operation which allows us to determine a function from a knowledge of its Fourier series, *regardless* of whether or not the series converges. This is the problem that will concern us in this chapter. The operation in question will be referred to as *summation* of the series. Summation must not be confused with the operation of finding the sum of a series which is known beforehand to converge, since summation can also be applied to divergent series.

The problem of summation can also be posed for arbitrary series of numbers or functions. Naturally, a properly defined summation operation must always have the property that it gives the sum of the series in the usual sense, if the series converges.

2. The Method of Arithmetic Means

Consider the series

$$u_0 + u_1 + u_2 + \cdots + u_n + \cdots, \tag{2.1}$$

and let

$$s_n = u_0 + u_1 + \cdots u_n,$$

$$\sigma_n = \frac{s_0 + s_1 + \cdots + s_{n-1}}{n} \qquad (n = 1, 2, \ldots).$$

It may happen that whereas the series (2.1) diverges, the quantities σ_n (the arithmetic means of the partial sums of the series) converge to a definite limit as $n \to \infty$. For example, the series

$$1 - 1 + 1 - 1 + \cdots$$

diverges, but in this case $s_0 = 1$, $s_1 = 0$, $s_2 = 1$, $s_3 = 0, \ldots$, so that $\sigma_n = \frac{1}{2}$ for even n and $\sigma_n = \frac{1}{2} + (2n)^{-1}$ for odd n, and hence

$$\lim_{n\to\infty} \sigma_n = \tfrac{1}{2}.$$

More generally, if

$$\lim_{n\to\infty} \sigma_n = \sigma,$$

we say that the series (2.1) is *summable by the method of arithmetic means*[1] to the value σ.

We now ask whether this method of summation satisfies the requirement mentioned at the end of Sec. 1, i.e., does the number σ give the sum of the series (2.1) in the usual sense, if the series converges? The answer is in the affirmative, as the following theorem shows:

THEOREM. *If the series* (2.1) *converges and its sum equals* σ, *then the series is summable by the method of arithmetic means to the same number* σ.

Proof. Since by hypothesis

$$\lim_{n\to\infty} s_n = \sigma,$$

then for any $\varepsilon > 0$, there exists a number m such that

$$|s_n - \sigma| < \frac{\varepsilon}{2}, \tag{2.2}$$

[1] Also known as *Cesàro's method of summation.* (*Translator*)

provided that $n \geqslant m$. Now consider

$$\sigma_n - \sigma = \frac{s_0 + s_1 + \cdots + s_{n-1} - n\sigma}{n} = \frac{1}{n}\sum_{k=0}^{n-1}(s_k - \sigma).$$

For $n > m$

$$\sigma_n - \sigma = \frac{1}{n}\sum_{k=0}^{m-1}(s_k - \sigma) + \frac{1}{n}\sum_{k=m}^{n-1}(s_k - \sigma),$$

and hence

$$|\sigma_n - \sigma| \leqslant \frac{1}{n}\sum_{k=0}^{m-1}|s_k - \sigma| + \frac{1}{n}\sum_{k=m}^{n-1}|s_k - \sigma|.$$

Since m is fixed

$$\frac{1}{n}\sum_{k=0}^{m-1}|s_k - \sigma| < \frac{\varepsilon}{2}$$

for all sufficiently large n. On the other hand, by (2.2)

$$\frac{1}{n}\sum_{k=m}^{n-1}|s_k - \sigma| < \frac{n-m-1}{n}\frac{\varepsilon}{2} < \frac{\varepsilon}{2}.$$

Therefore

$$|\sigma_n - \sigma| < \varepsilon$$

for all sufficiently large n, which proves the theorem.

Of course, the method of arithmetic means is also applicable to series of functions, in particular, to Fourier series.

3. The Integral Formula for the Arithmetic Mean of the Partial Sums of a Fourier Series

Suppose that

$$f(x) \sim \frac{a_0}{2} + \sum_{k=1}^{\infty}(a_k \cos kx + b_k \sin kx),$$

$$s_n(x) = \frac{a_0}{2} + \sum_{k=1}^{n}(a_k \cos kx + b_k \sin kx).$$

Then for the arithmetic mean of the partial sums, i.e., for

$$\sigma_n(x) = \frac{s_0(x) + s_1(x) + \cdots + s_{n-1}(x)}{n},$$

we obtain

$$\sigma_n(x) = \frac{a_0}{2} + \sum_{k=1}^{n-1} \frac{n-k}{n}(a_k \cos kx + b_k \sin kx). \tag{3.1}$$

We can also write an integral formula for $\sigma_n(x)$. In fact, as we know from (4.1) of Ch. 3

$$s_n(x) = \frac{1}{\pi}\int_{-\pi}^{\pi} f(x+u)\frac{\sin(n+\frac{1}{2})u}{2\sin(u/2)}\,du,$$

and hence

$$\sigma_n(x) = \frac{1}{\pi n}\int_{-\pi}^{\pi} \frac{f(x+u)}{2\sin(u/2)}\sum_{k=0}^{n-1}\sin(k+\tfrac{1}{2})u\,du.$$

We now calculate the sum of sines appearing in this expression. Since

$$2\sin\frac{u}{2}\sin(k+\tfrac{1}{2})u = \cos ku - \cos(k+1)u,$$

we have

$$2\sin\frac{u}{2}\sum_{k=0}^{n-1}\sin(k+\tfrac{1}{2})u = \sum_{k=0}^{n-1}(\cos ku - \cos(k+1)u)$$

$$= 1 - \cos nu = 2\sin^2\frac{nu}{2},$$

so that

$$\sum_{k=0}^{n-1}\sin(k+\tfrac{1}{2})u = \frac{\sin^2(nu/2)}{\sin(u/2)} \qquad (u \neq 0). \tag{3.2}$$

Therefore

$$\sigma_n(x) = \frac{1}{\pi n}\int_{-\pi}^{\pi} f(x+u)\frac{\sin^2(nu/2)}{2\sin^2(u/2)}\,du, \tag{3.3}$$

which is the desired integral formula.

We note the following consequence of (3.3). Suppose that $f(x) = 1$ for all x. Then $s_n(x) = 1$ $(n = 0, 1, 2, \ldots)$, and hence $\sigma_n(x) = 1$ $(n = 1, 2, \ldots)$ so that (3.3) implies

$$1 = \frac{1}{\pi n}\int_{-\pi}^{\pi}\frac{\sin^2(nu/2)}{2\sin^2(u/2)}\,du \qquad (n = 1, 2, \ldots). \tag{3.4}$$

4. Summation of Fourier Series by the Method of Arithmetic Means

THEOREM 1. *The Fourier series of an absolutely integrable function $f(x)$ of period 2π is summable by the method of arithmetic means to $f(x)$*

at every point of continuity and to the value

$$\frac{f(x+0)+f(x-0)}{2}$$

at every point of jump discontinuity.

Proof. Since

$$\frac{f(x+0)+f(x-0)}{2}=f(x)$$

at continuity points, it is sufficient to prove the relation

$$\lim_{n\to\infty}\sigma_n(x)=\frac{f(x+0)+f(x-0)}{2},$$

and to show this, it is in turn sufficient to prove that

$$\lim_{n\to\infty}\frac{1}{\pi n}\int_0^{\pi}f(x+u)\frac{\sin^2(nu/2)}{2\sin^2(u/2)}\,du=\frac{f(x+0)}{2},\tag{4.1}$$

$$\lim_{n\to\infty}\frac{1}{\pi n}\int_{-\pi}^{0}f(x+u)\frac{\sin^2(nu/2)}{2\sin^2(u/2)}\,du=\frac{f(x-0)}{2}\tag{4.2}$$

[see (3.3)]. Both of these formulas are proved in the same way, and hence we shall consider only the first of them. Since the integrand of (3.4) is an even function, we have

$$\frac{1}{2}=\frac{1}{\pi n}\int_0^{\pi}\frac{\sin^2(nu/2)}{2\sin^2(u/2)}\,du\tag{4.3}$$

so that

$$\frac{f(x+0)}{2}=\frac{1}{\pi n}\int_0^{\pi}f(x+0)\frac{\sin^2(nu/2)}{2\sin^2(u/2)}\,du.$$

Thus, it follows from (4.1) that we have to prove the formula

$$\lim_{n\to\infty}\frac{1}{\pi n}\int_0^{\pi}[f(x+u)-f(x+0)]\frac{\sin^2(nu/2)}{2\sin^2(u/2)}\,du=0.\tag{4.4}$$

Let $\varepsilon>0$ be arbitrary. Then, since

$$\lim_{\substack{u\to x\\ u>x}}f(x+u)=f(x+0),$$

it follows that

$$|f(x+u)-f(x+0)|<\varepsilon\tag{4.5}$$

for $0 < u \leqslant \delta$, if $\delta > 0$ is sufficiently small. We now divide the integral appearing in (4.4) into two integrals:

$$\frac{1}{\pi n}\int_0^{\delta} [f(x+u) - f(x+0)] \frac{\sin^2 (nu/2)}{2\sin^2 (u/2)}\, du$$
$$+ \frac{1}{\pi n}\int_{\delta}^{\pi} [f(x+u) - f(x+0)] \frac{\sin^2 (nu/2)}{2\sin^2 (u/2)}\, du = I_1 + I_2. \tag{4.6}$$

Then (4.5) implies that

$$|I_1| \leqslant \frac{\varepsilon}{\pi n}\int_0^{\delta} \frac{\sin^2 (nu/2)}{2\sin^2 (u/2)}\, du < \frac{\varepsilon}{\pi n}\int_0^{\pi} \frac{\sin^2 (nu/2)}{2\sin^2 (u/2)}\, du$$

whence, by (4.3)

$$|I_1| < \frac{\varepsilon}{2} \tag{4.7}$$

for any n. On the other hand

$$|I_2| \leqslant \frac{1}{2\pi n \sin^2 (\delta/2)}\int_{\delta}^{\pi} |f(x+u) - f(x+0)|\, du,$$

and therefore, for all sufficiently large n

$$|I_2| < \frac{\varepsilon}{2}. \tag{4.8}$$

The formula (4.4) follows from (4.6), (4.7) and (4.8).

THEOREM 2. *The Fourier series of an absolutely integrable function $f(x)$ of period 2π is uniformly summable by the method of arithmetic means to $f(x)$ on every interval $[\alpha, \beta]$ lying entirely within an interval of continuity $[a, b]$ of $f(x)$.*

Uniform summability on $[\alpha, \beta]$ means that for any $\varepsilon > 0$ and any x in $[\alpha, \beta]$, there exists a number N such that

$$|f(x) - \sigma_n(x)| \leqslant \varepsilon,$$

provided that $n \geqslant N$.

Proof. Let x lie in the interval $[\alpha, \beta]$. Then, using (3.3) and (3.4), we can write

$$\sigma_n(x) - f(x) = \frac{1}{\pi n}\int_{-\pi}^{\pi} [f(x+u) - f(x)] \frac{\sin^2 (nu/2)}{2\sin^2 (u/2)}\, du = J + j, \tag{4.9}$$

where J denotes the integral from 0 to π, and j denotes the integral from $-\pi$ to 0. Let $\delta > 0$ be so small that

$$|f(x+u) - f(x)| \leqslant \frac{\varepsilon}{2} \tag{4.10}$$

for any x in $[\alpha, \beta]$, provided that $|u| \leqslant \delta$. [It was to guarantee the existence of such a δ that we required that the interval $[\alpha, \beta]$ should lie entirely within a larger interval $[a, b]$ of continuity of $f(x)$.]

Now let M denote the maximum of $|f(x)|$ on $[\alpha, \beta]$, and consider

$$\begin{aligned} J = \frac{1}{\pi n}\int_0^\delta [f(x+u) - f(x)]\frac{\sin^2(nu/2)}{2\sin^2(u/2)}\,du \\ + \frac{1}{\pi n}\int_\delta^\pi [f(x+u) - f(x)]\frac{\sin^2(nu/2)}{2\sin^2(u/2)}\,du = I_1 + I_2. \end{aligned} \tag{4.11}$$

Then, by (4.10), the inequality

$$|I_1| \leqslant \frac{\varepsilon}{4}$$

holds for any x in $[\alpha, \beta]$. [See the proof of the inequality (4.7).] On the other hand, for any x in $[\alpha, \beta]$, we have

$$\begin{aligned} |I_2| &\leqslant \frac{1}{2\pi n \sin^2(\delta/2)}\int_\delta^\pi |f(x+u) - f(x)|\,du \\ &\leqslant \frac{1}{2\pi n \sin^2(\delta/2)}\left(\int_{-\pi}^\pi |f(x+u)|\,du + \pi M\right). \end{aligned}$$

Here, the term in parentheses is a constant, and hence there exists an N such that the inequality

$$|I_2| \leqslant \frac{\varepsilon}{4}$$

holds for all $n \geqslant N$. But then, by (4.11)

$$|J| \leqslant \frac{\varepsilon}{2}$$

for all $n \geqslant N$ and any x in $[\alpha, \beta]$.

A similar inequality can be proved for the integral j, by using the same method. The proof of the theorem is then completed by using (4.9).

Theorem 2 has the following remarkable consequence:

THEOREM 3. *The Fourier series of a continuous function $f(x)$ of period 2π is uniformly summable to $f(x)$ by the method of arithmetic means.*

To emphasize the power of the theorems just proved, we recall once more that the Fourier series of even a continuous function $f(x)$ may be *divergent*, so that the partial sums of the Fourier series may be bad approximations to $f(x)$. However, if $f(x)$ is continuous, the arithmetic means of the partial sums, i.e., the sums of the form (3.1) will be *uniform* approximations to $f(x)$.

The theorems just proved can be used to make certain questions concerning the convergence of Fourier series more precise. Thus, Theorem 1 and the result of Sec. 2 imply

THEOREM 4. *If the Fourier series of an absolutely integrable function $f(x)$ converges at a point of continuity of $f(x)$ (or at a point of jump discontinuity), then its sum must equal $f(x)$ (or $\frac{1}{2}[f(x + 0) + f(x - 0)]$).*

THEOREM 5. *If the Fourier series of an absolutely integrable function $f(x)$ converges everywhere, except possibly at a finite number of points, then its sum equals $f(x)$, except possibly at certain points.*

Theorem 5 reduces to Theorem 4 if we recall that an integrable function, as we understand the term, can be discontinuous only at a finite number of points.

Finally, Theorem 1 also implies that given the Fourier series of an absolutely integrable function $f(x)$, the method of arithmetic means can be used to reconstruct $f(x)$ at all its continuity points, i.e., everywhere except possibly at a finite number of points. Therefore, we have the following important proposition, which generalizes Theorem 3 of Ch. 5, Sec. 3:

THEOREM 6. *Any absolutely integrable function is completely defined (except for its values at a finite number of points) by its trigonometric Fourier series, whether or not the series converges.*

5. Abel's Method of Summation[2]

Consider the series

$$u_0 + u_1 + u_2 + \cdots + u_n + \cdots \tag{5.1}$$

and also the series

$$u_0 + u_1 r + u_2 r^2 + \cdots + u_n r^n + \cdots \tag{5.2}$$

We assume that the series (5.2) converges for $0 < r < 1$ [which will always be the case if the terms of the series (5.1) are bounded] and that the limit

$$\lim_{r \to 1} \sigma(r) = \sigma$$

exists, where $\sigma(r)$ is the sum of (5.2). In this case, we say that the series (5.1) is *summable by Abel's method* to the value σ.

---

[2] Equivalently, the "method of convergence factors" = Russian "метод степпеных множителей." (*Translator*)

Abel's method can be used to sum certain divergent series. For example, the series

$$1 - 1 + 1 - 1 + \cdots$$

already encountered in Sec. 2 is summable by Abel's method to the value $\sigma = \frac{1}{2}$, as well as by the method of arithmetic means. In fact, in this case

$$\sigma(r) = 1 - r + r^2 - r^3 + \cdots = \frac{1}{1 + r},$$

and therefore

$$\lim_{r \to 1} \sigma(r) = \tfrac{1}{2}.$$

We now ask whether Abel's method of summation gives the sum of the series (5.1) in the usual sense, if the series converges. The answer is in the affirmative, as the following theorem shows:

THEOREM. *If the series* (5.1) *converges and its sum equals* σ, *then the series is summable by Abel's method to the same number* σ.

Proof. If the series (5.1) converges, then by the Lemma of Ch. 4, Sec. 6, the series (5.2) converges and its sum $\sigma(r)$ is continuous on the interval $0 \leqslant r \leqslant 1$. This means that

$$\lim_{r \to 1} \sigma(r) = \sigma(1) = \sigma,$$

which proves the theorem.

6. Poisson's Kernel

We now calculate the sum of the series

$$\frac{1}{2} + \sum_{n=1}^{\infty} r^n \cos n\varphi \qquad (0 \leqslant r < 1).$$

To do this, we consider the series

$$\frac{1}{2} + \sum_{n=1}^{\infty} z^n, \quad z = r(\cos \varphi + i \sin \varphi).$$

Since $|z| = r < 1$

$$\frac{1}{2} + \sum_{n=1}^{\infty} z^n = \frac{1}{2} + \frac{z}{1 - z} = \frac{1 + z}{2(1 - z)} = \frac{1 + r\cos\varphi + ir\sin\varphi}{2(1 - r\cos\varphi - ir\sin\varphi)}$$

$$= \frac{(1 + r\cos\varphi + ir\sin\varphi)(1 - r\cos\varphi + ir\sin\varphi)}{2[(1 - r\cos\varphi)^2 + r^2\sin^2\varphi]}$$

$$= \frac{1 - r^2 + 2ir\sin\varphi}{2(1 - 2r\cos\varphi + r^2)}.$$

On the other hand

$$\frac{1}{2} + \sum_{n=1}^{\infty} z^n = \frac{1}{2} + \sum_{n=1}^{\infty} r^n (\cos n\varphi + i \sin n\varphi).$$

Therefore

$$\frac{1}{2} + \sum_{n=1}^{\infty} r^n \cos n\varphi = \frac{1}{2}\frac{1 - r^2}{1 - 2r\cos\varphi + r^2} \qquad (0 \leqslant r < 1), \tag{6.1}$$

and at the same time we have also obtained the formula

$$\sum_{n=1}^{\infty} r^n \sin n\varphi = \frac{r\sin\varphi}{1 - 2r\cos\varphi + r^2} \qquad (0 \leqslant r < 1). \tag{6.2}$$

The function

$$\frac{1 - r^2}{1 - 2r\cos\varphi + r^2}$$

of the variables r and φ is called *Poisson's kernel*. It should be pointed out that Poisson's kernel is a positive quantity, since

$$1 - r^2 > 0, \quad 1 - 2r\cos\varphi + r^2 = (1 - r)^2 + 4r\sin^2\frac{\varphi}{2} > 0, \tag{6.3}$$

for $0 \leqslant r < 1$.

7. Application of Abel's Method to the Summation of Fourier Series

Let $f(x)$ be an absolutely integrable function, and let

$$f(x) \sim \frac{a_0}{2} + \sum_{n=1}^{\infty} (a_n \cos nx + b_n \sin nx). \tag{7.1}$$

Consider the series

$$f(x, r) = \frac{a_0}{2} + \sum_{n=1}^{\infty} r^n(a_n \cos nx + b_n \sin nx), \tag{7.2}$$

where $0 \leqslant r < 1$. This series converges, since $a_n \to 0$ and $b_n \to 0$ as $n \to \infty$, and therefore

$$|a_n| \leqslant M, \quad |b_n| \leqslant M \qquad (n = 1, 2, \ldots;\ M = \text{const}),$$

$$|r^n(a_n \cos nx + b_n \sin nx)| \leqslant 2Mr^n,$$

where $2Mr^n$ is the general term of a convergent series, since $r < 1$. If

$\lim_{r\to 1} f(x, r)$ exists, this means that the series (7.2) is summable by Abel's method.

For convenience in studying the properties of this kind of summation, we first represent the function $f(x, r)$ as an integral. Recalling that

$$a_n = \frac{1}{\pi}\int_{-\pi}^{\pi} f(t)\cos nt\,dt \qquad (n = 0, 1, 2, \ldots)$$

$$b_n = \frac{1}{\pi}\int_{-\pi}^{\pi} f(t)\sin nt\,dt \qquad (n = 1, 2, \ldots),$$

we can write

$$f(x, r) = \frac{1}{2\pi}\int_{-\pi}^{\pi} f(t)\,dt + \frac{1}{\pi}\sum_{n=1}^{\infty} r^n \int_{-\pi}^{\pi} f(t)\cos n(t - x)\,dt. \tag{7.3}$$

But for fixed $r < 1$ and x, the series

$$\frac{1}{2} + \sum_{n=1}^{\infty} r^n \cos n(t - x)$$

converges uniformly in t (since its terms do not exceed in absolute value the corresponding terms of the series

$$\frac{1}{2} + \sum_{n=1}^{\infty} r^n,$$

which is known to converge) and can therefore be integrated term by term. Thus, the series

$$\frac{f(t)}{2} + \sum_{n=1}^{\infty} r^n f(t)\cos n(t - x)$$

can also be integrated term by term, and instead of (7.3), we can write

$$f(x, r) = \frac{1}{\pi}\int_{-\pi}^{\pi} f(t)\left[\frac{1}{2} + \sum_{n=1}^{\infty} r^n \cos n(t - x)\right] dt$$

or

$$f(x, r) = \frac{1}{2\pi}\int_{-\pi}^{\pi} f(t)\frac{1 - r^2}{1 - 2r\cos(t - x) + r^2}\,dt \qquad (0 \leqslant r < 1) \tag{7.4}$$

if we use (6.1). In this way, we have written $f(x, r)$ as an integral known as *Poisson's integral*. It should be noted that if $f(x) \equiv 1$, then $a_0/2 = 1$, $a_n = 0$, $b_n = 0$, and hence $f(x, r) \equiv 1$, so that (7.4) becomes

$$1 = \frac{1}{2\pi}\int_{-\pi}^{\pi} \frac{1 - r^2}{1 - 2r\cos(t - x) + r^2}\,dt \qquad (0 \leqslant r < 1). \tag{7.5}$$

THEOREM 1. *Let $f(x)$ be an absolutely integrable function of period 2π. Then*

$$\lim_{r \to 1} f(x, r) = f(x)$$

at every point where $f(x)$ is continuous, and

$$\lim_{r \to 1} f(x, r) = \frac{f(x+0) + f(x-0)}{2}$$

at every point where $f(x)$ has a jump discontinuity.

In other words, the Fourier series of $f(x)$ is summable by Abel's method to the value $f(x)$ at every point of continuity of $f(x)$ and to the value $\frac{1}{2}[f(x+0) + f(x-0)]$ at every point of jump discontinuity of $f(x)$.

Proof. Set $t - x = u$ in (7.4). Then

$$f(x, r) = \frac{1}{2\pi} \int_{-\pi-x}^{\pi-x} f(x+u) \frac{1 - r^2}{1 - 2r \cos u + r^2} \, du$$

or, since the integrand is periodic

$$f(x, r) = \frac{1}{2\pi} \int_{-\pi}^{\pi} f(x+u) \frac{1 - r^2}{1 - 2r \cos u + r^2} \, du. \tag{7.6}$$

Similarly, we can write

$$1 = \frac{1}{2\pi} \int_{-\pi}^{\pi} \frac{1 - r^2}{1 - 2r \cos u + r^2} \, du \tag{7.7}$$

instead of (7.5). Since

$$\frac{f(x+0) + f(x-0)}{2} = f(x)$$

at the continuity points of $f(x)$, it is sufficient to prove that

$$\lim_{r \to 1} f(x, r) = \frac{f(x+0) + f(x-0)}{2}$$

at every point where the right-hand and left-hand limits exist, and to show this, it is in turn sufficient to prove the formulas

$$\lim_{r \to 1} \frac{1}{2\pi} \int_{0}^{\pi} f(x+u) \frac{1 - r^2}{1 - 2r \cos u + r^2} \, du = \frac{f(x+0)}{2}, \tag{7.8}$$

$$\lim_{r \to 1} \frac{1}{2\pi} \int_{-\pi}^{0} f(x+u) \frac{1 - r^2}{1 - 2r \cos u + r^2} \, du = \frac{f(x-0)}{2}.$$

Both of these formulas are proved in the same way, and hence we shall consider only the first of them.

Since the integrand in (7.7) is even (in u), the relation

$$1 = \frac{1}{\pi}\int_0^{\pi} \frac{1 - r^2}{1 - 2r\cos u + r^2}\, du \tag{7.9}$$

holds, so that

$$\frac{f(x+0)}{2} = \frac{1}{2\pi}\int_0^{\pi} f(x+0)\frac{1 - r^2}{1 - 2r\cos u + r^2}\, du.$$

Therefore, instead of (7.8), we can prove the formula

$$\lim_{r\to 1} \frac{1}{2\pi}\int_0^{\pi} [f(x+u) - f(x+0)]\frac{1 - r^2}{1 - 2r\cos u + r^2}\, du = 0. \tag{7.10}$$

Let $\varepsilon > 0$ be arbitrary. Then, since

$$\lim_{\substack{u\to x\\ u>x}} f(x+u) = f(x+0),$$

it follows that

$$|f(x+u) - f(x+0)| \leqslant \varepsilon \tag{7.11}$$

for $0 < u \leqslant \delta$, if $\delta > 0$ is sufficiently small. We now write the integral in (7.10) as the sum of two integrals

$$\frac{1}{2\pi}\int_0^{\delta} [f(x+u) - f(x+0)]\frac{1 - r^2}{1 - 2r\cos u + r^2}\, du \tag{7.12}$$

$$+ \frac{1}{2\pi}\int_\delta^{\pi} [f(x+u) - f(x+0)]\frac{1 - r^2}{1 - 2r\cos u + r^2}\, du = I_1 + I_2.$$

Since Poisson's kernel is positive, it follows from (7.11) that

$$|I_1| \leqslant \frac{\varepsilon}{2\pi}\int_0^{\delta}\frac{1 - r^2}{1 - 2r\cos u + r^2}\, du \leqslant \frac{\varepsilon}{2\pi}\int_0^{\pi}\frac{1 - r^2}{1 - 2r\cos u + r^2}\, du$$

or, if we use (7.9)

$$|I_1| \leqslant \frac{\varepsilon}{2} \tag{7.13}$$

for any r $(0 \leqslant r < 1)$. On the other hand

$$|I_2| \leqslant \frac{1 - r^2}{8\pi r\sin^2(\delta/2)}\int_\delta^{\pi} |f(x+u) - f(x+0)|\, du \tag{7.14}$$

[see (6.3)]. Noting that

$$\lim_{r\to 1}\frac{1 - r^2}{r} = 0$$

for all r near enough to 1, we find that

$$|I_2| \leqslant \frac{\varepsilon}{2}. \tag{7.15}$$

The formula (7.10) now follows from (7.12), (7.13) and (7.15).

THEOREM 2. *The Fourier series of an absolutely integrable function $f(x)$ of period 2π is uniformly summable by Abel's method to $f(x)$ on every interval $[\alpha, \beta]$ lying entirely within an interval of continuity $[a, b]$ of $f(x)$.*

Uniform summability on $[\alpha, \beta]$ means that for any $\varepsilon > 0$ and any x in $[\alpha, \beta]$, there exists a number r_0 $(0 < r_0 < 1)$ such that the inequality $r_0 < r < 1$ implies the inequality

$$|f(x) - f(x, r)| \leqslant \varepsilon. \tag{7.16}$$

Proof. Because of (7.7), we can write

$$f(x, r) - f(x)$$
$$= \frac{1}{2\pi} \int_{-\pi}^{\pi} [f(x + u) - f(x)] \frac{1 - r^2}{1 - 2r \cos u + r^2} du = J + j, \tag{7.17}$$

where J is the integral from 0 to π, and j is the integral from $-\pi$ to 0. We can choose $\delta > 0$ to be so small that

$$|f(x + u) - f(x)| \leqslant \frac{\varepsilon}{2} \tag{7.18}$$

for any x in $[\alpha, \beta]$, provided that $|u| \leqslant \delta$. Now consider

$$J = \frac{1}{2\pi} \int_0^{\delta} [f(x + u) - f(x)] \frac{1 - r^2}{1 - 2r \cos u + r^2} du$$
$$+ \frac{1}{2\pi} \int_{\delta}^{\pi} [f(x + u) - f(x)] \frac{1 - r^2}{1 - 2r \cos u + r^2} du = I_1 + I_2.$$

By (7.18), we have

$$|I_1| \leqslant \frac{\varepsilon}{4}$$

for any x in $[\alpha, \beta]$ [cf. the proof of the inequality (7.13)]. On the other hand, for any x in $[\alpha, \beta]$ we have

$$|I_2| \leqslant \frac{1 - r^2}{8\pi r \sin^2 (\delta/2)} \int_{\delta}^{\pi} |f(x + u) - f(x)| du$$
$$\leqslant \frac{1 - r^2}{8\pi r \sin^2 (\delta/2)} \left(\int_{-\pi}^{\pi} |f(x + u)| du + M\pi \right),$$

[cf. (7.14)], where $|f(x)| \leqslant M = \text{const}$, for $\alpha \leqslant x \leqslant \beta$. [Recall that $f(x)$ is continuous on $[\alpha, \beta]$!] Since the term in parentheses is a constant, there exists a number r_0 $(0 < r_0 < 1)$ such that the inequality $r_0 < r < 1$ implies that

$$|I_2| \leqslant \frac{\varepsilon}{4}$$

for any x in $[\alpha, \beta]$. But then

$$|J| \leqslant \frac{\varepsilon}{2}$$

for $r_0 < r < 1$, $\alpha \leqslant x \leqslant \beta$.

A similar inequality can be proved for the integral j, and then (7.16) follows from (7.17), which proves the theorem.

Theorem 2 implies the following result:

THEOREM 3. *If $f(x)$ is a continuous function of period 2π, then $f(x, r) \to f(x)$ as $r \to 1$, uniformly for all x.*

In other words, the Fourier series of a continuous function $f(x)$ of period 2π is uniformly summable by Abel's method to $f(x)$.

We now cite without proof the following remarkable theorem:

THEOREM 4. *If an absolutely integrable function $f(x)$ of period 2π has a derivative $f^{(m)}(x)$ of order m at the point x, then the series obtained by differentiating the Fourier series of $f(x)$ m times term by term is summable by Abel's method to the value $f^{(m)}(x)$.*

It should be noted that when we differentiate a Fourier series term by term, we do not in general obtain the Fourier series of the derivative, even if the derivative exists for $-\pi < x < \pi$. Moreover, as a rule, term by term differentiation gives series whose coefficients do not approach zero (and such series can be shown to diverge) or even approach infinity. Thus, for example, we know that

$$\frac{x}{2} = \sum_{n=1}^{\infty} (-1)^{n+1} \frac{\sin nx}{n} \qquad (-\pi < x < \pi).$$

Term by term differentiation of this series gives

$$\sum_{n=1}^{\infty} (-1)^{n+1} \cos nx, \tag{7.19}$$

and another differentiation gives

$$-\sum_{n=1}^{\infty} (-1)^{n+1} n \sin nx. \tag{7.20}$$

According to Theorem 4, the series (7.19) is summable to $\frac{1}{2}$ for $-\pi < x < \pi$, and the series (7.20) is summable to 0.

Thus, while from the standpoint of ordinary convergence, the legitimacy of term by term differentiation of a series can only be guaranteed by rather strong requirements, Theorem 4 shows that from the standpoint of Abel's method of summation, term by term differentiation of a Fourier series (a number of times equal to the differentiability of the function itself) is always legitimate, even if the resulting series turns out to be divergent.

PROBLEMS

1. Sum the following series by the method of arithmetic means:

a) $\frac{1}{2} + \sum_{n=1}^{\infty} \cos nx$, b) $\sum_{n=1}^{\infty} \sin nx$.

Hint. For a) use formula (3.2) of Ch. 6, for b) use formula (2.1) of Ch. 4.

2. Show that Abel's method of summation gives the same answer as the method of arithmetic means for the series in Prob. 1.

3. Show that for $0 \leqslant r < 1$, the two series

$$\frac{1}{2} + \sum_{n=1}^{\infty} r^n \cos n\varphi, \qquad \sum_{n=1}^{\infty} r^n \sin n\varphi$$

can be differentiated term by term any number of times with respect to r and φ.

4. Sum the following series by Abel's method of summation:

a) $1 - 2 + 3 - 4 + \cdots$,

b) $p - 2p^2 + 3p^3 - 4p^4 + \cdots$, $|p| < 1$.

Hint.

$$1 - 2r + 3r^2 - 4r^3 + \cdots = -(1 - r + r^2 - r^3 + \cdots)' = \frac{1}{(1 + r)^2},$$

$$p - 2p^2r + 3p^3r^2 - 4p^4r^3 + \cdots = \frac{1}{(1 + pr)^2}.$$

5. Calculate the sum of the convergent series

$$\sum_{n=1}^{\infty} np^n \cos nx, \qquad |p| < 1.$$

Why is this series convergent?

Hint.

$$p \cos x + 2rp^2 \cos 2x + 3r^2p^3 \cos 3x + \cdots$$

$$= \frac{\partial}{\partial r}\left(\frac{1}{2} + rp \cos x + r^2p^2 \cos 2x + r^3p^3 \cos 3x + \cdots\right)$$

$$= \frac{\partial}{\partial r}\left(\frac{1}{2}\,\frac{1 - r^2p^2}{1 - 2rp \cos x + r^2p^2}\right).$$

6. Let $f(x)$ be square integrable, with Fourier series

$$\frac{a_0}{2} + \sum_{k=1}^{\infty} (a_k \cos kx + b_k \sin kx),$$

and let $\sigma_n(x)$ be the arithmetic mean of the partial sums of the Fourier series of $f(x)$, as in Sec. 3. Prove that

$$\frac{1}{\pi}\int_{-\pi}^{\pi} [\sigma_n(x) - f(x)]^2\,dx = \frac{1}{n^2}\sum_{k=1}^{n-1} k^2(a_k^2 + b_k^2) + \sum_{k=n}^{\infty} (a_k^2 + b_k^2).$$

7. Sum each of the following series by the method of arithmetic means and also by Abel's method:

a) $1 + 0 - 1 + 1 + 0 - 1 + \cdots;$

b) $1 + 0 + 0 - 1 + 0 + 0 + 1 + 0 + 0 - 1 + \cdots;$

c) $1 - 2 + 3 - 4 + 5 - 6 + 7 - \cdots$ (cf. Prob. 4a);

d) $1 - 2^2 + 3^2 - 4^2 + 5^2 - 6^2 + \cdots;$

e) $1\cdot 2 - 2\cdot 3 + 3\cdot 4 - 4\cdot 5 + 5\cdot 6 - \cdots.$

8. With the notation of Sec. 2, show that if $u_0 + u_1 + u_2 + \cdots + u_n + \cdots$ is summable by the method of arithmetic means, then $s_n/n \to 0$.

9. Let $u_0 + u_1 + u_2 + \cdots + u_n + \cdots$ be summable by the method of arithmetic means, and let $t_n = u_1 + 2u_2 + \cdots + nu_n$. Show that

a) The series $u_0 + u_1 + u_2 + \cdots + u_n + \cdots$ is convergent if and only if $t_n/n \to 0$;

b) If $nu_n \to 0$, then the series $u_0 + u_1 + u_2 + \cdots + u_n + \cdots$ is convergent.

10. With the notation of Sec. 2, show that

a) $$\sum_{n=0}^{\infty} u_n x^n = (1 - x)\sum_{n=0}^{\infty} s_n x^n;$$

b) $$\sum_{n=0}^{\infty} s_n x^n = (1 - x)\sum_{n=0}^{\infty} (n + 1)\sigma_{n+1} x^n.$$

11. Let $\sum_{n=0}^{\infty} u_n$ be summable to the value S by the method of arithmetic means, and let

$$\varphi(x) = \sum_{n=0}^{\infty} u_n x^n \qquad (0 \leqslant x < 1).$$

Show that

$$\varphi(x) - S = (1 - x)^2 \sum_{n=0}^{\infty} (n + 1)(\sigma_{n+1} - S)x^n.$$

Now let $\varepsilon > 0$ be given, so that by assumption, there is an integer N such that if $n \geqslant N$, then $|\sigma_{n+1} - S| < \varepsilon$. Show that

$$|\varphi(x) - S| \leqslant |1 - x|^2 \sum_{n=0}^{N} (n + 1)|\sigma_{n+1} - S| + \varepsilon|1 - x|^2 \sum_{n=N+1}^{\infty} (n + 1)x^n,$$

and hence that

$$\lim_{x\to 1} |\varphi(x) - S| < \varepsilon, \quad \text{i.e.,} \quad \lim_{x\to 1} \varphi(x) = S.$$

Comment. It follows that a series which is summable by the method of arithmetic means is summable to the same value by Abel's method.

7

DOUBLE FOURIER SERIES. THE FOURIER INTEGRAL

1. Orthogonal Systems in Two Variables

Let R be a rectangle in the xy-plane described by the inequalities $a \leqslant x \leqslant b$, $c \leqslant y \leqslant d$, and let

$$\varphi_n(x, y) \qquad (n = 0, 1, 2, \ldots), \tag{1.1}$$

be a system of continuous[1] functions defined on R, none of which vanishes identically. The system (1.1) is said to be *orthogonal* if

$$\int_R\int \varphi_n(x, y)\varphi_m(x, y)\, dx\, dy = 0$$

provided that $n \neq m$. The number

$$\|\varphi_n\| = \sqrt{\int_R\int \varphi_n^2(x, y)\, dx\, dy} \tag{1.2}$$

is called the *norm* of the function $\varphi_n(x, y)$. The system (1.1) is said to be *normalized* if

$$\|\varphi_n\| = 1 \qquad (n = 0, 1, 2, \ldots)$$

or equivalently

$$\int_R\int \varphi_n^2(x, y)\, dx\, dy = 1 \qquad (n = 0, 1, 2, \ldots).$$

[1] Instead of continuous functions, we can also consider square integrable functions, as in Ch. 2.

Every orthogonal system can be normalized, i.e., constants μ_n ($n = 0, 1, 2, \ldots$) can always be chosen such that the new system of functions

$$\mu_n\varphi_n(x, y) \qquad (n = 0, 1, 2, \ldots),$$

which is obviously still an orthogonal system, is also normalized. In fact, it is sufficient to set

$$\mu_n = \frac{1}{\|\varphi_n\|}.$$

Just as in the case of one variable (see Ch. 1), we can associate a Fourier series with every absolutely integrable function $f(x, y)$ defined on R, i.e.,

$$f(x, y) \sim c_0\varphi_0(x, y) + c_1\varphi_1(x, y) + c_2\varphi_2(x, y) + \cdots + c_n\varphi_n(x, y) + \cdots, \tag{1.3}$$

where

$$c_n = \frac{\int_R\int f(x, y)\varphi_n(x, y)\,dx\,dy}{\int_R\int \varphi_n^2(x, y)\,dx\,dy} = \frac{\int_R\int f(x, y)\varphi_n(x, y)\,dx\,dy}{\|\varphi_n\|^2}. \tag{1.4}$$

In the case where the *equality* holds in (1.3) and the series on the right converges uniformly, we find the expression (1.4) by multiplying (1.3) by each of the (continuous) functions $\varphi_n(x, y)$ and integrating term by term. The quantities c_n given by (1.4) are called the *Fourier coefficients* of $f(x, y)$.

In approximating any square integrable function $f(x, y)$ by a linear combination of functions of the system (1.1), we find that the Fourier coefficients give the least mean square error, in the way described for functions of one variable in Ch. 2, Sec. 5. Moreover, we also have *Bessel's inequality*

$$\int_R\int f^2(x, y)\,dx\,dy \geqslant \sum_{n=0}^{\infty} c_n^2\|\varphi_n\|^2, \tag{1.5}$$

where if the equality sign holds for any square integrable function, the system (1.1) is said to be *complete*. All the properties of complete systems proved for functions of one variable in Ch. 2, Secs. 7 and 8, are still valid. The completeness criterion of Ch. 2, Sec. 9 is also valid (properly rephrased for the two-dimensional case, of course).

The reader who has carefully read this section and Ch. 2 will see clearly how to generalize all that has been said about orthogonal systems to the case of functions of any number of variables.

2. The Basic Trigonometric System in Two Variables. Double Trigonometric Fourier Series

The functions

$$\begin{aligned}&1, \cos mx, \sin mx, \cos ny, \sin ny, \ldots,\\ &\cos mx \cos ny, \sin mx \cos ny,\\ &\cos mx \sin ny, \sin mx \sin ny, \ldots \qquad (m = 1, 2, \ldots;\ n = 1, 2, \ldots)\end{aligned} \tag{2.1}$$

form the *basic trigonometric system* in two variables. Each of these functions is of period 2π both in x and y. The functions of the system (2.1) are orthogonal on the square $K\ (-\pi \leqslant x \leqslant \pi, -\pi \leqslant y \leqslant \pi)$, as well as on any square of the form $(a \leqslant x \leqslant a + 2\pi, b \leqslant y \leqslant b + 2\pi)$. In fact

$$\iint_K 1 \cdot \cos mx\, dx\, dy = \int_{-\pi}^{\pi} dy \int_{-\pi}^{\pi} \cos mx\, dx = 0$$

and similarly

$$\begin{aligned}\iint_K 1 \cdot \sin mx\, dx\, dy &= \iint_K 1 \cdot \cos ny\, dx\, dy\\ &= \iint_K 1 \cdot \sin ny\, dx\, dy = 0.\end{aligned}$$

Moreover

$$\begin{aligned}&\iint_K (\cos mx \cos ny)(\cos rx \cos sy)\, dy\\ &\qquad = \int_{-\pi}^{\pi} \cos mx \cos rx \left(\int_{-\pi}^{\pi} \cos ny \cos sy\, dy\right) dx\\ &\qquad = \int_{-\pi}^{\pi} \cos mx \cos rx\, dx \int_{-\pi}^{\pi} \cos ny \cos sy\, dy = 0,\end{aligned}$$

if $m \neq r$ or $n \neq s$. The orthogonality of any pair of *different* functions of the system (2.1) is proved similarly. A calculation of the norms gives

$$\begin{aligned}\|1\| = 2\pi;\ \|\cos mx\| &= \|\sin mx\|\\ &= \|\cos ny\| = \|\sin ny\| = \sqrt{2}\,\pi;\end{aligned}$$

$$\begin{aligned}\|\cos mx \cos ny\| &= \|\sin mx \cos ny\|\\ &= \|\cos mx \sin ny\| = \|\sin mx \sin ny\| = \pi.\end{aligned}$$

For the Fourier coefficients of the function $f(x, y)$ defined on K, we obtain

$$A_{00} = \frac{\iint_K f(x, y)\, dx\, dy}{\|1\|^2} = \frac{1}{4\pi^2} \iint_K f(x, y)\, dx\, dy,$$

$$A_{m0} = \frac{\iint_K f(x, y) \cos mx \, dx \, dy}{\|\cos mx\|^2}$$

$$= \frac{1}{2\pi^2} \iint_K f(x, y) \cos mx \, dx \, dy \qquad (m = 1, 2, \ldots),$$

$$A_{0n} = \frac{\iint_K f(x, y) \cos ny \, dx \, dy}{\|\cos ny\|^2}$$

$$= \frac{1}{2\pi^2} \iint_K f(x, y) \cos ny \, dx \, dy \qquad (n = 1, 2, \ldots),$$

$$B_{m0} = \frac{\iint_K f(x, y) \sin mx \, dx \, dy}{\|\sin mx\|^2}$$

$$= \frac{1}{2\pi^2} \iint_K f(x, y) \sin mx \, dx \, dy \qquad (m = 1, 2, \ldots),$$

$$B_{0n} = \frac{\iint_K f(x, y) \sin ny \, dx \, dy}{\|\sin ny\|^2}$$

$$= \frac{1}{2\pi^2} \iint_K f(x, y) \sin ny \, dx \, dy \qquad (n = 1, 2, \ldots),$$

and similarly,

$$\begin{aligned} a_{mn} &= \frac{1}{\pi^2} \iint_K f(x, y) \cos mx \cos ny \, dx \, dy, \\ b_{mn} &= \frac{1}{\pi^2} \iint_K f(x, y) \sin mx \cos ny \, dx \, dy, \\ c_{mn} &= \frac{1}{\pi^2} \iint_K f(x, y) \cos mx \sin ny \, dx \, dy, \\ d_{mn} &= \frac{1}{\pi^2} \iint_K f(x, y) \sin mx \sin ny \, dx \, dy, \end{aligned} \qquad (2.2)$$

for $m = 1, 2, \ldots$ and $n = 1, 2, \ldots$.

Instead of A_{00}, one usually writes $\frac{1}{4}a_{00}$, and then a_{00} can be found by using the first of the formulas (2.2), with $m = 0, n = 0$. Similarly, if instead of A_{m0}, A_{0n}, B_{m0}, and B_{0n}, one writes $\frac{1}{2}a_{m0}$, $\frac{1}{2}a_{0n}$, $\frac{1}{2}b_{m0}$, and $\frac{1}{2}c_{0n}$, respectively, then a_{m0}, a_{0n}, b_{m0}, and c_{0n} can likewise be found by using the appropriate formulas (2.2).

With this notation, the Fourier series of $f(x, y)$ can be written in the form

$$f(x, y) \sim \sum_{m,n=0}^{\infty} \lambda_{mn}[a_{mn} \cos mx \cos ny + b_{mn} \sin mx \cos ny + c_{mn} \cos mx \sin ny + d_{mn} \sin mx \sin ny], \tag{2.3}$$

where

$$\lambda_{mn} = \begin{cases} \frac{1}{4} & \text{for } m = n = 0, \\ \frac{1}{2} & \text{for } m > 0, n = 0 \text{ or } m = 0, n > 0, \\ 1 & \text{for } m > 0, n > 0, \end{cases}$$

and the coefficients a_{mn}, b_{mn}, c_{mn}, and d_{mn} are calculated by the formulas (2.2) for $m = 0, 1, 2, \ldots$ and $n = 0, 1, 2, \ldots$.

The Fourier series of $f(x, y)$ can be written more compactly in the complex form

$$f(x, y) \sim \sum_{m,n=-\infty}^{\infty} c_{mn} e^{i(mx+ny)}, \tag{2.4}$$

where

$$c_{mn} = \frac{1}{4\pi^2} \int_K\int f(x, y) e^{-i(mx+ny)}\, dx\, dy \qquad (m = 0, \pm 1, \pm 2, \ldots;\ n = 0, \pm 1, \pm 2, \ldots). \tag{2.5}$$

We leave the proof of this result to the reader [it is recommended to go from (2.4) to (2.3)].

As in Ch. 5, it can be shown that the system (2.1) is complete. This means that

$$\int_K\int f^2(x, y)\, dx\, dy = 4\pi^2 A_{00}^2 + \sum_{m=1}^{\infty} 2\pi^2 A_{m0}^2 + \sum_{n=1}^{\infty} 2\pi^2 A_{0n}^2 + \sum_{m=1}^{\infty} 2\pi^2 B_{m0}^2 + \sum_{n=1}^{\infty} 2\pi^2 B_{0n}^2 + \sum_{m,n=1}^{\infty} \pi^2 (a_{mn}^2 + b_{mn}^2 + c_{mn}^2 + d_{mn}^2),$$

so that

$$\frac{1}{\pi^2} \int_K\int f^2(x, y)\, dx\, dy = \sum_{m,n=0}^{\infty} \lambda_{mn} (a_{mn}^2 + b_{mn}^2 + c_{mn}^2 + d_{mn}^2). \tag{2.6}$$

Formula (2.6) expresses *Parseval's theorem* for the case of two variables.

The analogs of all the results implied by the completeness of the trigonometric system (proved in Ch. 5 for the case of one variable) remain valid, provided that we suitably modify the way in which the results are stated.

3. The Integral Formula for the Partial Sums of a Double Trigonometric Fourier Series. A Convergence Criterion

Suppose that we have an expansion of the form (2.3), where it is assumed that $f(x, y)$ is of period 2π both in x and in y. If $f(x, y)$ is defined only on K, we extend it periodically (with respect to x and y) onto the whole xy-plane. Now let

$$s_{mn}(x, y) = \sum_{\mu=0}^{m} \sum_{\nu=0}^{n} \lambda_{\mu\nu}[a_{\mu\nu} \cos \mu x \cos \nu y + b_{\mu\nu} \sin \mu x \cos \nu y + c_{\mu\nu} \cos \mu x \sin \nu y + d_{\mu\nu} \sin \mu x \sin \nu y].$$

The quantities $s_{mn}(x, y)$ ($m = 0, 1, 2, \ldots$; $n = 0, 1, 2, \ldots$) are called the partial sums of the double Fourier series. According to (2.2)

$$\begin{aligned} s_{mn}(x, y) &= \frac{1}{\pi^2} \sum_{\mu=0}^{m} \sum_{\nu=0}^{n} \lambda_{\mu\nu} \int\!\!\int_K f(s, t) \cos \mu(s - x) \cos \nu(t - y)\, ds\, dt \\ &= \frac{1}{\pi^2} \int\!\!\int_K f(s, t) \left[\frac{1}{2} + \sum_{\mu=1}^{m} \cos \mu(s - x)\right]\left[\frac{1}{2} + \sum_{\nu=1}^{n} \cos \nu(t - y)\right] ds\, dt. \end{aligned}$$

Recalling the formula for the sum of cosines (see Ch. 3, Sec. 3), we have

$$s_{mn}(x, y) = \frac{1}{\pi^2} \int\!\!\int_K f(s, t) \frac{\sin [(m + \frac{1}{2})(s - x)] \sin [(n + \frac{1}{2})(t - y)]}{4 \sin [(s - x)/2] \sin [(t - y)/2]}\, ds\, dt.$$

Setting $s - x = u$, $t - y = v$ and using the periodicity of the integrand, we obtain

$$s_{mn}(x, y) = \frac{1}{\pi^2} \int\!\!\int_K f(x + u, y + v) \frac{\sin [(m + \frac{1}{2})u] \sin [(n + \frac{1}{2})v]}{4 \sin (u/2) \sin (v/2)}\, du\, dv.$$

This formula is completely analogous to the corresponding formula (4.1) of Ch. 3, proved for the case of one variable.

The following result can be proved by a method similar to that used in Ch. 3, Secs. 6 and 7:

THEOREM. *Let $f(x, y)$ be a continuous function defined on a square K, with bounded partial derivatives $\partial f/\partial x$ and $\partial f/\partial y$. Then, the Fourier series of $f(x, y)$ converges to $f(x, y)$ at every interior point of K in a neighborhood of which the mixed partial derivative $\partial^2 f/\partial x \partial y$ exists. If $f(x, y)$ is of period 2π in x and in y and has continuous partial derivatives $\partial f/\partial x$, $\partial f/\partial y$, $\partial^2 f/\partial x \partial y$, then the Fourier series of $f(x, y)$ converges to $f(x, y)$ everywhere.*

For the reader who is not accustomed to dealing with *double* series, we make the following remark, to avoid confusion: The formula

$$f(x, y) = \sum_{m,n=0}^{\infty} \lambda_{mn}[a_{mn} \cos mx \cos ny + b_{mn} \sin mx \cos ny + c_{mn} \cos mx \sin ny + d_{mn} \sin mx \sin ny]$$

means that

$$\lim_{\substack{m\to\infty \\ n\to\infty}} s_{mn}(x, y) = f(x, y),$$

or more precisely, that given any $\varepsilon > 0$, there exists a number N such that the inequality

$$|f(x, y) - s_{mn}(x, y)| \leqslant \varepsilon$$

holds for $m \geqslant N$, $n \geqslant N$. All that has been said above concerning the square K $(-\pi \leqslant x \leqslant \pi, -\pi \leqslant y \leqslant \pi)$ is also applicable to every square Q $(a \leqslant x \leqslant a + 2\pi, b \leqslant x \leqslant b + 2\pi)$.

Example 1. *Expand the function* $f(x, y) = xy$ *for* $-\pi < x < \pi$, $-\pi < y < \pi$ *in double Fourier series.* The formulas (2.2) give

$$a_{mn} = b_{mn} = c_{mn} = 0,$$
$$d_{mn} = (-1)^{m+n} \frac{4}{mn}.$$

Using the preceding theorem, we can write

$$xy = 4 \sum_{m,n=1}^{\infty} (-1)^{m+n} \frac{\sin mx \sin ny}{mn} \qquad (-\pi < x < \pi, -\pi < y < \pi).$$

Example 2. *Expand the same function in the square* $0 < x < 2\pi, 0 < y < 2\pi$. The formulas (2.2) give

$$a_{00} = 4\pi^2, \qquad \text{with the remaining } a_{mn} = 0,$$
$$b_{m0} = -\frac{4\pi}{m}, \qquad \text{with the remaining } b_{mn} = 0,$$
$$c_{0n} = -\frac{4\pi}{n}, \qquad \text{with the remaining } c_{mn} = 0,$$
$$d_{mn} = \frac{4}{mn} \qquad (m = 1, 2, \ldots; n = 1, 2, \ldots).$$

Therefore we have

$$xy = \pi^2 - 2\pi \sum_{m=1}^{\infty} \frac{\sin mx}{m}$$
$$- 2\pi \sum_{n=1}^{\infty} \frac{\sin ny}{n} + 4 \sum_{m,n=1}^{\infty} \frac{\sin mx \sin ny}{mn} \qquad (0 < x < 2\pi, 0 < y < 2\pi).$$

4. Double Fourier Series for a Function with Different Periods in x and y

A problem which frequently arises in the applications is that of expanding in a double trigonometric series a function $f(x, y)$ defined on a rectangle $R\,(-l \leqslant x \leqslant l, -h \leqslant y \leqslant h)$, or perhaps a function defined for all x and y, with period $2l$ in x and period $2h$ in y. This problem can be reduced to the problem already considered by making the substitutions $u = \pi x/l$, $v = \pi y/h$, since then the function

$$f\left(\frac{lu}{\pi}, \frac{hv}{\pi}\right) = \varphi(u, v)$$

has period 2π in both u and v.

If we have

$$\varphi(u, v) \sim \sum_{m,n=0}^{\infty} \lambda_{mn}[a_{mn} \cos mu \cos nv + b_{mn} \sin mu \cos nv + c_{mn} \cos mu \sin nv + d_{mn} \sin mu \sin nv],$$

then, returning to the original variables x and y, we obtain

$$f(x, y) \sim \sum_{m,n=0}^{\infty} \lambda_{mn} \left[a_{mn} \cos \frac{\pi mx}{l} \cos \frac{\pi ny}{h} + b_{mn} \sin \frac{\pi mx}{l} \cos \frac{\pi ny}{h} + c_{mn} \cos \frac{\pi mx}{l} \sin \frac{\pi ny}{h} + d_{mn} \sin \frac{\pi mx}{l} \sin \frac{\pi ny}{h}\right],$$

where

$$a_{mn} = \frac{1}{lh} \int_R\int f(x, y) \cos \frac{\pi mx}{l} \cos \frac{\pi ny}{h}\, dx\, dy,$$

and so forth. In this case, the complex form of the Fourier series becomes

$$f(x, y) \sim \sum_{m,n=-\infty}^{\infty} c_{mn} e^{i\pi[(mx/l)+(ny/h)]}$$

where

$$c_{mn} = \frac{1}{4lh} \int_R\int f(x, y) e^{-i\pi[(mx/l)+(ny/h)]}\, dx\, dy$$

$$(m = 0, \pm 1, \pm 2, \ldots;\ n = 0, \pm 1, \pm 2, \ldots).$$

All the results of Secs. 2 and 3 remain valid in the present case, provided we appropriately change the way in which the results are stated.

5. The Fourier Integral as a Limiting Case of the Fourier Series

Let $f(x)$ be a function defined for all real x, and let $f(x)$ be piecewise smooth (continuous or discontinuous) on every finite interval $[-l, l]$. Then on every such interval, $f(x)$ can be expanded as a Fourier series

$$f(x) = \frac{a_0}{2} + \sum_{n=1}^{\infty}\left(a_n \cos\frac{\pi n x}{l} + b_n \sin\frac{\pi n x}{l}\right), \tag{5.1}$$

where

$$a_n = \frac{1}{l}\int_{-l}^{l} f(u)\cos\frac{\pi n u}{l}\,du \qquad (n = 0, 1, 2, \ldots)$$

$$b_n = \frac{1}{l}\int_{-l}^{l} f(u)\sin\frac{\pi n u}{l}\,du \qquad (n = 1, 2, \ldots)$$

(see Ch. 1, Sec. 15). [At points of discontinuity, we have to write $\frac{1}{2}[f(x + 0) + f(x - 0)]$ instead of $f(x)$ in (5.1).] Substituting the expressions for a_n and b_n in (5.1), we obtain

$$f(x) = \frac{1}{2l}\int_{-l}^{l} f(u)\,du + \sum_{n=1}^{\infty}\frac{1}{l}\int_{-l}^{l} f(u)\cos\frac{\pi n}{l}(u - x)\,du. \tag{5.2}$$

We now assume that $f(x)$ is absolutely integrable on the whole x-axis, i.e., we suppose that the integral

$$\int_{-\infty}^{\infty} |f(x)|\,dx$$

exists. Then as $l \to \infty$ (x fixed), (5.2) becomes

$$f(x) = \lim_{l\to\infty}\sum_{n=1}^{\infty}\frac{1}{l}\int_{-l}^{l} f(u)\cos\frac{\pi n}{l}(u - x)\,du. \tag{5.3}$$

To investigate the limit on the right, we set

$$\lambda_1 = \frac{\pi}{l}, \quad \lambda_2 = \frac{2\pi}{l}, \ldots, \quad \lambda_n = \frac{n\pi}{l}, \ldots,$$

$$\Delta\lambda_n = \lambda_{n+1} - \lambda_n = \frac{\pi}{l},$$

after which the sum in (5.3) takes the form

$$\frac{1}{\pi}\sum_{n=1}^{\infty}\Delta\lambda_n\int_{-l}^{l} f(u)\cos\lambda_n(u - x)\,du. \tag{5.4}$$

This reminds one of the sum defining the integral of the function

$$\frac{1}{\pi}\int_{-\infty}^{\infty} f(u)\cos\lambda(u - x)\,du$$

of the variable λ over the interval $[0, +\infty]$. Therefore, it is natural to expect that as $l \to \infty$, (5.4) goes into an improper double integral, i.e., it can be anticipated that we have the formula

$$f(x) = \frac{1}{\pi}\int_0^\infty d\lambda \int_{-\infty}^\infty f(u) \cos \lambda(u - x)\, du. \tag{5.5}$$

[instead of (5.3)]. Of course, this argument is not rigorous, but at least we now know what formula to expect. It will be proved below that (5.5) is actually valid, with our assumptions concerning $f(x)$ (and even with somewhat weaker assumptions). In this regard, we point out once more that in (5.5) we have to write $\frac{1}{2}[f(x + 0) + f(x - 0)]$ instead of $f(x)$ at points where $f(x)$ is discontinuous.

The inner integral in (5.5) is called the *Fourier integral*, and the entire formula is called the *Fourier integral theorem*. Using the formula for the cosine of a difference, we can write

$$f(x) = \int_0^\infty [a(\lambda) \cos \lambda x + b(\lambda) \sin \lambda x]\, d\lambda \tag{5.6}$$

instead of (5.5), where

$$a(\lambda) = \frac{1}{\pi}\int_{-\infty}^\infty f(u) \cos \lambda u\, du, \quad b(\lambda) = \frac{1}{\pi}\int_{-\infty}^\infty f(u) \sin \lambda u\, du. \tag{5.7}$$

The reader will immediately notice a resemblance between (5.6) and a Fourier series: The sum has been replaced by an integral and the formula involves a continuously varying parameter λ instead of the integral parameter n. Moreover, the coefficients $a(\lambda)$ and $b(\lambda)$ are quite like Fourier coefficients.

6. Improper Integrals Depending on a Parameter

Consider the integral

$$\int_a^\infty F(x, \lambda)\, dx, \tag{6.1}$$

and suppose that it is convergent for $\alpha \leqslant \lambda \leqslant \beta$. We shall say that it is *uniformly convergent* for $\alpha \leqslant \lambda \leqslant \beta$ if for every $\varepsilon > 0$, there exists a number L such that

$$\left|\int_l^\infty F(x, \lambda)\, dx\right| \leqslant \varepsilon, \tag{6.2}$$

for all $l \geqslant L$ and for all λ $(\alpha \leqslant \lambda \leqslant \beta)$.

We begin by proving the following result:

A necessary and sufficient condition for the integral (6.1) *to be uniformly convergent is that for every sequence of numbers*

$$x_0 = a < x_1 < x_2 < \cdots < x_n < \cdots,$$
$$\lim_{n\to\infty} x_n = \infty, \tag{6.3}$$

the series

$$\int_a^\infty F(x, \lambda)\, dx = \int_{x_0}^{x_1} F(x, \lambda)\, dx + \int_{x_1}^{x_2} F(x, \lambda)\, dx + \cdots \\ + \int_{x_n}^{x_{n+1}} F(x, \lambda)\, dx + \cdots, \tag{6.4}$$

whose terms are functions of λ, *should be uniformly convergent for* $\alpha \leqslant \lambda \leqslant \beta$.

The proof goes as follows: If (6.2) holds, then for $x_n \geqslant L$ we have

$$\begin{aligned} \left|\int_a^\infty F(x, \lambda)\, dx - \sum_{k=1}^{n} \int_{x_{k-1}}^{x_k} F(x, \lambda)\, dx\right| \\ &= \left|\int_a^\infty F(x, \lambda)\, dx - \int_a^{x_n} F(x, \lambda)\, dx\right| \\ &= \left|\int_{x_n}^\infty F(x, \lambda)\, dx\right| \leqslant \varepsilon \qquad (\alpha \leqslant \lambda \leqslant \beta), \end{aligned} \tag{6.5}$$

which means that the series (6.4) is uniformly convergent. Conversely, suppose that the series (6.4) is uniformly convergent for every sequence of the form (6.3). Then, if the integral (6.1) were not uniformly convergent, this would imply the existence for at least one $\varepsilon > 0$ of arbitrarily large numbers $x_1 < x_2 < \cdots < x_n < \cdots$, such that

$$\left|\int_{x_n}^\infty F(x, \lambda)\, dx\right| > \varepsilon \qquad (n = 1, 2, \ldots)$$

for each x_i $(i = 1, 2, \ldots)$ and some λ. But this is impossible, since if we choose these values of x_i as the sequence (6.3), then (6.5) must hold for sufficiently large n. This contradiction shows that the integral (6.1) is uniformly convergent.

This result implies

THEOREM 1. *If* $F(x, \lambda)$ *is continuous, regarded as a function of two variables (or if* $F(x, \lambda)$ *has discontinuities at a finite number of values of* x *in each finite interval, while remaining integrable with respect to* x *and continuous in* λ*) and if the integral* (6.1) *is uniformly convergent for* $\alpha \leqslant \lambda \leqslant \beta$, *then the integral is a continuous function of* λ.

Proof. Every term of the series (6.4) is a continuous function of λ (by a property of integrals with finite limits), and since this series converges uniformly, its sum, i.e., the integral (6.1), is a continuous function.

THEOREM 2. *With the hypotheses of Theorem* 1, *we have*

$$\int_a^\beta d\lambda \int_a^\infty F(x, \lambda)\, dx = \int_a^\infty dx \int_a^\beta F(x, \lambda)\, d\lambda.$$

Proof. Because of the uniform convergence of the series (6.4), it can be integrated term by term, and the order of integration can be interchanged in every term (this is legitimate for *finite* intervals). Thus, we obtain

$$\int_\alpha^\beta d\lambda \int_a^\infty F(x, \lambda)\, dx = \int_{x_0}^{x_1} dx \int_\alpha^\beta F(x, \lambda)\, d\lambda + \int_{x_1}^{x_2} dx \int_\alpha^\beta F(x, \lambda)\, d\lambda + \cdots$$

$$= \lim_{n\to\infty} \int_a^{x_n} dx \int_\alpha^\beta F(x, \lambda)\, d\lambda = \int_a^\infty dx \int_\alpha^\beta F(x, \lambda)\, d\lambda.$$

THEOREM 3. *Suppose that $F(x, \lambda)$ is continuous, regarded as a function of two variables, and has a continuous partial derivative $\partial F(x, \lambda)/\partial\lambda$. Suppose further that the integrals*

$$\int_a^\infty F(x, \lambda)\, dx, \qquad \int_a^\infty \frac{\partial F(x, \lambda)}{\partial \lambda}\, dx$$

exist and that the second of them is uniformly convergent for $\alpha \leqslant \lambda \leqslant \beta$. Then we have

$$\frac{\partial}{\partial\lambda}\int_a^\infty F(x, \lambda)\, dx = \int_a^\infty \frac{\partial F(x, \lambda)}{\partial \lambda}\, dx \qquad (\alpha \leqslant \lambda \leqslant \beta). \tag{6.6}$$

Proof. The series

$$\int_{x_0}^{x_1} \frac{\partial F(x, \lambda)}{\partial \lambda}\, dx + \int_{x_1}^{x_2} \frac{\partial F(x, \lambda)}{\partial \lambda}\, dx + \cdots = \int_a^\infty \frac{\partial F(x, \lambda)\, dx}{\partial \lambda}$$

whose terms are the derivatives of the corresponding terms of the series (6.4) (since for *finite* intervals, differentiation under the sign of integration is legitimate), is uniformly convergent. But then we are justified in differentiating (6.4) term by term (see Theorem 2 of Ch. 1, Sec. 4).

The next theorem gives a very useful criterion for the uniform convergence of the integral (6.1):

THEOREM 4. *If for $\alpha \leqslant \lambda \leqslant \beta$*

$$|F(x, \lambda)| \leqslant f(x)$$

and the integral

$$\int_a^\infty |f(x)|\, dx$$

exists, then the integral (6.1) *is uniformly convergent.* (The requirements on $F(x, \lambda)$, regarded as a function of x, are the same as in Theorem 1.)

Proof. The absolute values of the terms of the series (6.4) do not exceed the terms of the convergent numerical series

$$\int_a^\infty |f(x)|\,dx = \int_{x_0}^{x_1} |f(x)|\,dx + \int_{x_1}^{x_2} |f(x)|\,dx + \cdots,$$

which proves the theorem.

Instead of (6.1), we can consider integrals of the form

$$\int_{-\infty}^b F(x, \lambda)\,dx, \qquad \int_{-\infty}^\infty F(x, \lambda)\,dx, \tag{6.7}$$

with the condition (6.2) replaced by the condition

$$\left|\int_{-\infty}^{-l} F(x, \lambda)\,dx\right| \leqslant \varepsilon$$

for the first integral, and by the two conditions

$$\left|\int_{-\infty}^{-l} F(x, \lambda)\,dx\right| \leqslant \varepsilon, \qquad \left|\int_l^\infty F(x, \lambda)\,dx\right| \leqslant \varepsilon$$

for the second integral. The theorems just proved remain valid for integrals of this form. In fact, the proofs reduce to the case already considered if we make the substitution $x = -y$ in the first integral, and divide the second integral into the two integrals

$$\int_{-\infty}^0 F(x, \lambda)\,dx, \qquad \int_0^\infty F(x, \lambda)\,dx.$$

7. Two Lemmas

We now refine the results of Ch. 3, Sec. 2:

LEMMA 1. *If $f(x)$ is absolutely integrable on $a \leqslant x < \infty$, then*

$$\lim_{l\to\infty} \int_a^\infty f(u) \sin lu\,du = 0 \tag{7.1}$$

(*where l takes arbitrary values*).

Proof. For any given $\varepsilon > 0$

$$\left|\int_b^\infty f(u) \sin lu\,du\right| \leqslant \frac{\varepsilon}{2}$$

for sufficiently large b. Moreover, by the theorem of Ch. 3, Sec. 2

$$\left|\int_a^b f(u) \sin lu\,du\right| \leqslant \frac{\varepsilon}{2}$$

for sufficiently large l. Therefore

$$\left| \int_a^{\infty} f(u) \sin lu \, du \right| \leqslant \varepsilon$$

for all sufficiently large l, which proves (7.1).

Remark. Instead of the integral from a to ∞, we can consider the integrals from $-\infty$ to a and from $-\infty$ to ∞. Also, instead of $\sin lu$, we can consider $\cos lu$. The proofs are essentially the same.

LEMMA 2. *If the function $f(x)$ is absolutely integrable on the whole x-axis, and if $f(x)$ has left-hand and right-hand limits at the point x, then*

$$\lim_{l \to \infty} \frac{1}{\pi} \int_{-\infty}^{\infty} f(x + u) \frac{\sin lu}{u} \, du = \frac{f(x + 0) + f(x - 0)}{2}. \tag{7.2}$$

Proof. Given any $\varepsilon > 0$, the inequality

$$\frac{1}{\pi} \int_{-\delta}^{\delta} |f(x + u)| \, du < \frac{\varepsilon}{2}$$

holds for sufficiently small $\delta > 0$. The function $(1/u)f(x + u)$ is absolutely integrable for $-\infty < u \leqslant \delta$ and $\delta \leqslant u < \infty$. Therefore, by Lemma 1

$$\lim_{l \to \infty} \frac{1}{\pi} \int_{\delta}^{\infty} f(x + u) \frac{\sin lu}{u} \, du = \lim_{l \to \infty} \frac{1}{\pi} \int_{-\infty}^{-\delta} f(x + u) \frac{\sin lu}{u} \, du = 0. \tag{7.3}$$

Now consider the equality

$$\lim_{m \to \infty} \frac{1}{\pi} \int_{-\pi}^{\pi} f(x + u) \frac{\sin mu}{2 \sin (u/2)} \, du = \frac{f(x + 0) + f(x - 0)}{2},$$

proved in Ch. 3, Sec. 7, where $m = n + \frac{1}{2}$ for integral n. This can be written as

$$\lim_{m \to \infty} \frac{1}{\pi} \int_{-\delta}^{\delta} f(x + u) \frac{\sin mu}{2 \sin (u/2)} \, du = \frac{f(x + 0) + f(x - 0)}{2}, \tag{7.4}$$

since the integrals on the intervals $[-\pi, -\delta]$, $[\delta, \pi]$ go to zero as $m \to \infty$, because of the absolute integrability of the function

$$\frac{f(x + u)}{2 \sin (u/2)}$$

on these intervals (see Ch. 3, Sec. 2).

Next, we observe that the integral in the left-hand side of (7.4) differs from the integral

$$\frac{1}{\pi} \int_{-\delta}^{\delta} f(x + u) \frac{\sin mu}{u} \, du \tag{7.5}$$

by the quantity

$$\frac{1}{\pi}\int_{-\delta}^{\delta} f(x+u)\left[\frac{1}{2\sin(u/2)}-\frac{1}{u}\right]\sin mu\,du, \tag{7.6}$$

where the function in brackets is continuous, if it is regarded as being equal to zero at the origin. (This can be seen by applying L'Hospital's rule.) But by Ch. 3, Sec. 2, the integral (7.6) converges to zero as $m \to \infty$. Therefore, instead of (7.4), we can write

$$\lim_{m\to\infty}\frac{1}{\pi}\int_{-\delta}^{\delta} f(x+u)\frac{\sin mu}{u}\,du = \frac{f(x+0)+f(x-0)}{2}. \tag{7.7}$$

Now let $m \leqslant l < m + 1$, so that $l = m + \theta$, where $0 \leqslant \theta < 1$. Applying the mean value theorem, we obtain

$$\frac{\sin lu - \sin mu}{u} = (l-m)\cos hu = \theta\cos hu,$$

where h lies between m and l. Therefore

$$\left|\frac{1}{\pi}\int_{-\delta}^{\delta} f(x+u)\frac{\sin lu}{u}\,du - \frac{1}{\pi}\int_{-\delta}^{\delta} f(x+u)\frac{\sin mu}{u}\,du\right|$$

$$= \frac{1}{\pi}\left|\int_{-\delta}^{\delta} f(x+u)\theta\cos hu\,du\right| \leqslant \frac{1}{\pi}\int_{-\delta}^{\delta}|f(x+u)|\,du < \frac{\varepsilon}{2} \tag{7.8}$$

for any l. If l is large, then m is also large, and therefore by (7.7), we obtain

$$\left|\frac{f(x+0)+f(x-0)}{2} - \frac{1}{\pi}\int_{-\delta}^{\delta} f(x+u)\frac{\sin mu}{u}\,du\right| < \frac{\varepsilon}{2}$$

for the values of m corresponding to large values of l. Together with (7.8), this inequality gives

$$\left|\frac{f(x+0)+f(x-0)}{2} - \frac{1}{\pi}\int_{-\delta}^{\delta} f(x+u)\frac{\sin lu}{u}\,du\right| < \varepsilon$$

for all sufficiently large l. By (7.3), instead of (7.9), we can write

$$\left|\frac{f(x+0)+f(x-0)}{2} - \frac{1}{\pi}\int_{-\infty}^{\infty} f(x+u)\frac{\sin lu}{u}\,du\right| < \varepsilon,$$

for sufficiently large l. This proves (7.2).

8. Proof of the Fourier Integral Theorem

Suppose that $f(x)$ is absolutely integrable on the whole x-axis. Then, by the very definition of an improper integral, we have

$$\frac{1}{\pi}\int_0^\infty d\lambda \int_{-\infty}^\infty f(u)\cos\lambda(u-x)\,du = \lim_{l\to\infty}\frac{1}{\pi}\int_0^l d\lambda\int_{-\infty}^\infty f(u)\cos\lambda(u-x)\,du, \tag{8.1}$$

i.e., the existence of the integral in the left-hand side of (8.1) is equivalent to the existence of the limit in the right-hand side. But the integral

$$\int_{-\infty}^\infty f(u)\cos\lambda(u-x)\,du$$

is uniformly convergent for $-\infty < \lambda < \infty$, since

$$|f(u)\cos\lambda(u-x)| \leqslant |f(u)|,$$

and $f(u)$ is absolutely integrable on the whole x-axis (see Theorem 4 of Sec. 6). It follows by Theorem 2 of Sec. 6 that

$$\int_0^l d\lambda\int_{-\infty}^\infty f(u)\cos\lambda(u-x)\,du = \int_{-\infty}^\infty du\int_0^l f(u)\cos\lambda(u-x)\,d\lambda$$
$$= \int_{-\infty}^\infty f(u)\frac{\sin l(u-x)}{u-x}\,du = \int_{-\infty}^\infty f(x+u)\frac{\sin lu}{u}\,du,$$

where we have first made the substitution $u - x = v$ and then changed v back to u. By (8.1), we have

$$\frac{1}{\pi}\int_0^\infty d\lambda\int_{-\infty}^\infty f(u)\cos\lambda(u-x)\,du = \lim_{l\to\infty}\frac{1}{\pi}\int_{-\infty}^\infty f(x+u)\frac{\sin lu}{u}\,du. \tag{8.2}$$

Now, if the function $f(x)$ has a left-hand and a right-hand derivative at the point x, then by Lemma 2 of Sec. 7, the limit in (8.2) exists and is equal to $\frac{1}{2}[f(x+0)+f(x-0)]$. Therefore, the integral in the left-hand side of (8.2) exists and we have

$$\frac{1}{\pi}\int_0^\infty d\lambda\int_{-\infty}^\infty f(u)\cos\lambda(u-x)\,du = \frac{f(x+0)+f(x-0)}{2}. \tag{8.3}$$

At the continuity points of $f(x)$, the right-hand side of (8.3) coincides with $f(x)$. Thus, we have the following results:

1. *If $f(x)$ is absolutely integrable on the whole x-axis, the Fourier integral theorem holds at every point where $f(x)$ has both a left-hand and a right-hand derivative.*

2. *If $f(x)$ is absolutely integrable on the whole x-axis and if $f(x)$ is piecewise smooth on every finite interval, then the Fourier integral theorem holds for all x.*

9. Different Forms of the Fourier Integral Theorem

Assume that $f(x)$ is absolutely integrable on the whole x-axis and consider the integral

$$\int_{-\infty}^{\infty} f(u) \sin \lambda(u - x)\, du,$$

which converges uniformly for $-\infty < \lambda < \infty$, since

$$|f(u) \sin \lambda(u - x)| \leqslant |f(u)|$$

(see Theorem 4 of Sec. 6). Therefore the integral represents a continuous function of λ, which is obviously *odd*. But then[2]

$$\lim_{l\to\infty} \int_{-l}^{l} d\lambda \int_{-\infty}^{\infty} f(u) \sin \lambda(u - x)\, du$$
$$= \int_{-\infty}^{\infty} d\lambda \int_{-\infty}^{\infty} f(u) \sin \lambda(u - x)\, du = 0.$$

On the other hand, the integral

$$\int_{-\infty}^{\infty} f(u) \cos \lambda(u - x)\, du$$

represents an *even* function of λ. Therefore, instead of (5.5), we can write

$$\begin{aligned} f(x) &= \frac{1}{2\pi} \int_{-\infty}^{\infty} d\lambda \int_{-\infty}^{\infty} f(u)[\cos \lambda(u - x) + i \sin \lambda(u - x)]\, du \\ &= \frac{1}{2\pi} \int_{-\infty}^{\infty} d\lambda \int_{-\infty}^{\infty} f(u) e^{i\lambda(u-x)}\, du. \end{aligned} \tag{9.1}$$

which is the *complex form* of the Fourier integral theorem.

Next, we write (5.5) in the form

$$\begin{aligned} f(x) = \frac{1}{\pi} \int_{0}^{\infty} \cos \lambda x \left(\int_{-\infty}^{\infty} f(u) \cos \lambda u\, du \right) d\lambda \\ + \frac{1}{\pi} \int_{0}^{\infty} \sin \lambda x \left(\int_{-\infty}^{\infty} f(u) \sin \lambda u\, du \right) d\lambda. \end{aligned} \tag{9.2}$$

---

[2] If the integral with respect to λ in the right-hand side of this formula does not exist *in the usual sense*, then we interpret it as

$$\lim_{l\to\infty} \int_{-l}^{l}$$

a limit which in this case obviously exists and equals 0 (the *Cauchy principal value* of the integral).

If $f(x)$ is *even*, then

$$\int_{-\infty}^{\infty} f(u) \cos \lambda u \, du = 2 \int_{0}^{\infty} f(u) \cos \lambda u \, du,$$

$$\int_{-\infty}^{\infty} f(u) \sin \lambda u \, du = 0,$$

and (9.2) becomes

$$f(x) = \frac{2}{\pi} \int_0^\infty \cos \lambda x \left(\int_0^\infty f(u) \cos \lambda u \, du \right) d\lambda. \tag{9.3}$$

Similarly, if $f(x)$ is odd, we obtain

$$f(x) = \frac{2}{\pi} \int_0^\infty \sin \lambda x \left(\int_0^\infty f(u) \sin \lambda u \, du \right) d\lambda. \tag{9.4}$$

If $f(x)$ is defined only on $[0, \infty]$, then (9.3) gives the *even extension* of $f(x)$ onto the whole x-axis, while (9.4) gives the *odd extension* of $f(x)$. Thus, both formulas are applicable for positive x, but for negative x they give different values of $f(x)$. We note that if $f(x)$ is continuous, then (9.3) is always valid at $x = 0$, whereas (9.4) is valid only if $f(0) = 0$. [This is because when $x = 0$, the integral (9.4) takes the value $\frac{1}{2}[f(+0) + f(-0)]$, and this value is always zero for the odd extension of $f(x)$.]

*10. The Fourier Transform

Given the function $f(x)$, the new function

$$F(x) = \frac{1}{\sqrt{2\pi}} \int_{-\infty}^{\infty} f(u) e^{ixu} \, du \tag{10.1}$$

is called the *Fourier transform* of $f(x)$. If the Fourier integral theorem holds for $f(x)$, then by (9.1) we have

$$f(x) = \frac{1}{\sqrt{2\pi}} \int_{-\infty}^{\infty} F(u) e^{-ixu} \, du, \tag{10.2}$$

i.e., $f(x)$ is the (*inverse*) *Fourier transform* of $F(x)$. Thus, the function (10.1) can be regarded as the solution of the *integral equation* (10.2), where $f(x)$ is a given function.

We now note some properties of the Fourier transform (10.1).

1. *If $f(x)$ is absolutely integrable on the whole x-axis, then the function $F(x)$ is continuous for all x and converges to zero as $|x| \to \infty$.*

Proof. The continuity of $F(x)$ follows from the uniform convergence (in x) of the integral (10.1), since

$$|e^{ixu}| = 1, \qquad |f(u)e^{ixu}| = |f(u)|,$$

and the integral

$$\int_{-\infty}^{\infty} |f(u)|\, du$$

exists. (All the considerations of Sec. 6 continue to hold in the case where $f(x)$ is a complex-valued function.) Moreover

$$\lim_{|x|\to\infty} F(x)$$
$$= \frac{1}{\sqrt{2\pi}}\left[\lim_{|x|\to\infty}\int_{-\infty}^{\infty} f(u)\cos xu\, du + i\lim_{|x|\to\infty}\int_{-\infty}^{\infty} f(u)\sin xu\, du\right] = 0,$$

by the remark to Lemma 1 of Sec. 7.

2. *If the function $x^n f(x)$ is absolutely integrable on the whole x-axis (n is a positive integer), then $F(x)$ is differentiable n times, where*

$$F^{(k)}(x) = \frac{i^k}{\sqrt{2\pi}}\int_{-\infty}^{\infty} f(u)u^k e^{ixu}\, du \qquad (k = 1, 2, \ldots, n), \quad (10.3)$$

and all these derivatives converge to zero as $|x| \to \infty$.

Proof. The formula (10.3) can be obtained by differentiating (10.1) behind the integral sign, since each time we obtain an integral which converges uniformly in x. This follows from the relations

$$|f(u)u^k e^{ixu}| = |f(u)u^k| \qquad (k = 1, 2, \ldots, n),$$

where the functions on the right are absolutely integrable. (See Theorem 3 of Sec. 6.) To prove that the derivatives $F^{(k)}(x)$ converge to zero as $|x| \to \infty$, we again use the remark to Lemma 1 of Sec. 7.

3. *If $f(x)$ is continuous and converges to zero as $|x| \to \infty$ and if $f'(x)$ is absolutely integrable on the whole x-axis, then*

$$\frac{1}{\sqrt{2\pi}}\int_{-\infty}^{\infty} f'(u)e^{ixu}\, du = \frac{x}{i}F(x).$$

4. *If $f(x)$ is absolutely integrable on the whole x-axis and if*

$$\int_0^x f(u)\, du \to 0$$

as $|x| \to \infty$, then

$$\frac{1}{\sqrt{2\pi}}\int_{-\infty}^{\infty}\left(\int_0^u f(t)\, dt\right) e^{ixu}\, du = \frac{i}{x}F(x).$$

To prove the last two formulas, we use integration by parts. These formulas show that differentiating the original function $f(x)$ corresponds to multiplying its Fourier transform $F(x)$ by x/i, while integrating $f(x)$ corresponds to dividing $F(x)$ by x/i. This idea of reducing complicated mathematical operations on the original function to simple algebraic operations on its transform (and then taking the inverse transform of the final result) is the basis for the *operational calculus*, a very important branch of applied mathematics.

Next, we consider transforms of a somewhat different form. The function

$$F(\lambda) = \sqrt{\frac{2}{\pi}} \int_0^\infty f(u) \cos \lambda u \, du \tag{10.4}$$

is called the (*Fourier*) *cosine transform* of the function $f(x)$. If the Fourier integral theorem holds for $f(x)$, then it follows from (9.3) that

$$f(x) = \sqrt{\frac{2}{\pi}} \int_0^\infty F(\lambda) \cos x\lambda \, d\lambda, \tag{10.5}$$

i.e., $f(x)$ is itself the Fourier cosine transform of $F(\lambda)$. In other words, the functions f and F are cosine transforms of each other. Similarly, the function

$$\Phi(\lambda) = \sqrt{\frac{2}{\pi}} \int_0^\infty f(u) \sin \lambda u \, du \tag{10.6}$$

is called the (*Fourier*) *sine transform* of $f(x)$, and (9.4) gives

$$f(x) = \sqrt{\frac{2}{\pi}} \int_0^\infty \Phi(\lambda) \sin x\lambda \, d\lambda, \tag{10.7}$$

i.e., just as in the case of cosine transforms, f and Φ are sine transforms of each other.

The function (10.4) can be regarded as the solution of the integral equation (10.5) [where $f(x)$ is a given function], and the function (10.6) can be regarded as the solution of the integral equation (10.7).

We now illustrate the use of Fourier cosine and sine transforms by evaluating some integrals.

Example 1. *Let* $f(x) = e^{-ax}$ $(a > 0, x \geqslant 0)$. This function is integrable for $0 \leqslant x < \infty$ and has a derivative everywhere. Integrating by parts, we find

$$F(\lambda) = \sqrt{\frac{2}{\pi}} \int_0^\infty e^{-au} \cos \lambda u \, du = \sqrt{\frac{2}{\pi}} \frac{a}{a^2 + \lambda^2},$$

$$\Phi(\lambda) = \sqrt{\frac{2}{\pi}} \int_0^\infty e^{-au} \sin \lambda u \, du = \sqrt{\frac{2}{\pi}} \frac{\lambda}{a^2 + \lambda^2}.$$

Then (10.5) and (10.7) give

$$e^{-ax} = \frac{2a}{\pi}\int_0^\infty \frac{\cos x\lambda\, d\lambda}{a^2+\lambda^2} \qquad (x \geqslant 0),$$

$$e^{-ax} = \frac{2}{\pi}\int_0^\infty \frac{\lambda \sin x\lambda\, d\lambda}{a^2+\lambda^2} \qquad (x > 0).$$

Example 2. If

$$f(x) = \begin{cases} 1 & \text{for } 0 \leqslant x < a, \\ \frac{1}{2} & \text{for } x = a, \\ 0 & \text{for } x > a, \end{cases}$$

then obviously

$$F(\lambda) = \sqrt{\frac{2}{\pi}}\int_0^a \cos \lambda u\, du = \sqrt{\frac{2}{\pi}}\frac{\sin a\lambda}{\lambda},$$

and by (10.5)

$$f(x) = \frac{2}{\pi}\int_0^\infty \frac{\sin a\lambda \cos x\lambda\, d\lambda}{\lambda} = \begin{cases} 1 & \text{for } 0 \leqslant x < a, \\ \frac{1}{2} & \text{for } x = a, \\ 0 & \text{for } x > a. \end{cases}$$

In particular, for $x = a$

$$\frac{1}{2} = \frac{1}{\pi}\int_0^\infty \frac{\sin 2a\lambda}{\lambda}\, d\lambda,$$

and setting $a = \frac{1}{2}$, we obtain

$$\frac{\pi}{2} = \int_0^\infty \frac{\sin \lambda}{\lambda}\, d\lambda.$$

*11. The Spectral Function

It is easy to see that equation (9.1) can be written in the form

$$f(x) = \frac{1}{2\pi}\int_{-\infty}^\infty d\lambda \int_{-\infty}^\infty f(u)e^{i\lambda(x-u)}\, du \tag{11.1}$$

since

$$\begin{aligned} e^{i\lambda(x-u)} &= \cos \lambda(x-u) + i \sin \lambda(x-u) \\ &= \cos \lambda(u-x) - i \sin \lambda(x-u), \end{aligned}$$

and since the integral containing the sine vanishes (see Sec. 9). Now we set[3]

$$A(\lambda) = \frac{1}{2\pi}\int_{-\infty}^\infty f(u)e^{-i\lambda u}\, du. \tag{11.2}$$

[3] Sometimes the factor $1/2\pi$ is omitted in the formula for $A(\lambda)$; of course, the factor then reappears in (11.3).

This function (which is in general complex) is of great importance in electrical engineering, and is called the *spectral function* (synonymously, the *spectral density* or *spectrum*) of $f(x)$. According to (11.1) and (11.2)

$$f(x) = \int_{-\infty}^{\infty} A(\lambda)e^{i\lambda x}\, d\lambda; \tag{11.3}$$

(11.2) is the analog of formula (14.7) of Ch. 1 (which gives the values of the *complex Fourier coefficients*), and (11.3) is the analog of formula (14.6) of Ch. 1.

Example. Find the spectral function of the function

$$f(x) = \begin{cases} 1 & \text{for } |x| < a, \\ 0 & \text{for } |x| > a, \end{cases}$$

shown in Fig. 42.

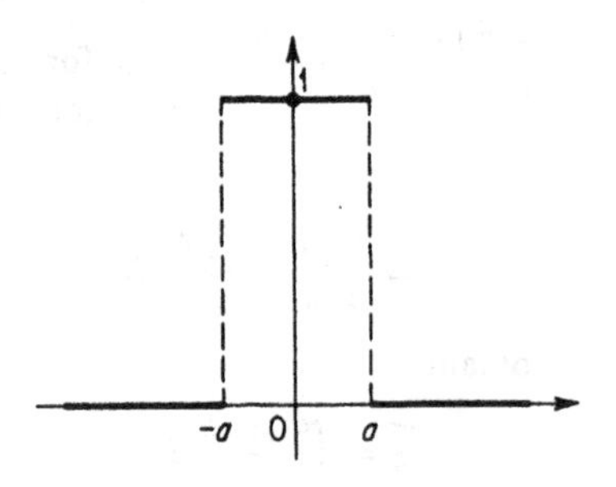

FIGURE 42

According to (11.2)

$$A(\lambda) = \frac{1}{2\pi}\int_{-a}^{a} e^{-i\lambda u}\, du = \frac{1}{2\pi}\left[\frac{e^{-i\lambda u}}{-i\lambda}\right]_{u=-a}^{u=a}$$

$$= \frac{1}{2\pi}\frac{e^{i\lambda a} - e^{-i\lambda a}}{i\lambda} = \frac{1}{\pi}\frac{\sin a\lambda}{\lambda},$$

and therefore in this case $A(\lambda)$ turns out to be real. The graph of $A(\lambda)$ is shown in Fig. 43 (for $a = 1$).

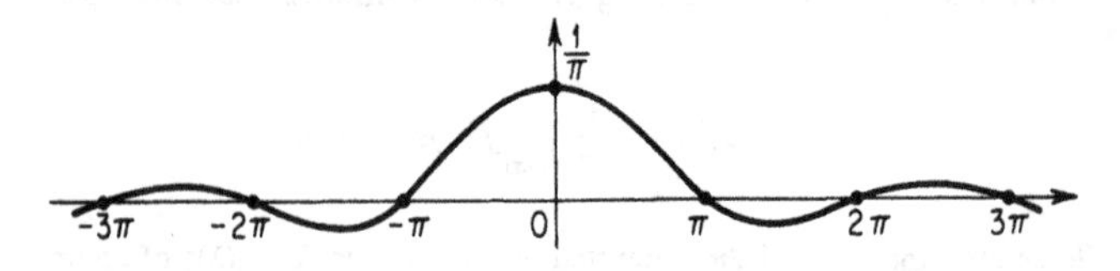

FIGURE 43

PROBLEMS

1. Show that if $f(x, y) = g(x)h(y)$, where

$$g(x) \sim \sum_{m=-\infty}^{\infty} a_m e^{imx}, \qquad h(y) \sim \sum_{n=-\infty}^{\infty} b_n e^{iny},$$

then

$$f(x, y) \sim \sum_{m,\, n=-\infty}^{\infty} a_m b_n e^{i(mx+ny)}.$$

2. Show that if $f(x, y) = g(x + y)$, where

$$g(x) \sim \sum_{m=-\infty}^{\infty} a_m e^{imx},$$

then

$$f(x, y) \sim \sum_{m=-\infty}^{\infty} e^{im(x+y)}.$$

3. Expand the following functions in double Fourier series in the region $-\pi \leqslant x, y \leqslant \pi$:

a) $f(x, y) = x + y$; b) $f(x, y) = (x + y)^2$;
c) $f(x, y) = |\sin (x + y)|$.

4. Find the Fourier sine transform (10.6) of each of the following functions:

a) $f(x) = e^{-x} \cos x \ (0 \leqslant x < \infty)$; b) $f(x) = \begin{cases} \sin x & \text{for } 0 \leqslant x < \pi, \\ 0 & \text{for } x > \pi; \end{cases}$

c) $f(x) = \begin{cases} x & \text{for } 0 \leqslant x < 1, \\ 0 & \text{for } x > 1; \end{cases}$ d) $f(x) = xe^{-x} \ (0 \leqslant x < \infty)$.

5. Find the Fourier cosine transform (10.4) of each of the functions of the preceding problem.

6. Solve the following integral equations for $f(x)$ $(0 \leqslant x < \infty)$:

a) $\displaystyle\int_0^\infty f(x) \sin \lambda x \, dx = g(\lambda)$, where $g(\lambda) = \begin{cases} 1 & \text{for } 0 \leqslant \lambda < \pi, \\ 0 & \text{for } \lambda > \pi; \end{cases}$

b) $\displaystyle\int_0^\infty f(x) \cos \lambda x \, dx = e^{-\lambda}$;

c) $\displaystyle\int_0^\infty f(x) \sin \lambda x \, dx = \lambda e^{-a\lambda}$.

7. Let $f(x)$ and $g(x)$ be absolutely integrable on $-\infty < x < \infty$, and let

$$h(x) = \int_{-\infty}^{\infty} f(x - y) g(y) \, dy.$$

Let $F(\lambda)$, $G(\lambda)$ and $H(\lambda)$ be the Fourier transforms [see equation (10.1)] of $f(x)$, $g(x)$ and $h(x)$, respectively. Show that

a) $\displaystyle\int_{-\infty}^{\infty} |h(x)| \, dx \leqslant \left(\int_{-\infty}^{\infty} |f(x)| \, dx\right)\left(\int_{-\infty}^{\infty} |g(y)| \, dy\right)$;

b) $H(\lambda) = \sqrt{2\pi}\, F(\lambda)G(\lambda)$;

c) $\int_{-\infty}^{\infty} f(x - y)g(y)\, dy = \int_{-\infty}^{\infty} g(x - y)f(y)\, dy$.

The function h is called the *convolution* of f and g, written $h = f*g$.

8. Find $h = f*g$ (cf. preceding problem) where

a) $f(x) = e^{-x}, \quad g(x) = \begin{cases} x & \text{for } 0 \leqslant x < 1, \\ 0 & \text{for } x > 1; \end{cases}$

b) $f(x) = g(x) = e^{-x}$;

c) $f(x) = \begin{cases} 1 & \text{for } 0 \leqslant x < 1, \\ 0 & \text{for } x > 1; \end{cases} \qquad g(x) = \begin{cases} x & \text{for } 0 \leqslant x < 1, \\ 0 & \text{for } x > 1; \end{cases}$

d) $f(x) = g(x) = \begin{cases} x & \text{for } 0 \leqslant x < 1, \\ 0 & \text{for } x > 1. \end{cases}$

(In every case, $f(x) = g(x) = 0$ for $x < 0$.)

9. Solve the following integral equations:

a) $\int_0^x f(x - y)e^{-y}\, dy = \begin{cases} 0 & \text{for } x < 0, \\ x^2e^{-x} & \text{for } x \geqslant 0; \end{cases}$

b) $\int_{-\infty}^{\infty} e^{-|x-y|}f(y)\, dy = e^{-|x|}(1 + |x|)$.

8

BESSEL FUNCTIONS AND FOURIER-BESSEL SERIES

1. Bessel's Equation

By *Bessel's equation* is meant the second order differential equation

$$x^2y'' + xy' + (x^2 - p^2)y = 0, \tag{1.1}$$

or equivalently

$$y'' + \frac{1}{x}y' + \left(1 - \frac{p^2}{x^2}\right)y = 0,$$

where p is a constant [which we call the *order* of (1.1)], and the prime denotes differentiation with respect to x. With the exception of certain special values of p, the solution of (1.1) cannot be expressed in terms of elementary functions (in finite form); this leads to the so-called *Bessel functions*, which have many applications in engineering and physics. Tables of Bessel functions are available for use in practical calculations.

Since Bessel's equation is linear, it has a general solution of the form

$$y = C_1y_1 + C_2y_2, \tag{1.2}$$

where y_1, y_2 are any two linearly independent particular solutions of Bessel's equation, and C_1, C_2 are arbitrary constants. Thus, to find the general solution of (1.1), it is sufficient to find any two linearly independent solutions of (1.1).

2. Bessel Functions of the First Kind of Nonnegative Order

Suppose that $p \geqslant 0$. To simplify subsequent calculations, we make the substitution

$$y = x^p z \tag{2.1}$$

in the equation (1.1). Since

$$y' = px^{p-1}z + x^p z', \; y'' = p(p-1)x^{p-2}z + 2px^{p-1}z' + x^p z'',$$

the function z satisfies the equation

$$z'' + \frac{2p+1}{x} z' + z = 0. \tag{2.2}$$

We now look for a solution of this equation in the form of a power series

$$z = c_0 + c_1 x + c_2 x^2 + \cdots + c_n x^n + \cdots.$$

Elementary calculations give

$$z' = c_1 + 2c_2 x + 3c_3 x^2 + 4c_4 x^3 + \cdots + (n+2)c_{n+2}x^{n+1} + \cdots,$$

$$\frac{z'}{x} = \frac{c_1}{x} + 2c_2 + 3c_3 x + 4c_4 x^2 + \cdots + (n+2)c_{n+2}x^n + \cdots,$$

$$z'' = 2c_2 + 2\cdot 3c_3 x + 3\cdot 4c_4 x^2 + \cdots + (n+1)(n+2)c_{n+2}x^n + \cdots.$$

Substituting these series in (2.2), we obtain

$$\begin{aligned}\frac{2p+1}{x} c_1 &+ [2c_2 + (2p+1)2c_2 + c_0] + [2\cdot 3c_3 + (2p+1)\cdot 3c_3 + c_1]x \\ &+ [3\cdot 4c_4 + (2p+1)4c_4 + c_2]x^2 + \cdots \\ &+ [(n+1)(n+2)c_{n+2} + (2p+1)(n+2)c_{n+2} + c_n]x^n \\ &+ \cdots = 0.\end{aligned}$$

If this equation is to be satisfied, all the coefficients of the different powers of x must vanish, i.e.,

$$c_1 = 0, \tag{2.3}$$

$$(n+1)(n+2)c_{n+2} + (2p+1)(n+2)c_{n+2} + c_n = 0 \qquad (n = 0, 1, 2, \ldots).$$

Therefore, we have

$$c_{n+2} = -\frac{c_n}{(n+2)(n+2p+2)} \qquad (n = 0, 1, 2, \ldots). \tag{2.4}$$

It follows from (2.3) and (2.4) that

$$c_1 = c_3 = c_5 = \cdots = c_{2m-1} = \cdots = 0,$$

$$c_2 = -\frac{c_0}{2(2p+2)},$$

$$c_4 = -\frac{c_2}{4(2p+4)} = \frac{c_0}{2\cdot 4\cdot(2p+2)(2p+4)},$$

$$c_6 = \frac{c_4}{6(2p+6)} = -\frac{c_0}{2\cdot 4\cdot 6\cdot(2p+2)(2p+4)(2p+6)},$$

and in general

$$c_{2m} = (-1)^m \frac{c_0}{2\cdot 4\cdot 6\cdots 2m(2p+2)(2p+4)(2p+6)\cdots(2p+2m)}$$

$$= (-1)^m \frac{c_0}{2^{2m}\cdot 1\cdot 2\cdot 3\cdots m(p+1)(p+2)(p+3)\cdots(p+m)}.$$

Thus, the series

$$z = c_0\left[1 - \frac{x^2}{2(2p+2)} + \frac{x^4}{2\cdot 4(2p+2)(2p+4)} - \cdots\right]$$

$$= c_0\left\{1 + \sum_{m=1}^{\infty} \frac{(-1)^m x^{2m}}{2^{2m}\cdot 1\cdot 2\cdots m(p+1)(p+2)\cdots(p+m)}\right\},$$

where c_0 is an arbitrary constant, gives a formal solution of the equation (2.2). By using the familiar ratio test, it is easy to show that this series converges for all x. Since term by term differentiation of a power series is always legitimate (inside the interval of convergence), z is actually a solution of (2.2). But then, the function

$$y = x^p z = c_0 x^p\left[1 - \frac{x^2}{2(2p+2)} + \frac{x^4}{2\cdot 4\cdot(2p+2)(2p+4)} - \cdots\right]$$
$$= c_0\left\{1 + \sum_{m=1}^{\infty} \frac{(-1)^m x^{p+2m}}{2^{2m}\cdot 1\cdot 2\cdots m(p+1)(p+2)\cdots(p+m)}\right\} \tag{2.5}$$

where c_0 is arbitrary, is a solution of (1.1).

It is customary to set

$$c_0 = \frac{1}{2^p\Gamma(p+1)}, \tag{2.6}$$

where Γ is the well-known *gamma function* of mathematical analysis, which has the following properties:

1) $\Gamma(1) = 1$,
2) $\Gamma(p+1) = p\Gamma(p)$ for any p,
3) $\Gamma(p+1) = p!$ for positive integral p.

(We shall discuss the gamma function in more detail in the next section.)

If the constant c_0 is defined by (2.6), then the series (2.5) gives the *Bessel function of the first kind of order p* (for the time being, $p \geqslant 0$), denoted by $J_p(x)$:

$$J_p(x) = \frac{x^p}{2^p\Gamma(p+1)}\left[1 - \frac{x^2}{2(2p+2)} + \frac{x^4}{2\cdot 4(2p+2)(2p+4)} - \cdots\right] \tag{2.7}$$
$$= \left\{\frac{(x/2)^p}{\Gamma(p+1)} + \sum_{m=1}^{\infty}\frac{(-1)^m(x/2)^{p+2m}}{1\cdot 2\cdots m(p+1)(p+2)\cdots(p+m)\Gamma(p+1)}\right\}.$$

But according to the properties of the gamma function, we have

$$1\cdot 2\cdots m = m! = \Gamma(m+1),$$

$$\begin{aligned}(p+1)(p+2)\cdots(p+m)\Gamma(p+1) &= (p+2)(p+3)\cdots(p+m)\Gamma(p+2)\\ &= (p+3)(p+4)\cdots(p+m)\Gamma(p+3)\\ &= (p+m)\Gamma(p+m) = \Gamma(p+m+1),\end{aligned}$$

and therefore

$$J_p(x) = \sum_{m=0}^{\infty}\frac{(-1)^m(x/2)^{p+2m}}{\Gamma(m+1)\Gamma(p+m+1)} \tag{2.8}$$

(see also Property 1 of the gamma function).

In particular, for $p = 0$

$$J_0(x) = 1 - \frac{x^2}{2^2} + \frac{x^4}{2^2\cdot 4^2} - \frac{x^6}{2^2\cdot 4^2\cdot 6^2} + \cdots \tag{2.9}$$
$$= \sum_{m=0}^{\infty}\frac{(-1)^m(x/2)^{2m}}{(m!)^2}$$

where we must set $0! = 1$. For $p = 1$

$$J_1(x) = \frac{x}{2}\left[1 - \frac{x^2}{2\cdot 4} + \frac{x^4}{2\cdot 4\cdot 4\cdot 6} - \frac{x^6}{2\cdot 4\cdot 6\cdot 4\cdot 6\cdot 8} + \cdots\right]$$
$$= \sum_{m=0}^{\infty}\frac{(-1)^m(x/2)^{2m+1}}{m!(m+1)!}.$$

In general, for positive integral p

$$J_p(x) = \frac{x^p}{2^p p!}\left[1 - \frac{x^2}{2(2p+2)} + \frac{x^4}{2\cdot 4\cdot(2p+2)(2p+4)} - \cdots\right] \tag{2.10}$$
$$= \sum_{m=0}^{\infty}\frac{(-1)^m(x/2)^{2m+p}}{m!(p+m)!}.$$

Formulas (2.9) and (2.10) show that if $p = 0$ or if p is an even integer, then $J_p(x)$ is an even function (since it contains only even powers of x). On the other hand, if p is an odd integer, then $J_p(x)$ is an odd function (since it contains only odd powers of x). The graphs of the functions $y = J_0(x)$ and $y = J_1(x)$ are shown in Fig. 44.

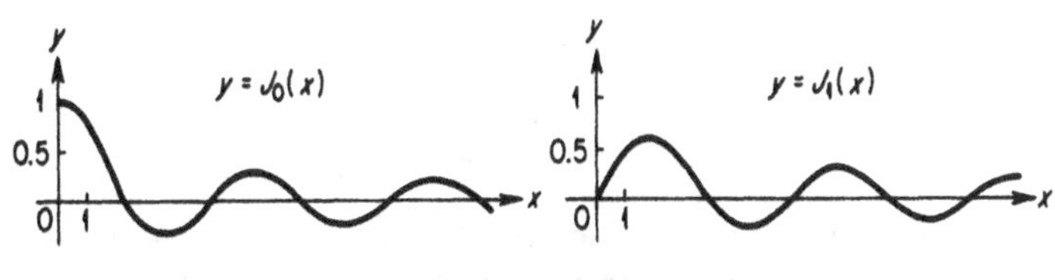

FIGURE 44

Remark. It should be noted that if $x < 0$ and if p is not an integer, then in general $J_p(x)$ will take complex values [see (2.7)]. In order to avoid complex values, we shall consider $J_p(x)$ only for $x \geqslant 0$ (when p is not an integer).

3. The Gamma Function

For $p > 0$, the gamma function is usually defined by the formula

$$\Gamma(p) = \int_0^\infty e^{-x}x^{p-1}\,dx. \tag{3.1}$$

(This improper integral has meaning only for $p > 0$.) We now verify that (3.1) actually has the properties given in Sec. 2.

1)
$$\Gamma(1) = \int_0^\infty e^{-x}\,dx = [-e^{-x}]_{x=0}^{x=\infty} = 1,$$

2)
$$\Gamma(p + 1) = \int_0^\infty e^{-x}\,x^p\,dx.$$

Integrating by parts, we obtain

$$\Gamma(p + 1) = [-e^{-x}x^p]_{x=0}^{x=\infty} + p\int_0^\infty e^{-x}x^{p-1}\,dx.$$

The first term on the right vanishes (provided, of course, that $p > 0$), and the integral is just $\Gamma(p)$. This proves Property 2.

3) If p is a positive integer, then using Property 2 we obtain

$$\Gamma(p + 1) = p\Gamma(p) = p(p - 1)\Gamma(p - 1) = \cdots = p(p - 1)\cdots 2\cdot 1\cdot \Gamma(1),$$

or, in view of Property 1

$$\Gamma(p + 1) = p!$$

Thus, for $p > 0$, the function defined by (3.1) actually has the required properties.

To extend the function $\Gamma(p)$ to all real values of p, we start with the formula

$$\Gamma(p + 1) = p\Gamma(p)$$

or

$$\Gamma(p) = \frac{\Gamma(p + 1)}{p}. \tag{3.2}$$

If $-1 < p < 0$, the right-hand side of (3.2) has meaning, since then $0 < p + 1 < 1$. Therefore, (3.2) can be used to define $\Gamma(p)$ for $-1 < p < 0$. Incidentally, we note that as $p \to 0$, the numerator in the right-hand side of (3.2) approaches 1 while the denominator approaches 0, and hence we have

$$\Gamma(0) = \infty.$$

Now let $-2 < p < -1$. Then $-1 < p + 1 < 0$, and the right-hand side of (3.2) again has meaning. Therefore, (3.2) can be used to define $\Gamma(p)$ for $-2 < p < -1$. If $p \to -1$, then it follows from (3.2) that $\Gamma(p) \to \infty$, and hence we have

$$\Gamma(-1) = \infty.$$

Next, we consider the values $-3 < p < -2$, and so forth. Thus, step by step, we can define $\Gamma(p)$ for all negative values of p, with $\Gamma(p) = \infty$ for $p = 0, -1, -2, \ldots$. In other words, (3.1) can be used to define $\Gamma(p)$ for $p > 0$ and then (3.2) can be used to define $\Gamma(p)$ for all real p. Moreover, by its very construction, $\Gamma(p)$ has the three properties listed above.

4. Bessel Functions of the First Kind of Negative Order

Since p^2 appears in equation (1.1), it is natural to expect that the considerations of Sec. 2 will be applicable to $-p$ as well as p and will also lead to a solution of (1.1). Replacing p by $-p$ in (2.8) gives

$$J_{-p}(x) = \sum_{m=0}^{\infty} \frac{(-1)^m (x/2)^{-p+2m}}{\Gamma(m + 1)\Gamma(-p + m + 1)}. \tag{4.1}$$

We note that for integral p and $m = 0, 1, 2, \ldots, p - 1$, the quantity $-p + m + 1$ takes negative values and the value zero. Therefore, $\Gamma(-p + m + 1) = \infty$ for these values of m, and the corresponding terms in the series (4.1) are taken to be zero. Thus, for integral p

$$J_{-p}(x) = \sum_{m=0}^{\infty} \frac{(-1)^m (x/2)^{-p+2m}}{\Gamma(m + 1)\Gamma(-p + m + 1)},$$

or, if we set $m = p + k$

$$J_{-p}(x) = (-1)^p \sum_{k=0}^{\infty} \frac{(-1)^k (x/2)^{p+2k}}{\Gamma(k+1)\Gamma(p+k+1)} \tag{4.2}$$

$$= (-1)^p J_p(x).$$

If p is not an integer, then none of the denominators in (4.1) becomes infinite. The ratio test shows that the series (4.1) converges for all $x \neq 0$ if p is not an integer and for all x if p is an integer [see (4.2)].

The function $J_{-p}(x)$ is again called a *Bessel function of the first kind*, this time *of order* $-p$. It is easily verified by direct substitution that $J_{-p}(x)$ is actually a solution of (1.1). We leave this verification to the reader.[1]

For what follows, it is useful to observe that the formulas (2.8) and (4.1) can be combined in one formula

$$J_p(x) = \sum_{m=0}^{\infty} \frac{(-1)^m (x/2)^{p+2m}}{\Gamma(m+1)\Gamma(p+m+1)}, \tag{4.3}$$

where the number p can be either positive or negative. The remark made at the end of Sec. 2 applies to the case of fractional values of p of either sign.

5. The General Solution of Bessel's Equation

Consider first the case where the number $p > 0$ is not an integer. Then, the functions $J_p(x)$ and $J_{-p}(x)$ cannot be linearly dependent, i.e., there cannot exist a constant C such that

$$J_p(x) = CJ_{-p}(x).$$

To see this, we observe that for $x = 0$, the function $J_p(x)$ vanishes, whereas $J_{-p}(x)$ becomes infinite [see (2.8) and (4.1)]. Thus, if p is not an integer, the general solution of Bessel's equation (1.1) has the form

$$y = C_1 J_p(x) + C_2 J_{-p}(x), \tag{5.1}$$

where C_1 and C_2 are arbitrary constants [cf. (1.2)].

If $p \geqslant 0$ is an integer, then by (4.2) the functions $J_p(x)$ and $J_{-p}(x)$ are linearly dependent, and hence in this case, (5.1) does not give the general solution. Therefore, for integral p we have to find another particular solution of (1.1), which is linearly independent of $J_p(x)$. This particular solution $Y_p(x)$ is the so-called *Bessel function of the second kind*, to be

[1] It should be noted that for some values of p (e.g., for integral p), part of the argument given in Sec. 2 ceases to be legitimate when applied to $-p$, since it leads to fractions with zero denominators [see e.g. (2.4)]. Nevertheless, the final formula (2.8) has meaning when p is changed to $-p$, and in fact gives a solution of (1.1), as noted.

discussed in the next section. Thus, when p is an integer, the general solution of (1.1) has the form

$$y = C_1 J_p(x) + C_2 Y_p(x).$$

6. Bessel Functions of the Second Kind

For fractional p, the Bessel function of the second kind is obtained from (5.1) by a special choice of the constants C_1 and C_2, i.e., we set

$$Y_p(x) = J_p(x)\cot p\pi - J_{-p}(x)\csc p\pi = \frac{J_p(x)\cos p\pi - J_{-p}(x)}{\sin p\pi}. \tag{6.1}$$

For integral p, (6.1) is indeterminate; in fact, the numerator reduces to $(-1)^p J_p(x) - J_{-p}(x)$ which vanishes according to (4.2), and the denominator also vanishes. This suggests the following question: Can the indeterminacy be "removed" by taking the limit of the ratio (6.1) as p approaches integral values, and will this limit give the required solution for integral p? As we now show, the answer is in the affirmative.

According to L'Hospital's rule

$$\begin{aligned} Y_n(x) &= \lim_{p\to n} \frac{(\partial/\partial p)[J_p(x)\cos p\pi - J_{-p}(x)]}{(\partial/\partial p)\sin p\pi} \\ &= \lim_{p\to n} \frac{\cos p\pi(\partial/\partial p)J_p(x) - \pi J_p(x)\sin p\pi - (\partial/\partial p)J_{-p}(x)}{\pi\cos p\pi} \\ &= \left[\frac{(\partial/\partial p)J_p(x)\cdot(-1)^n - (\partial/\partial p)J_{-p}(x)}{\pi(-1)^n}\right]_{p=n}. \end{aligned}$$

We substitute the series (2.8) and (4.1) in this expression, differentiate them with respect to p, and then set the arbitrary index p equal to the integral index n. After some manipulations, the details of which we omit (since they involve special properties of the gamma function and are rather tedious), we obtain

$$\begin{aligned} Y_n(x) &= \frac{2}{\pi} J_n(x)\left(\ln\frac{x}{2} + C\right) - \frac{1}{\pi}\sum_{m=0}^{n-1} \frac{(n-m-1)!}{m!}\left(\frac{x}{2}\right)^{-n+2m} \\ &\quad - \frac{1}{\pi}\sum_{m=0}^{\infty} \frac{(-1)^m (x/2)^{n+2m}}{m!(n+m)!}\left(\sum_{k=1}^{n+m}\frac{1}{k} + \sum_{k=1}^{m}\frac{1}{k}\right), \end{aligned}$$

where $C = 0.5772156649015 32\cdots$ is the so-called *Euler's constant.* In particular, when $n = 0$

$$\begin{aligned} Y_0(x) &= \frac{2}{\pi} J_0(x)\left(\ln\frac{x}{2} + C\right) \\ &\quad - \frac{2}{\pi}\sum_{m=1}^{\infty} \frac{(-1)^m}{(m!)^2}\left(\frac{x}{2}\right)^{2m}\left(1 + \frac{1}{2} + \frac{1}{3} + \cdots + \frac{1}{m}\right). \end{aligned}$$

Substitution of the function $Y_n(x)$ into the equation (1.1) with $p = n$ shows that $Y_n(x)$ is actually a solution of (1.1). Moreover, the functions $J_n(x)$ and $Y_n(x)$ cannot be linearly dependent, since for $x = 0$, $J_n(x)$ has a finite value, whereas $Y_n(x)$ becomes infinite. Therefore, $Y_n(x)$ is the required particular solution of (1.1) (cf. the end of Sec. 5). The graph of the function $y = Y_0(x)$ is shown in Fig. 45.

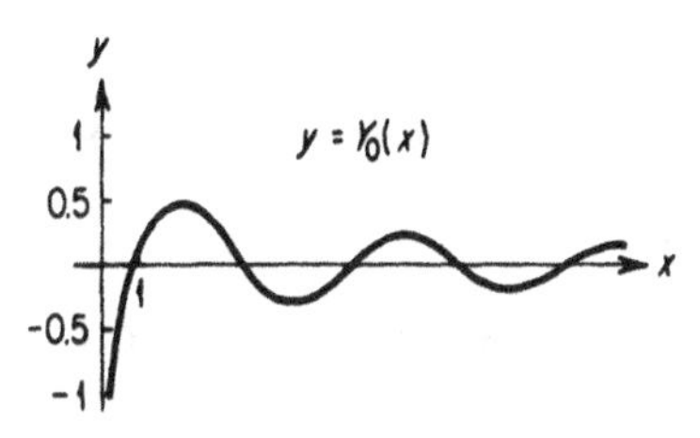

FIGURE 45

7. Relations between Bessel Functions of Different Orders

For *any* p, we have the formulas

$$\frac{d}{dx}[x^p J_p(x)] = x^p J_{p-1}(x), \tag{7.1}$$

$$\frac{d}{dx}[x^{-p} J_p(x)] = -x^{-p} J_{p+1}(x). \tag{7.2}$$

Similar formulas hold for the corresponding Bessel functions of the second kind.

Proof. By (4.3), we have

$$\frac{d}{dx}[x^p J_p(x)] = \frac{d}{dx}\sum_{m=0}^{\infty}\frac{(-1)^m x^{2p+2m}}{2^{p+2m}\Gamma(m+1)\Gamma(p+m+1)}$$

$$= \sum_{m=0}^{\infty}\frac{(-1)^m x^{2p+2m-1}}{2^{p+2m-1}\Gamma(m+1)\Gamma(p+m)}$$

$$= x^p J_{p-1}(x),$$

for any p, which proves (7.1). Formula (7.2) is proved in the same way. To prove the corresponding formulas for Bessel functions of the second kind, we first replace p by $-p$, obtaining

$$\frac{d}{dx}[x^p J_{-p}(x)] = -x^p J_{-p+1}(x). \tag{7.3}$$

Assuming that p is not an integer, we multiply (7.1) by $\cot p\pi$ and subtract (7.3) multiplied by $\csc p\pi$. The result is

$$\frac{d}{dx}\left[x^p \frac{J_p(x)\cos p\pi - J_{-p}(x)}{\sin p\pi}\right] = x^p \frac{J_{p-1}(x)\cos p\pi + J_{-p+1}(x)}{\sin p\pi}$$

$$= x^p \frac{J_{p-1}(x)\cos(p-1)\pi - J_{-p+1}(x)}{\sin(p-1)\pi}$$

since

$$\cos(p-1)\pi = -\cos p\pi, \quad \sin(p-1)\pi = -\sin p\pi.$$

Thus, for fractional p

$$\frac{d}{dx}[x^p Y_p(x)] = x^p Y_{p-1}(x) \tag{7.4}$$

[see (6.1)].

Next we replace p by $-p$ in (7.1), obtaining

$$\frac{d}{dx}[x^{-p}J_{-p}(x)] = x^{-p}J_{-p-1}(x). \tag{7.5}$$

Then we multiply (7.2) by $\cot p\pi$ and subtract (7.5) multiplied by $\csc p\pi$. The result is

$$\frac{d}{dx}\left[x^{-p} \frac{J_p(x)\cos p\pi - J_{-p}(x)}{\sin p\pi}\right]$$

$$= x^{-p} \frac{-J_{p+1}(x)\cos p\pi - J_{-p-1}(x)}{\sin p\pi}$$

$$= -x^{-p} \frac{J_{p+1}(x)\cos(p+1)\pi - J_{-p-1}(x)}{\sin(p+1)\pi}$$

since

$$\cos(p+1)\pi = -\cos p\pi, \quad \sin(p+1)\pi = -\sin p\pi.$$

Therefore, for fractional p [see (6.1)]

$$\frac{d}{dx}[x^{-p} Y_p(x)] = -x^{-p} Y_{p+1}(x). \tag{7.6}$$

The formulas (7.5) and (7.6) are the analogs of (7.1) and (7.2), and also hold for integral p (as can be seen by letting p approach integral values).

From (7.1) and (7.2), we obtain the formulas

$$xJ'_p(x) + pJ_p(x) = xJ_{p-1}(x), \tag{7.7}$$

$$xJ'_p(x) - pJ_p(x) = -xJ_{p+1}(x), \tag{7.8}$$

$$J_{p-1}(x) - J_{p+1}(x) = 2J'_p(x), \tag{7.9}$$

$$J_{p-1}(x) + J_{p+1}(x) = \frac{2p}{x} J_p(x) \tag{7.10}$$

and similar formulas for Bessel functions of the second kind. In fact, it follows from (7.1) that

$$x^p J'_p(x) + p x^{p-1} J_p(x) = x^p J_{p-1}(x),$$

from which (7.7) is obtained by dividing by x^{p-1}. In the same way, we obtain (7.8) from (7.2). To obtain (7.9), we add (7.7) and (7.8) and divide the result by x. Finally, (7.10) is obtained from (7.7) and (7.8) by subtraction and subsequent division by x.

The utility of these formulas will be repeatedly demonstrated below, and hence we shall not discuss them now. However, it should be noted that (7.10) shows how to calculate the values of the function $J_p(x)$ for any positive or negative integral p from a knowledge of the functions $J_0(x)$ and $J_1(x)$. A similar remark applies to Bessel functions of the second kind.

8. Bessel Functions of the First Kind of Half-Integral Order

Referring to (2.7), we first consider

$$J_{1/2}(x) = \frac{\sqrt{x}}{\sqrt{2}\,\Gamma(\frac{3}{2})}\left[1 - \frac{x^2}{2\cdot 3} + \frac{x^4}{2\cdot 4\cdot 3\cdot 5} - \frac{x^6}{2\cdot 4\cdot 6\cdot 3\cdot 5\cdot 7} + \cdots\right]$$

$$= \frac{1}{\sqrt{2x}\,\Gamma(\frac{3}{2})}\left[x - \frac{x^3}{3!} + \frac{x^5}{5!} - \frac{x^7}{7!} + \cdots\right]$$

$$= \frac{1}{\sqrt{2x}\,\Gamma(\frac{3}{2})} \sin x.$$

According to (3.1)

$$\Gamma\left(\frac{3}{2}\right) = \frac{1}{2}\Gamma\left(\frac{1}{2}\right) = \frac{1}{2}\int_0^\infty e^{-x}\frac{dx}{\sqrt{x}} = \int_0^\infty e^{-t^2}\,dt,$$

where the integral equals $\sqrt{\pi}/2$.[2] Therefore

$$J_{1/2}(x) = \sqrt{2/\pi x}\,\sin x, \tag{8.1}$$

[2] To see this, write

$$\int_0^\infty e^{-t^2}\,dt \int_0^\infty e^{-u^2}\,du = \int_0^\infty\int_0^\infty e^{-(t^2+u^2)}\,dt\,du = \int_0^{\pi/2}\int_0^\infty e^{-r^2} r\,dr\,d\theta = \frac{\pi}{4}.$$

(*Translator*)

and similarly, we find

$$J_{-1/2}(x) = \sqrt{2/\pi x}\cos x. \tag{8.2}$$

Thus, the functions $J_{1/2}(x)$ and $J_{-1/2}(x)$ can be expressed in terms of elementary functions. But then it is an immediate consequence of formula (7.10) that any function $J_p(x)$ where $p = n + \frac{1}{2}$ and n is integral, can be expressed in terms of elementary functions. For example, setting $p = \frac{1}{2}$ in (7.10), we obtain

$$J_{-1/2}(x) + J_{3/2}(x) = \frac{1}{x}J_{1/2}(x),$$

whence

$$J_{3/2}(x) = \frac{1}{x}J_{1/2}(x) - J_{-1/2}(x) = \sqrt{2/\pi x^3}\sin x - \sqrt{2/\pi x}\cos x.$$

Similarly, setting $p = \frac{3}{2}$ in (7.10), we can find $J_{5/2}(x)$, etc.

9. Asymptotic Formulas for the Bessel Functions

In this section, we derive formulas that permit us to readily determine the behavior of Bessel functions for large values of x (so-called *asymptotic* formulas). First of all, we transform equation (1.1) by making the substitution

$$y = \frac{z}{\sqrt{x}}. \tag{9.1}$$

The resulting equation for the function z is

$$z'' + \left(1 - \frac{p^2 - \frac{1}{4}}{x^2}\right)z = 0. \tag{9.2}$$

If we set

$$m = \tfrac{1}{4} - p^2, \qquad \frac{m}{x^2} = \rho, \tag{9.3}$$

then (9.2) becomes

$$z'' + (1 + \rho)z = 0. \tag{9.4}$$

For large x, the function $\rho = \rho(x)$ becomes very small. Therefore, it is natural to expect that for large x, the solution of the equation (9.4) will not differ much from the solution of the equation

$$z'' + z = 0,$$

i.e., from the function

$$z = A \sin (x + \omega) \qquad (A = \text{const}, \ \omega = \text{const}).$$

Now, let z be a solution of (9.4) which is not identically zero. Then, in view of what has just been said, it is natural to assume that there exist functions $\alpha = \alpha(x)$ and $\delta = \delta(x)$ such that

$$z = \alpha \sin (x + \delta), \tag{9.5}$$

where α and δ converge to definite finite limits as $x \to \infty$. To *prove* the existence of such functions, we consider the equation

$$z' = \alpha \cos (x + \delta) \tag{9.6}$$

as well as (9.5), and we then regard (9.5) and (9.6) as a system of equations where the left-hand sides are known and the functions α and δ are unknowns.

It follows from (9.4) and (9.5) that

$$z'' = -(1 + \rho)\alpha \sin (x + \delta),$$

and from (9.6) that

$$z'' = \alpha' \cos (x + \delta) - \alpha(1 + \delta') \sin (x + \delta),$$

which imply at once that

$$\tan (x + \delta) = \frac{\alpha'}{\alpha(\delta' - \rho)}. \tag{9.7}$$

Differentiating (9.5) gives

$$z' = \alpha' \sin (x + \delta) + \alpha(1 + \delta') \cos (x + \delta).$$

Comparing this with (9.6), we easily obtain

$$\tan (x + \delta) = -\frac{\alpha\delta'}{\alpha'}. \tag{9.8}$$

Multiplying (9.7) and (9.8), we find

$$\tan^2 (x + \delta) = -\frac{\delta'}{\delta' - \rho},$$

whence

$$\delta' = \rho \sin^2 (x + \delta). \tag{9.9}$$

Then (9.8) implies

$$\frac{\alpha'}{\alpha} = -\frac{\delta'}{\tan (x + \delta)} = -\rho \sin (x + \delta) \cos (x + \delta). \tag{9.10}$$

We observe that the denominator α of the ratio α'/α cannot vanish. In fact, if α vanishes, then it follows from (9.5) and (9.6) that z and z' vanish simultaneously, i.e., the solution z satisfies zero initial conditions at some point $x = x_0$. However, because of the uniqueness of a solution satisfying given initial conditions, this implies that z is identically zero, contrary to hypothesis.

The required function δ is found from the differential equation (9.9), and the initial condition for δ can be found from the initial conditions for z by using (9.5) and (9.6) and eliminating α (e.g., by division). Then, from a knowledge of δ, we can easily find α from (9.10), and the initial condition for α is obtained from the initial conditions for z and δ by using (9.5) and (9.6).

Thus, all that remains is to analyze the asymptotic behavior of α and δ. Since

$$\delta(x) = \delta(b) - \int_x^b \delta'(t)\,dt$$

it follows from (9.3) and (9.9) that

$$\delta(x) = \delta(b) - m\int_x^b \frac{\sin^2(t+\delta)}{t^2}\,dt.$$

If we take the limit as $b \to \infty$, the resulting improper integral obviously converges (since the integrand does not exceed $1/t^2$). Therefore, the limit of $\delta(b)$ as $b \to \infty$ also exists. If we set

$$\lim_{b\to\infty} \delta(b) = \omega,$$

then

$$\delta(x) = \omega - m\int_x^\infty \frac{\sin^2(t+\delta)}{t^2}\,dt. \tag{9.11}$$

But

$$0 < \int_x^\infty \frac{\sin^2(t+\delta)}{t^2}\,dt < \int_x^\infty \frac{dt}{t^2} = \left[-\frac{1}{t}\right]_{t=x}^{t=\infty} = \frac{1}{x},$$

and hence

$$0 < mx\int_x^\infty \frac{\sin^2(t+\delta)}{t^2}\,dt < m.$$

In other words, the function

$$\eta(x) = -mx\int_x^\infty \frac{\sin^2(t+\delta)}{t^2}\,dt$$

is bounded for any x, and (9.11) can be written in the form

$$\delta(x) = \omega + \frac{\eta(x)}{x}. \tag{9.12}$$

Moreover

$$\frac{\alpha'}{\alpha} = (\ln \alpha)',$$

and hence

$$\ln \alpha(x) = \ln \alpha(b) - \int_x^b \frac{\alpha'(t)}{\alpha(t)}\, dt$$

or by (9.10)

$$\ln \alpha(x) = \ln \alpha(b) + m \int_x^b \frac{\sin (t + \delta) \cos (t + \delta)}{t^2}\, dt.$$

Just as before, we pass to the limit as $b \to \infty$, and note that the resulting improper integral converges. This implies the existence of a finite limit as $b \to \infty$ for $\ln \alpha(b)$ and hence for $\alpha(b)$. We set

$$\lim_{b \to \infty} \alpha(b) = A,$$

where $A \neq 0$, since otherwise $\ln \alpha(b) \to \infty$ as $b \to \infty$. Then we have

$$\ln \alpha(x) = \ln A + m \int_x^\infty \frac{\sin (t + \delta) \cos (t + \delta)}{t^2}\, dt.$$

Just as we proved the boundedness of the function $\eta(x)$, we can also prove the boundedness of the function

$$\varphi(x) = mx \int_x^\infty \frac{\sin (t + \delta) \cos (t + \delta)}{t^2}\, dt.$$

Then

$$\ln \alpha(x) = \ln A + \frac{\varphi(x)}{x},$$

whence

$$\alpha(x) = A \exp [\varphi(x)/x].$$

According to Taylor's theorem, for any t we have

$$e^t = 1 + te^{\theta t} \qquad (0 < \theta < 1),$$

and hence, writing $t = \varphi(x)/x$, we obtain

$$\exp [\varphi(x)/x] = 1 + \frac{\varphi(x)}{x} \exp [\theta\varphi(x)/x].$$

The function $\exp[\theta\varphi(x)/x]$ is obviously bounded as $x \to \infty$. Therefore, we can write

$$\exp[\varphi(x)/x] = 1 + \frac{\xi(x)}{x}$$

or

$$\alpha(x) = A\left(1 + \frac{\xi(x)}{x}\right), \tag{9.13}$$

where $\xi(x)$ remains bounded as $x \to \infty$.

The formulas (9.12) and (9.13) confirm our conjecture about the way the functions α and δ behave as $x \to \infty$ and about the behavior of the solution z of the equation (9.4). [Cf. (9.5).] Substituting (9.12) and (9.13) into (9.5) gives

$$z = A\left(1 + \frac{\xi(x)}{x}\right)\sin\left(x + \omega + \frac{\eta(x)}{x}\right). \tag{9.14}$$

Next, we transform the last factor in (9.14). According to Taylor's theorem

$$\sin(a + t) = \sin a + t\cos(a + \theta t) \qquad (0 < \theta < 1).$$

Setting $a = x + \omega$, $t = \eta(x)/x$, we obtain

$$\sin\left(x + \omega + \frac{\eta(x)}{x}\right) = \sin(x + \omega) + \frac{\zeta(x)}{x},$$

where

$$\zeta(x) = \eta(x)\cos\left(x + \omega + \frac{\theta\eta(x)}{x}\right)$$

is a function which is bounded for all x. Therefore, it follows from (9.14) that

$$\begin{aligned} z &= A\left(1 + \frac{\xi(x)}{x}\right)\left[\sin(x + \omega) + \frac{\zeta(x)}{x}\right] \\ &= A\sin(x + \omega) + A\,\frac{\xi(x)\sin(x + \omega) + (1 + \xi(x)/x)\zeta(x)}{x} \end{aligned}$$

or

$$z = A\sin(x + \omega) + \frac{r(x)}{x}, \tag{9.15}$$

where $r(x)$ is bounded as $x \to \infty$. Thus, we have found an asymptotic formula for the solution of (9.2).

To go to the case of Bessel's equation, we use (9.1) and obtain

$$y = \frac{A}{\sqrt{x}} \sin (x + \omega) + \frac{r(x)}{x\sqrt{x}}, \tag{9.16}$$

where $A = \text{const}$, $\omega = \text{const}$ and $r(x)$ is bounded as $x \to \infty$. This formula shows that for large x, any solution of Bessel's equation differs very little from the damped sinusoid

$$y = \frac{A}{\sqrt{x}} \sin (x + \omega).$$

In particular, the above considerations apply to the Bessel functions $J_p(x)$ and $Y_p(x)$. A more exact calculation, which we omit, shows that

$$\begin{aligned} J_p(x) &= \sqrt{2/\pi x} \sin \left(x - \frac{p\pi}{2} + \frac{\pi}{4}\right) + \frac{r_p(x)}{x\sqrt{x}}, \\ Y_p(x) &= \sqrt{2/\pi x} \sin \left(x - \frac{p\pi}{2} - \frac{\pi}{4}\right) + \frac{\rho_p(x)}{x\sqrt{x}}, \end{aligned} \tag{9.17}$$

where the functions $r_p(x)$ and $\rho_p(x)$ remain bounded as $x \to \infty$.

For subsequent purposes, besides the formula (9.15), it is useful to have the corresponding formula for z'. Therefore, we substitute the expressions (9.12) and (9.13) for δ and α into (9.6). The result is

$$z' = A\left(1 + \frac{\xi(x)}{x}\right) \cos \left(x + \omega + \frac{\eta(x)}{x}\right).$$

If we transform this expression in the same way as we transformed (9.14) [which led to (9.15)], we obtain

$$z' = A \cos (x + \omega) + \frac{s(x)}{x}, \tag{9.18}$$

where the function $s(x)$ is bounded as $x \to \infty$.

10. Zeros of Bessel Functions and Related Functions

It follows at once from formula (9.15) that any solution of Bessel's equation has an infinite number of positive zeros and that these zeros are close to the zeros of the function $\sin (x + \omega)$, i.e., to the numbers of the form

$$k_n = n\pi - \omega,$$

where n is an integer. We now show that for sufficiently large n, there is just one zero near each k_n. Since, according to (9.1), the functions y and z have the same positive zeros, it is sufficient to prove this for the function

z. If for arbitrarily large n, there were a pair of zeros of the function z near the point k_n, then it would follow from Rolle's theorem that z' has a zero near z_n. But by (9.18) this is impossible, since near $z_n = n\pi - \omega$, the value of z' is near $A \cos n\pi$, provided that n is sufficiently large. Thus, for sufficiently large n, all the zeros of the function y lie near the numbers k_n, and there is only one zero near each k_n. It follows that the distance between consecutive zeros of the function y approaches π as the distance from the origin increases. In particular, these considerations apply to the functions $J_p(x)$ and $Y_p(x)$, where according to (9.17), the numbers k_n have the values

$$k_n = n\pi + \frac{p\pi}{2} - \frac{\pi}{4},$$

$$k_n = n\pi - \frac{p\pi}{2} + \frac{\pi}{4}$$

for $J_p(x)$ and $Y_p(x)$, respectively.

Below, we shall be concerned with the positive zeros of $J_p(x)$. (It should be noted that by (4.3), the positive and negative zeros of $J_p(x)$ are located symmetrically with respect to the origin.) It follows from (7.2) and Rolle's theorem that there is at least one zero of the function $x^{-p}J_{p+1}(x)$ between any two consecutive zeros of the function $x^{-p}J_p(x)$, i.e., there is at least one zero of the function $J_{p+1}(x)$ between any two consecutive zeros of the function $J_p(x)$. If we replace p by $p + 1$ in (7.1), we obtain

$$\frac{d}{dx}[x^{p+1}J_{p+1}(x)] = x^{p+1}J_p(x).$$

From this we conclude as before that there is at least one zero of the function $J_p(x)$ between any two consecutive zeros of the function $J_{p+1}(x)$. Thus, the zeros of $J_p(x)$ and $J_{p+1}(x)$ "separate each other." More precisely, *one and only one zero of* $J_{p+1}(x)$ *appears between any two consecutive positive zeros of* $J_p(x)$. Moreover, $J_p(x)$ and $J_{p+1}(x)$ cannot have any positive zeros in common. In fact, if $J_p(x)$ and $J_{p+1}(x)$ both vanished at $x_0 > 0$, then by (7.8), $J_p'(x)$ would also vanish at x_0. But this is impossible, since by the uniqueness theorem for solutions of a second order differential equation, $J_p(x_0) = 0$, $J_p'(x_0) = 0$ would imply that $J_p(x) \equiv 0$, which is obviously false.

Consider next the zeros of the function $J_p'(x)$. By Rolle's theorem, at least one zero of $J_p'(x)$ lies between any two consecutive zeros of $J_p(x)$. Therefore, like $J_p(x)$, the function $J_p'(x)$ has an infinite set of positive zeros.

Finally, we investigate the zeros of the function $xJ_p'(x) - HJ_p(x)$, where H is a constant, a function which is often encountered in the applications. As we have just seen, the functions $J_p(x)$ and $J_p'(x)$ cannot vanish simultaneously (for $x > 0$). It follows at once that $J_p(x)$ must change sign as we pass through a value of x for which $J_p(x)$ vanishes. Let $\lambda_1, \lambda_2, \ldots, \lambda_n, \ldots$ denote the positive zeros of $J_p(x)$ arranged in increasing order. For $0 < x < \lambda_1$, the function $J_p(x)$ does not change sign, and in fact for $p > -1$ (the only values of p of interest to us), $J_p(x)$ is positive for $0 < x < \lambda_1$. [See formula (4.3), where the first term, which determines the sign of $J_p(x)$ for x near zero, is positive.] As we pass through λ_1, $J_p(x)$ goes from positive to negative values, as we pass through λ_2, $J_p(x)$ goes from negative to positive values, etc., so that

$$J_p'(\lambda_1) < 0, \; J_p'(\lambda_2) > 0, \; J_p'(\lambda_3) < 0, \ldots$$

Clearly, we have

$$[xJ_p'(x) - HJ_p(x)]_{x=\lambda_n} = \lambda_n J_p'(\lambda_n).$$

Therefore, the function $xJ_p'(x) - HJ_p(x)$ is alternately negative and positive for $x = \lambda_1, \lambda_2, \lambda_3, \ldots$, and hence vanishes at least once between each pair of consecutive zeros $\lambda_1, \lambda_2, \lambda_3, \ldots$. This shows that $xJ_p'(x) - HJ_p(x)$ also has an infinite set of positive zeros.

It can be shown that the distance between consecutive zeros of both $J_p'(x)$ and $xJ_p'(x) - HJ_p(x)$ approaches π as the distance from the origin increases, just as in the case of the zeros of $J_p(x)$. However, we shall not prove this fact here.

11. Parametric Form of Bessel's Equation

Let the function $y(x)$ be any solution of Bessel's equation (1.1). Consider the function $y = y(\lambda x)$, and set $\lambda x = t$. Since obviously

$$t^2 \frac{d^2y}{dt^2} + t\frac{dy}{dt} + (t^2 - p^2)y = 0, \tag{11.1}$$

substituting

$$\frac{dy}{dt} = \frac{1}{\lambda}\frac{dy}{dx}, \quad \frac{d^2y}{dt^2} = \frac{1}{\lambda^2}\frac{d^2y}{dx^2}$$

into (11.1), and using $\lambda x = t$, we obtain

$$x^2\frac{d^2y}{dx^2} + x\frac{dy}{dx} + (\lambda^2x^2 - p^2)y = 0.$$

Thus, if the function $y(x)$ is a solution of Bessel's equation (1.1), the function $y(\lambda x)$ is a solution of the equation

$$x^2y'' + xy' + (\lambda^2x^2 - p^2)y = 0, \tag{11.2}$$

called the *parametric form of Bessel's equation*, with parameter λ.

12. Orthogonality of the Functions $J_p(\lambda x)$

Let λ and μ be two nonnegative numbers. Consider the functions $y = J_p(\lambda x)$ and $z = J_p(\mu x)$, where $p > -1$. According to Sec. 11, these functions obey the equations

$$x^2y'' + xy' + (\lambda^2x^2 - p^2)y = 0,$$
$$x^2z'' + xz' + (\mu^2x^2 - p^2)z = 0$$

or

$$xy'' + y' - \frac{p^2}{x}y = -\lambda^2xy,$$

$$xz'' + z' - \frac{p^2}{x}z = -\mu^2xz.$$

Subtract the first equation multiplied by z from the second equation multiplied by y. The result is

$$x(yz'' - zy'') + (yz' - zy') = (\lambda^2 - \mu^2)xyz,$$

or

$$x(yz' - zy')' + (yz' - zy') = (\lambda^2 - \mu^2)xyz,$$

so that finally

$$[x(yz' - zy')]' = (\lambda^2 - \mu^2)xyz. \tag{12.1}$$

We now integrate (12.1) from 0 to 1, obtaining

$$[x(yz' - zy')]_{x=0}^{x=1} = (\lambda^2 - \mu^2)\int_0^1 xyz\,dx. \tag{12.2}$$

Since by hypothesis $p > -1$, the integrand in (12.2) is actually integrable on the interval $[0, 1]$. In fact, since $y = J_p(\lambda x)$, $z = J_p(\mu x)$, it follows from (4.3) that

$$y = x^p\varphi(x), \qquad z = x^p\psi(x), \tag{12.3}$$

where $\varphi(x)$ and $\psi(x)$ are the sums of power series, and hence represent continuous functions with continuous derivatives. Therefore

$$|xyz| = |x^{2p+1}\varphi(x)\psi(x)| \leqslant Mx^{2p+1} \qquad (M = \text{const}),$$

and since $2p + 1 > -1$ by hypothesis, this implies the integrability of the right-hand side (and hence of the left-hand side) of (12.1). It follows from (12.3) that

$$[x(yz' - zy')]_{x=0} = 0$$

for $p > -1$. Therefore, instead of (12.2), we can write

$$[x(yz' - zy')]_{x=1} = (\lambda^2 - \mu^2)\int_0^1 xyz\,dx. \tag{12.4}$$

We now note that

$$[y]_{x=1} = J_p(\lambda), \qquad [z]_{x=1} = J_p(\mu),$$

while on the other hand

$$y' = \frac{d}{dx} J_p(\lambda x) = \lambda J_p'(\lambda x),$$

$$z' = \frac{d}{dx} J_p(\mu x) = \mu J_p'(\mu x)$$

so that

$$[y']_{x=1} = \lambda J_p'(\lambda), \qquad [z']_{x=1} = \mu J_p'(\mu).$$

Thus, (12.4) becomes

$$\mu J_p(\lambda) J_p'(\mu) - \lambda J_p(\mu) J_p'(\lambda) = (\lambda^2 - \mu^2) \int_0^1 x J_p(\lambda x) J_p(\mu x)\, dx. \tag{12.5}$$

So far, λ and μ have been arbitrary nonnegative numbers. We now impose certain restrictions on λ and μ. Consider the following three cases:

1) λ and μ are different zeros of the function $J_p(x)$, i.e., $J_p(\lambda) = 0$, $J_p(\mu) = 0$, $\lambda \neq \mu$. For such values of λ and μ, the left-hand side of (12.5) vanishes. Noting that $\lambda^2 - \mu^2 \neq 0$, we then obtain

$$\int_0^1 x J_p(\lambda x) J_p(\mu x)\, dx = 0. \tag{12.6}$$

If the factor x were absent in the integrand, the functions $J_p(\lambda x)$ and $J_p(\mu x)$ would be orthogonal in the usual sense. In the present case, we say that the functions $J_p(\lambda x)$ and $J_p(\mu x)$ are orthogonal *with weight* x. Of course, we might also say that the functions

$$z_1 = \sqrt{x} J_p(\lambda x), \quad z_2 = \sqrt{x} J_p(\mu x)$$

are orthogonal in the usual sense.

It should be noted that according to (8.1), for $p = 1/2$, z_1 and z_2 reduce to

$$z_1 = \sqrt{2/\pi} \sin \lambda x, \quad z_2 = \sqrt{2/\pi} \sin \mu x,$$

where λ and μ are numbers of the form $n\pi$. According to (8.2), for $p = -1/2$, we obtain the functions

$$z_1 = \sqrt{2/\pi} \cos \lambda x, \quad z_2 = \sqrt{2/\pi} \cos \mu x,$$

where in this case λ and μ are numbers of the form $(n + \frac{1}{2})\pi$. Thus, in these special cases, we obtain the trigonometric functions, i.e., in the first case we obtain the functions $\sin n\pi x$ (except for a factor), which are orthogonal on $[0, 1]$, and in the second case the functions $\cos (n + \frac{1}{2})\pi x$, which are also orthogonal on $[0, 1]$.

2) λ and μ are different zeros of the function $J'_p(x)$:

$$J'_p(\lambda) = 0, \ J'_p(\mu) = 0 \quad (\lambda \neq \mu).$$

In this case, the left-hand side of (12.5) also vanishes, and hence we still obtain the formula (12.6). Thus, here again the functions $J_p(\lambda x)$ and $J_p(\mu x)$ are orthogonal with weight x.

3) Finally, let λ and μ be two different zeros of the function $xJ'_p(x) - HJ_p(x)$:

$$\lambda J'_p(\lambda) - HJ_p(\lambda) = 0,$$

$$\mu J'_p(\mu) - HJ_p(\mu) = 0.$$

Multiply the first equation by $J_p(\mu)$ and subtract it from the second equation multiplied by $J_p(\lambda)$. The result is

$$\mu J_p(\lambda)J'_p(\mu) - \lambda J_p(\mu)J'_p(\lambda) = 0.$$

Therefore, in this case, the left-hand side of (12.5) also vanishes, and we again obtain the formula (12.6), i.e., the functions $J_p(\lambda x)$ and $J_p(\mu x)$ are orthogonal with weight x.

13. Evaluation of the Integral $\int_0^1 xJ_p^2(\lambda x)\,dx$

When λ and μ are different, (12.5) leads to

$$\int_0^1 xJ_p(\mu x)J_p(\lambda x)\,dx = \frac{\lambda J_p(\mu)J'_p(\lambda) - \mu J_p(\lambda)J'_p(\mu)}{\mu^2 - \lambda^2}.$$

When $\mu \to \lambda$, the fraction on the right becomes indeterminate, since the numerator and denominator both approach zero. To "remove" this indeterminacy, we use L'Hospital's rule, letting $\lambda = \text{const}$ and $\mu \to \lambda$. This gives

$$\int_0^1 xJ_p^2(\lambda x)\,dx = \lim_{\mu\to\lambda} \frac{\lambda J'_p(\mu)J'_p(\lambda) - \mu J_p(\lambda)J''_p(\mu) - J_p(\lambda)J'_p(\mu)}{2\mu}$$

$$= \frac{\lambda J'^2_p(\lambda) - \lambda J_p(\lambda)J''_p(\lambda) - J_p(\lambda)J'_p(\lambda)}{2\lambda} \tag{13.1}$$

$$= \frac{1}{2}\left[J'^2_p(\lambda) - J_p(\lambda)J''_p(\lambda) - \frac{J_p(\lambda)J'_p(\lambda)}{\lambda}\right].$$

But $J_p(\lambda)$, regarded as a function of λ, satisfies Bessel's equation, i.e.,

$$\lambda^2 J''_p(\lambda) + \lambda J'_p(\lambda) + (\lambda^2 - p^2)J_p(\lambda) = 0,$$

so that

$$-J_p(\lambda)J_p''(\lambda) - \frac{J_p(\lambda)J_p'(\lambda)}{\lambda} = \left(1 - \frac{p^2}{\lambda^2}\right) J_p^2(\lambda).$$

Therefore, it follows from (13.1) that

$$\int_0^1 xJ_p^2(\lambda x)\,dx = \frac{1}{2}\left[J_p'^2(\lambda) + \left(1 - \frac{p^2}{\lambda^2}\right) J_p^2(\lambda)\right]. \tag{13.2}$$

Thus, we can draw the following conclusions:

1) If λ is a zero of the function $J_p(\lambda)$, then

$$\int_0^1 xJ_p^2(\lambda x)\,dx = \tfrac{1}{2}J_p'^2(\lambda). \tag{13.3}$$

This formula can be given a different form. We first replace x by λ in (7.8), obtaining

$$\lambda J_p'(\lambda) - pJ_p(\lambda) = -\lambda J_{p+1}(\lambda).$$

Since in the present case $J_p(\lambda) = 0$, we have

$$J_p'(\lambda) = -J_{p+1}(\lambda),$$

so that

$$\int_0^1 xJ_p^2(\lambda x)\,dx = \tfrac{1}{2}J_{p+1}^2(\lambda). \tag{13.4}$$

2) If λ is a zero of the function $J_p'(\lambda)$, then

$$\int_0^1 xJ_p^2(\lambda x)\,dx = \frac{1}{2}\left(1 - \frac{p^2}{\lambda^2}\right) J_p^2(\lambda). \tag{13.5}$$

*14. Bounds for the Integral $\int_0^1 xJ_p^2(\lambda x)\,dx$

The following inequality, valid for sufficiently large λ, will be useful later

$$\frac{K}{\lambda} \leqslant \int_0^1 xJ_p^2(\lambda x)\,dx \leqslant \frac{M}{\lambda}. \tag{14.1}$$

Here $K > 0$ and $M > 0$ are constants (which can depend on p). Obviously we have

$$\int_0^1 xJ_p^2(\lambda x)\,dx = \frac{1}{\lambda^2}\int_0^\lambda tJ_p^2(t)\,dt. \tag{14.2}$$

According to the asymptotic formula (9.16)

$$|J_p(t)| \leqslant \frac{2A}{\sqrt{t}}$$

IER-BESSEL SERIES 219

ce

$$M \int_0^\lambda dt = M\lambda \qquad (M = \text{const}).$$

lies the right-hand inequality in (14.1).
by the same asymptotic formula (9.16)

$$= \left(A \sin (t + \omega) + \frac{r}{t}\right)^2$$

$$= A^2 \sin^2 (t + \omega) + \frac{2Ar \sin (t + \omega)}{t} + \frac{r^2}{t^2}$$

$$\geqslant A^2 \sin^2 (t + \omega) - \frac{L}{t} \qquad (L = \text{const})$$

for large t, i.e., for $t > \lambda_0$. But then

$$\int_0^\lambda tJ_p^2(t)\, dt > \int_{\lambda_0}^\lambda tJ_p^2(t)\, dt \geqslant \int_{\lambda_0}^\lambda \left(A^2 \sin^2 (t + \omega) - \frac{L}{t}\right) dt$$

$$= A^2 \int_{\lambda_0}^\lambda \sin^2 (t + \omega)\, dt - L(\ln \lambda - \ln \lambda_0) \geqslant K\lambda \quad (K = \text{const}, K > 0),$$

and in view of (14.2), this implies the left-hand side of (14.1).

15. Definition of Fourier-Bessel Series

The rest of this chapter is devoted to the study of Fourier expansions with respect to Bessel functions. Let $\lambda_1, \lambda_2, \ldots, \lambda_n, \ldots$ be the positive roots of the equation $J_p(x) = 0$ $(p > -1)$ arranged in increasing order. According to Sec. 12, the functions

$$J_p(\lambda_1 x), J_p(\lambda_2 x), \ldots, J_p(\lambda_n x), \ldots \tag{15.1}$$

form an orthogonal system on [0, 1], with weight x. To give the reader some idea of the appearance of the functions of the system (15.1), in Fig. 46 we

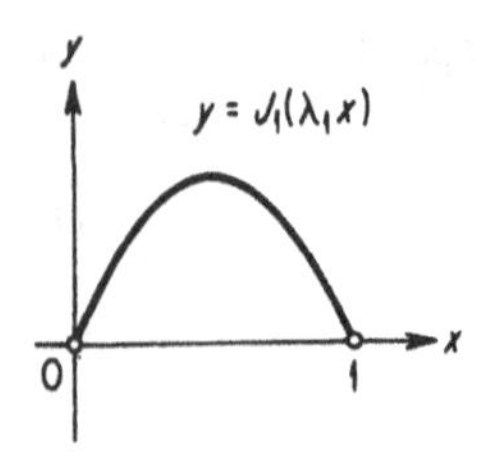

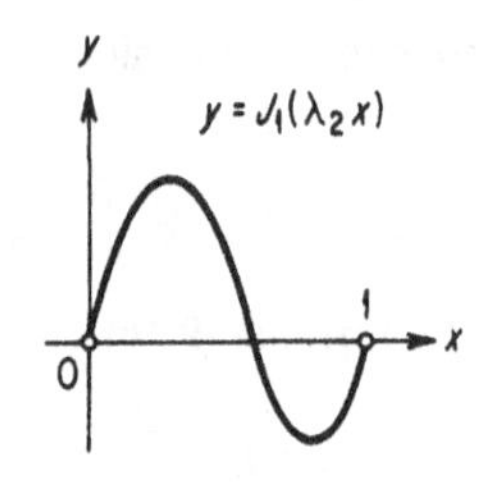

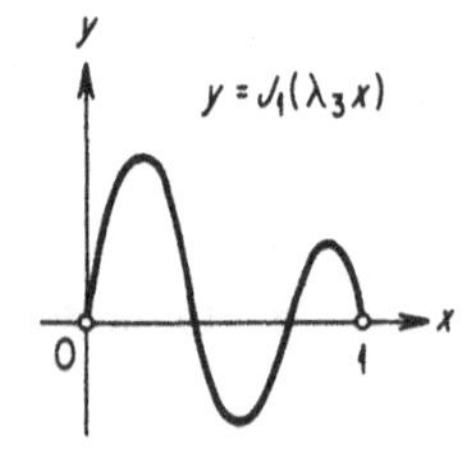

FIGURE 46

draw the graphs of the functions $J_1(\lambda_1 x)$, $J_1(\lambda_2 x)$, $J_1(\lambda_3 x)$ on
[0, 1]. The functions $J_1(\lambda_n x)$ $(n = 3, 4, \ldots)$ on [0, 1] become mo
more complicated, i.e., the number of "oscillations" increases.

For any function $f(x)$ which is absolutely integrable on [0, 1], we can form the Fourier series with respect to the system (15.1), or briefly, the *Fourier-Bessel* series

$$f(x) \sim c_1 J_p(\lambda_1 x) + c_2 J_p(\lambda_2 x) + \cdots, \tag{15.2}$$

where the constants

$$c_n = \frac{\int_0^1 x f(x) J_p(\lambda_n x)\,dx}{\int_0^1 x J_p^2(\lambda_n x)\,dx} = \frac{2}{J_{p+1}^2(\lambda_n)} \int_0^1 x f(x) J_p(\lambda_n x)\,dx \tag{15.3}$$

are called the *Fourier-Bessel coefficients* of $f(x)$. These coefficients can be obtained by using the following formal argument: Instead of (15.2), we write

$$f(x) = c_1 J_p(\lambda_1 x) + c_2 J_p(\lambda_2 x) + \cdots. \tag{15.4}$$

We multiply both sides of (15.4) by $x J_p(\lambda_n x)$ and integrate over the interval [0, 1], assuming that term by term integration is justified. Because of the orthogonality (with weight x) of the system (15.1), the result is

$$\int_0^1 x f(x) J_p(\lambda_n x)\,dx = c_n \int_0^1 x J_p^2(\lambda_n x)\,dx,$$

which implies (15.3). [See equation (13.4).]

If the equality (15.4) actually holds and if the series on the right converges uniformly, then term by term integration is known to be legitimate, and hence the coefficients c_n must be given by the formula (15.3). However, just as in the case of ordinary orthogonal systems (see Ch. 2, Sec. 2), we first use (15.3) to form the series (15.2), and only later examine the convergence of the series to $f(x)$.

16. Criteria for the Convergence of Fourier-Bessel Series

We now state without proof the most important criteria for the convergence of a Fourier-Bessel series to the function from which it is formed. These criteria are analogous to those with which we are familiar in the case of trigonometric Fourier series (see Ch. 3, Secs. 9 and 12). However, the proofs are much more complicated than in the case of Fourier-Bessel series, and hence we omit them.

THEOREM 1. *Let $f(x)$ be a piecewise smooth, continuous or discontinuous function on [0, 1]. Then the Fourier-Bessel series $(p \geqslant -\tfrac{1}{2})$ of*

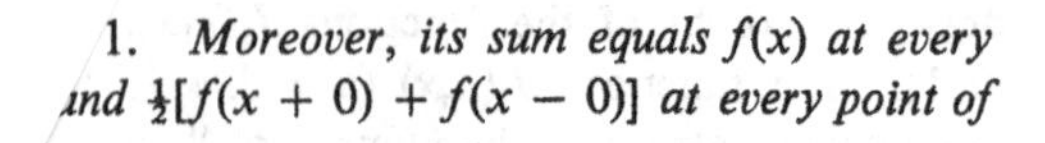

… 1. *Moreover, its sum equals* $f(x)$ *at every* … *and* $\frac{1}{2}[f(x+0)+f(x-0)]$ *at every point of*

…nverges to zero for $x = 1$, and converges to zero …ce all the functions of the system vanish for these …at for $p < 0$ all the functions of the system (15.1) … $x = 0$ [see (4.1)], so that it is meaningless to talk of …e series at $x = 0$.

R… …nstead of piecewise smoothness of $f(x)$ on [0, 1], it is sufficient t… …equire piecewise smoothness of $f(x)$ on every subinterval $[\delta, 1-\delta]$ where $\delta > 0$, in addition to the requirement that $f(x)$ [or even $\sqrt{x}f(x)$] be absolutely integrable on the whole interval [0, 1].

THEOREM 2. *Let* $f(x)$ *be continuous and have an absolutely integrable derivative on the interval* $[a, b]$, *where* $0 \leqslant a < b \leqslant 1$. *Then, the Fourier-Bessel series* ($p \geqslant -\frac{1}{2}$) *of* $f(x)$ *converges uniformly on every subinterval* $[a+\delta, b-\delta]$, *where* $\delta > 0$.

THEOREM 3. *Let* $f(x)$ *be absolutely integrable on* [0, 1], *let* $f(x)$ *be continuous and have an absolutely integrable derivative on the interval* $[a, 1]$, *where* $0 \leqslant a < 1$, *and let* $f(x)$ *satisfy the condition* $f(1) = 0$. *Then the Fourier-Bessel series* ($p \geqslant -\frac{1}{2}$) *of* $f(x)$ *converges uniformly on every subinterval* $[a+\delta, 1]$, *where* $\delta > 0$.

The condition $f(1) = 0$ is quite natural, since all the functions of the system (15.1) vanish for $x = 1$.

Remark. In Theorems 2 and 3, it is sufficient to require only absolute integrability of $\sqrt{x}f(x)$.

Example. *Expand the function* $f(x) = x^p$ ($p \geqslant -\frac{1}{2}$) *for* $0 < x < 1$ *in Fourier-Bessel series with respect to the system*

$$J_p(\lambda_1 x),\ J_p(\lambda_2 x), \ldots,\ J_p(\lambda_n x). \ldots$$

By formula (15.3)

$$c_n = \frac{2}{J_{p+1}^2(\lambda_n)} \int_0^1 x^{p+1} J_p(\lambda_n x)\, dx \qquad (n = 1, 2, \ldots).$$

But we have

$$\int_0^1 x^{p+1} J_p(\lambda_n x)\, dx = \frac{1}{\lambda_n^{p+2}} \int_0^{\lambda_n} t^{p+1} J_p(t)\, dt.$$

According to formula (7.1) (with p replaced by $p+1$)

$$\frac{d}{dt}[t^{p+1} J_{p+1}(t)] = t^{p+1} J_p(t).$$

Therefore

$$\int_0^{\lambda_n} t^{p+1}J_p(t)\,dt = \int_0^{\lambda_n} [t^{p+1}J_{p+1}(t)]'\,dt = [t^{p+1}J_{p+1}(t)]_{t=0}^{t=\lambda_n} = \lambda_n^{p+1}J_{p+1}(\lambda_n),$$

so that

$$\int_0^1 x^{p+1}J_p(\lambda_n x)\,dx = \frac{1}{\lambda_n}J_{p+1}(\lambda_n). \tag{16.1}$$

It follows that

$$c_n = \frac{2}{\lambda_n J_{p+1}(\lambda_n)} \qquad (n = 1, 2, \ldots).$$

Thus, by Theorem 1, we can write

$$x^p = 2\left(\frac{J_p(\lambda_1 x)}{\lambda_1 J_{p+1}(\lambda_1)} + \frac{J_p(\lambda_2 x)}{\lambda_2 J_{p+1}(\lambda_2)} + \cdots\right)$$

for $p \geqslant -\frac{1}{2}$ and $0 < x < 1$.

*17. Bessel's Inequality and Its Consequences

The orthogonality with weight x of the functions of the system (15.1) can be regarded as ordinary orthogonality of the functions

$$\sqrt{x}J_p(\lambda_1 x),\ \sqrt{x}J_p(\lambda_2 x), \ldots,\ \sqrt{x}J_p(\lambda_n x), \ldots. \tag{17.1}$$

Therefore, if we wish to make a series expansion of a function $f(x)$ with respect to the system (15.1), we can first make a series expansion of the function $\sqrt{x}f(x)$ with respect to the ordinary orthogonal system (17.1), obtaining

$$\sqrt{x}f(x) \sim c_1\sqrt{x}J_p(\lambda_1 x) + c_2\sqrt{x}J_p(\lambda_2 x) + \cdots,$$

and then go over to the expansion (15.2). (It is easily verified that the coefficients of both expansions are the same.) These considerations allow us to apply the results of Ch. 2 to Fourier-Bessel series.

Thus, assuming that $F(x) = \sqrt{x}f(x)$ is a square integrable function (which is certainly the case if $f(x)$ itself is square integrable) and applying Bessel's inequality (see Ch. 2, Sec. 6), we obtain

$$\int_0^1 F^2(x)\,dx \geqslant \sum_{n=1}^{\infty} c_n^2\|\sqrt{x}\,J_p(\lambda_n x)\|^2$$

or

$$\int_0^1 xf^2(x)\,dx \geqslant \sum_{n=0}^{\infty} c_n^2\int_0^1 xJ_p^2(\lambda_n x)\,dx.$$

Consequently, we have

$$\lim_{n\to\infty}\left[c_n^2\int_0^1 xJ_p^2(\lambda_n x)\,dx\right]=0.$$

But according to (14.1)

$$\int_0^1 xJ_p^2(\lambda_n x)\,dx \geqslant \frac{K}{\lambda_n} \qquad (K>0)$$

for all sufficiently large n, and therefore

$$\lim_{n\to\infty}\frac{c_n^2}{\lambda_n}=0$$

or

$$\lim_{n\to\infty}\frac{c_n}{\sqrt{\lambda_n}}=0. \tag{17.2}$$

Even more can be said: It follows from the asymptotic formula (9.16) that

$$|J_p(x)| < \frac{2A}{\sqrt{x}},$$

for all sufficiently large x, and hence for every fixed $x > 0$

$$|J_p(\lambda_n x)| < \frac{2A}{\sqrt{\lambda_n x}} \tag{17.3}$$

if n is sufficiently large. Then, using (17.2), we obtain

$$\lim_{n\to\infty}|c_n J_p(\lambda_n x)|=0 \tag{17.4}$$

for every fixed $x > 0$. Thus, *if $f(x)$ is square integrable, the general term of the Fourier-Bessel series of $f(x)$ always converges to zero* ($x > 0$). If $p > 0$, then obviously this will also be the case for $x = 0$ (since all the functions of the system (15.1) vanish for $x = 0$). Finally, it follows from (15.3) that

$$\int_0^1 xf(x)J_p(\lambda_n x)\,dx = c_n\int_0^1 xJ_p^2(\lambda_n x)\,dx,$$

or by (14.1)

$$\left|\int_0^1 xf(x)J_p(\lambda_n x)\,dx\right| \leqslant \frac{M|c_n|}{\lambda_n},$$

for sufficiently large n, so that by (17.2)

$$\lim_{n\to\infty}\int_0^1 xf(x)J_p(\lambda_n x)\,dx=0. \tag{17.5}$$

This is the analog of the property of the trigonometric integrals discussed in Ch. 3, Sec. 2.

Equations (17.2) and (17.5) are true not only for square integrable functions, but also for arbitrary absolutely integrable functions. (We omit the proof.)

*18. The Order of Magnitude of the Coefficients which Guarantees Uniform Convergence of a Fourier-Bessel Series

THEOREM 1. *If* $p \geqslant 0$ *and if*

$$|c_n| \leqslant \frac{c}{\lambda_n^{1+\varepsilon}}, \tag{18.1}$$

where $\varepsilon > 0$ *and* c *are constants, then the series*

$$c_1 J_p(\lambda_1 x) + c_2 J_p(\lambda_2 x) + \cdots + c_n J_p(\lambda_n x) + \cdots \tag{18.2}$$

converges absolutely and uniformly on $[0, 1]$.

Proof. For $p \geqslant 0$, the function $J_p(x)$ is bounded in the neighborhood of $x = 0$. By the asymptotic formula (9.16), $J_p(x)$ is bounded for large x. Therefore, $J_p(x)$ is bounded for *all* (positive) x, i.e.,

$$|c_n J_p(\lambda_n x)| \leqslant |c_n| L \qquad (L = \text{const}),$$

so that by (18.1)

$$|c_n J_p(\lambda_n x)| \leqslant \frac{cL}{\lambda_n^{1+\varepsilon}}.$$

But since $\lambda_{n+1} - \lambda_n \to \pi$ as $n \to \infty$ (see Sec. 9), it follows that for $n > m$ (where m is some fixed number),

$$\lambda_n > \lambda_m + (n - m) = n + h \qquad (h = \text{const}).$$

Therefore, if n is large, $\lambda_n \geqslant \frac{1}{2}n$ and

$$\frac{1}{\lambda_n} \leqslant \frac{2}{n}. \tag{18.3}$$

Consequently, for large n we have

$$|c_n J_p(\lambda_n x)| \leqslant \frac{H}{n^{1+\varepsilon}} \qquad (H = \text{const}).$$

Theorem 1 now follows from the fact that the right-hand side of this inequality is the general term of a convergent numerical series.

THEOREM 2. *If* $p \geqslant -\frac{1}{2}$ *and if*

$$|c_n| \leqslant \frac{c}{\lambda_n^{(1/2)+\varepsilon}}, \tag{18.4}$$

for all sufficiently large n, where $\varepsilon > 0$ *and c are constants, then the series*

$$c_1\sqrt{x}\,J_p(\lambda_1 x) + c_2\sqrt{x}\,J_p(\lambda_2 x) + \cdots + c_n\sqrt{x}\,J_p(\lambda_n x) + \cdots \tag{18.5}$$

converges absolutely and uniformly on the interval [0, 1]. *This implies that the series* (4.2) *converges absolutely and uniformly on every subinterval* $[\delta, 1]$, *where* $\delta > 0$.

Proof. For $p \geqslant -\frac{1}{2}$, the function $\sqrt{x}\,J_p(x)$ is bounded as $x \to 0$ [see equation (4.3)]. According to the asymptotic formula (9.16), $\sqrt{x}\,J_p(x)$ is bounded for large x. Therefore, $J_p(x)$ is bounded for all (positive) x, i.e.,

$$|\sqrt{x\lambda_n}\,J_p(\lambda_n x)| \leqslant L \qquad (L = \text{const})$$

and hence for any x in [0, 1] we have

$$|\sqrt{x}\,J_p(\lambda_n x)| \leqslant \frac{L}{\sqrt{\lambda_n}}. \tag{18.6}$$

It follows that

$$|c_n\sqrt{x}\,J_p(\lambda_n x)| \leqslant \frac{|c_n|\,L}{\sqrt{\lambda_n}} \leqslant \frac{cL}{\lambda_n^{1+\varepsilon}}$$

or by (18.3)

$$|c_n\sqrt{x}\,J_p(\lambda_n x)| \leqslant \frac{H}{n^{1+\varepsilon}} \qquad (H = \text{const}). \tag{18.7}$$

The right-hand side of (18.7) is the general term of a convergent numerical series. This proves the first part of the theorem. The second part of the theorem is a consequence of the following inequality, obtained from (18.7):

$$|c_n J_p(\lambda_n x)| \leqslant \frac{H}{\sqrt{x}\,n^{1+\varepsilon}} \leqslant \frac{H}{\sqrt{\delta}\,n^{1+\varepsilon}} \qquad (\delta \leqslant x \leqslant 1).$$

THEOREM 3. *If* $p > -1$ *and if*

$$|c_n| \leqslant \frac{c}{\lambda_n^{(1/2)+\varepsilon}},$$

for all sufficiently large n, where $\varepsilon > 0$ *and c are constants, then the series* (18.2) *converges absolutely and uniformly on every interval* $[\delta, 1]$, *where* $\delta > 0$.

Proof. Let $\delta \leqslant x \leqslant 1$. By the asymptotic formula (9.16),

$$|J_p(x)| \leqslant \frac{2A}{\sqrt{x}}.$$

for all sufficiently large x, i.e., for $x \geqslant x_0$. Moreover, the inequality $\lambda_n\delta \geqslant x_0$ holds for all sufficiently large n. Therefore if $x \geqslant \delta$, we have $\lambda_n x \geqslant x_0$ *a fortiori*, and hence for $\delta \leqslant x \leqslant 1$

$$|J_p(\lambda_n x)| \leqslant \frac{2A}{\sqrt{\lambda_n x}} \leqslant \frac{2A}{\sqrt{\lambda_n \delta}}$$

so that

$$|c_n J_p(\lambda_n x)| \leqslant \frac{2A}{\sqrt{\delta}} \frac{|c_n|}{\sqrt{\lambda_n}},$$

or

$$|c_n J_p(\lambda_n x)| \leqslant \frac{2Ac}{\sqrt{\delta}} \frac{1}{\lambda_n^{1+\varepsilon}}.$$

By (18.3), this implies

$$|c_n J_p(\lambda_n x)| \leqslant \frac{H}{n^{1+\varepsilon}} \qquad (H = \text{const})$$

for all sufficiently large x. Theorem 3 now follows from the fact that the right-hand side of this series is the general term of a convergent numerical series.

Remark. In the conditions (18.1) and (18.4) of Theorems 1, 2, and 3, we can write simply n instead of λ_n [in view of (18.3)].

Example 1. The series

$$J_1(\lambda_1 x) + \frac{J_1(\lambda_2 x)}{2^2} + \cdots + \frac{J_1(\lambda_n x)}{n^2} + \cdots$$

is absolutely and uniformly convergent on [0, 1], since here $p = 1$, $\varepsilon = 1$, and hence Theorem 1 is applicable.

Example 2. The series

$$J_{-1/4}(\lambda_1 x) + \frac{J_{-1/4}(\lambda_2 x)}{2} + \cdots + \frac{J_{-1/4}(\lambda_n x)}{n} + \cdots$$

is absolutely and uniformly convergent on every interval $[\delta, 1]$, $\delta > 0$, while the series

$$\sqrt{x}\, J_{-1/4}(\lambda_1 x) + \frac{\sqrt{x}\, J_{-1/4}(\lambda_2 x)}{2} + \cdots + \frac{\sqrt{x}\, J_{-1/4}(\lambda_n x)}{n} + \cdots$$

is absolutely and uniformly convergent on the whole interval [0, 1]. Here Theorem 2 is applicable.

Example 3. The series

$$J_{-3/4}(\lambda_1 x) + \frac{J_{-3/4}(\lambda_2 x)}{2} + \cdots + \frac{J_{-3/4}(\lambda_n x)}{n} + \cdots$$

is absolutely and uniformly convergent on every interval $[\delta, 1]$, $\delta > 0$. Here Theorem 3 is applicable.

*19. The Order of Magnitude of the Fourier-Bessel Coefficients of a Twice Differentiable Function

LEMMA. *Let $F(x)$ be a twice differentiable function defined on the interval $[0, 1]$, such that $F(0) = F'(0) = 0$, $F(1) = 0$, and such that $F''(x)$ is bounded (the second derivative may not exist at certain points). Then, if λ is a zero of the function $J_p(x)$, where $p > -1$, the inequality*

$$\left|\int_0^1 \sqrt{x}\, F(x) J_p(\lambda x)\, dx\right| \leqslant \frac{R}{\lambda^{5/2}} \qquad (R = \text{const}) \tag{19.1}$$

holds.

Proof. According to equations (9.1) and (9.2), the function

$$z(t) = \sqrt{t}\, J_p(t)$$

satisfies the equation

$$z''(t) + \left(1 - \frac{p^2 - \frac{1}{4}}{t^2}\right) z(t) = 0.$$

If we set $t = \lambda x$, then

$$z'(t) = \frac{1}{\lambda}\frac{dz}{dx},$$

$$z''(t) = \frac{1}{\lambda^2}\frac{d^2z}{dx^2},$$

so that

$$\frac{1}{\lambda^2}\frac{d^2z}{dx^2} + \left(1 - \frac{p^2 - \frac{1}{4}}{\lambda^2 x^2}\right) z = 0$$

or

$$\frac{d^2z}{dx^2} + \left(\lambda^2 - \frac{p^2 - \frac{1}{4}}{x^2}\right) z = 0. \tag{19.2}$$

Thus, the function $z = \sqrt{\lambda x}\, J_p(\lambda x)$ satisfies the equation (19.2). But then the function $z = \sqrt{x}\, J_p(\lambda x)$ also satisfies (19.2), since it differs from $\sqrt{\lambda x}\, J_p(\lambda x)$ only by a constant factor.

We now set $p^2 - \frac{1}{4} = m$ in (19.2) obtaining

$$z = \frac{1}{\lambda^2}\left(\frac{m}{x^2} z - z''\right).$$

It follows that

$$I = \int_0^1 \sqrt{x}\, F(x) J_p(\lambda x)\, dx = \int_0^1 F(x) z\, dx = \frac{1}{\lambda^2}\int_0^1 F(x)\left(\frac{m}{x^2} z - z''\right) dx.$$

Since

$$(F'z - Fz')' = F''z - Fz'',$$

we have

$$I = \frac{1}{\lambda^2}\int_0^1 \left[\left(F(x)\frac{m}{x^2} - F''(x)\right) z + (F'z - Fz')'\right] dx$$

$$= \frac{1}{\lambda^2}\int_0^1 \left(F(x)\frac{m}{x^2} - F''(x)\right) z\, dx + [F'z - Fz']_{x=0}^{x=1}.$$

But

$$[F'z - Fz']_{x=0}^{x=1} = [F'(1)z(1) - F(1)z'(1)] - [F'(0)z(0) - F(0)z'(0)] = 0,$$

in view of the following facts:

1) $z(1) = [\sqrt{x} J_p(\lambda x)]_{x=1} = J_p(\lambda) = 0$ and $F'(1)$ is finite;

2) $F(1) = 0$ by hypothesis and $z'(1)$ is finite;

3) By Taylor's formula, $F'(x) = xF''(\theta x)$ $(0 < \theta < 1)$, while $z = \sqrt{x}\, J_p(\lambda x) = x^{p+(1/2)}\varphi(x)$, where $\varphi(x)$ is continuous and differentiable, being the sum of a power series [see equation (4.3)]. Therefore

$$F'(0)z(0) = \lim_{x\to 0} F'(x)z(x) = \lim_{x\to 0} x^{p+(3/2)}\varphi(x)F''(\theta x) = 0$$

since F'' is bounded and $p > -1$;

4) By Taylor's formula, $F(x) = \frac{1}{2}x^2F''(\theta x)$ for $0 < \theta < 1$, while

$$z'(x) = (x^{p+(1/2)}\varphi(x))'$$

$$= (p + \tfrac{1}{2})x^{p-(1/2)}\varphi(x) + x^{p+(1/2)}\varphi'(x),$$

so that

$$F(0)z'(0) = \lim_{x\to 0} F(x)z'(x)$$

$$= \tfrac{1}{2}\lim_{x\to 0} [(p + \tfrac{1}{2})x^{p+(3/2)}\varphi(x) + x^{p+(5/2)}\varphi'(x)]F''(\theta x) = 0.$$

Using all these facts, we find that

$$I = \frac{1}{\lambda^2}\int_0^1 \left(F(x)\frac{m}{x^2} - F''(x)\right) z\,dx. \tag{19.3}$$

As we have already noted, it follows from Taylor's formula that

$$F(x) = \tfrac{1}{2}x^2F''(\theta x) \qquad (0 < \theta < 1)$$

near $x = 0$. This implies that the integrand of (19.3) is bounded. But then

$$\left|\int_0^1 \left(F(x)\frac{m}{x^2} - F''(x)\right) z\,dx\right| \leqslant L\int_0^1 |z|\,dx \qquad (L = \text{const}).$$

Thus, by the Schwarz inequality [see equation (4.1) of Ch. 2]

$$\left(\int_0^1 |z|\,dx\right)^2 \leqslant \int_0^1 z^2\,dx = \int_0^1 xJ_p^2(\lambda x)\,dx \leqslant \frac{M}{\lambda} \qquad (M = \text{const})$$

[see also equation (14.1)], and therefore

$$\int_0^1 |z|\,dx \leqslant \sqrt{M/\lambda}. \tag{19.4}$$

If we use (19.4), it follows from (19.3) that

$$|I| \leqslant \frac{L\sqrt{M}}{\lambda^{5/2}},$$

which proves the inequality (19.1).

THEOREM 1. *Let $f(x)$ be a bounded and twice differentiable function defined on the interval* $[0, 1]$, *such that* $f(0) = f'(0) = 0$, $f(1) = 0$, *and such that $f''(x)$ is bounded (the second derivative may not exist at certain points). Then the Fourier-Bessel coefficients of the function $f(x)$ satisfy the inequality*

$$|c_n| \leqslant \frac{C}{\lambda_n^{3/2}} \qquad (C = \text{const}). \tag{19.5}$$

Proof. If $f(x)$ satisfies the conditions of the theorem, so does the function $F(x) = \sqrt{x}f(x)$. Therefore, applying the lemma, we obtain

$$\left|\int_0^1 xf(x)J_p(\lambda_n x)\,dx\right| = \left|\int_0^1 \sqrt{x}\,F(x)J_p(\lambda_n x)\,dx\right|$$

$$\leqslant \frac{R}{\lambda_n^{5/2}} \qquad (R = \text{const}).$$

By equation (14.1)

$$\int_0^1 xJ_p^2(\lambda_n x)\,dx \geqslant \frac{K}{\lambda_n} \qquad (K \neq 0),$$

so that

$$|c_n| = \frac{\left|\int_0^1 xf(x)J_p(\lambda_n x)\,dx\right|}{\left|\int_0^1 xJ_p^2(\lambda_n x)\,dx\right|} \leqslant \frac{R}{K}\frac{1}{\lambda_n^{3/2}},$$

which proves the required inequality (19.5).

Remark. Theorem 1 remains true if we impose the requirements of the theorem on the function $F(x) = \sqrt{x}\,f(x)$ instead of on $f(x)$, since we actually applied the lemma to $F(x)$.

The following proposition supplementing Theorem 2 of Sec. 16 is a consequence of Theorem 1:

THEOREM 2. *Let $f(x)$ be a function which is continuous and twice differentiable on the interval $[0, 1]$, let $f(0) = f'(0) = 0$, $f(1) = 0$, and let $f''(x)$ be bounded (the second derivative may not exist at certain points). Then, the Fourier-Bessel series of $f(x)$ converges absolutely and uniformly on every subinterval $[\delta, 1]$ where $0 < \delta$ if $p > -1$, and on the whole interval $[0, 1]$ if $p \geqslant 0$.*

Proof. If $p > -1$, the assertion follows from the preceding theorem and Theorem 3 of Sec. 18. If $p \geqslant 0$, the uniform convergence on the whole interval $[0, 1]$ follows from the preceding theorem and Theorem 1 of Sec. 18.

Remark. If we use Theorem 2 of Sec. 18, then with these conditions on $f(x)$ and with $p \geqslant -\frac{1}{2}$, we obtain absolute and uniform convergence of the series (18.5) on the *whole* interval $[0, 1]$.

*20. The Order of Magnitude of the Fourier-Bessel Coefficients of a Function Which is Differentiable Several Times

THEOREM 1. *Let $f(x)$ be a function defined on the interval $[0, 1]$ such that $f(x)$ is differentiable $2s$ times $(s > 1)$ and such that*

1) $f(0) = f'(0) = \cdots = f^{(2s-1)}(0) = 0;$
2) *$f^{(2s)}(x)$ is bounded (this derivative may not exist at certain points);*
3) $f(1) = f'(1) = \cdots = f^{(2s-2)}(1) = 0.$

Then the following inequality is satisfied by the Fourier-Bessel coefficients of $f(x)$:

$$|c_n| \leqslant \frac{C}{\lambda^{2s-(1/2)}} \qquad (C = \text{const}). \tag{20.1}$$

Proof. It is easy to see that the function $F(x) = \sqrt{x}\,f(x)$ also satisfies the conditions of the theorem. In particular, $F(x)$ satisfies the conditions of the lemma of Sec. 19 and hence satisfies (19.3), i.e.,

$$I = \int_0^1 xf(x)J_p(\lambda_n x)\,dx = \int_0^1 \sqrt{x}\,F(x)J_p(\lambda_n x)\,dx$$

$$= \int_0^1 F(x)z\,dx = \frac{1}{\lambda_n^2}\int_0^1 \left(\frac{m}{x^2}F - F''\right)z\,dx,$$

where $m = p^2 - \frac{1}{4}$, $z = \sqrt{x}\,J_p(\lambda_n x)$. If F_1 denotes the function in parentheses in the last integral, then we have

$$I = \frac{1}{\lambda_n^2}\int_0^1 F_1 z dx.$$

Since the function F_1 satisfies all the conditions of the lemma, this time (19.3) gives

$$I = \frac{1}{\lambda_n^4}\int_0^1 F_2 z\,dx,$$

where we have written

$$F_2 = \frac{m}{x^2}F_1 - F_1''.$$

If $s > 2$, then F_2 again satisfies the conditions of the lemma, and in fact we can repeat the argument s times, finally obtaining

$$I = \frac{1}{\lambda_n^{2s}}\int_0^1 F_s z\,dx,$$

where

$$F_s = \frac{m}{x^2}F_{s-1} - F_{s-1}''$$

is a bounded function.

It follows that

$$\left|\int_0^1 F_s z\,dx\right| \leqslant L\int_0^1 |z|\,dx \qquad (L = \text{const}).$$

By (19.4)

$$\int_0^1 |z|\,dx \leqslant \sqrt{M/\lambda_n} \qquad (M = \text{const}),$$

so that we have

$$|I| \leqslant \frac{L\sqrt{M}}{\lambda_n^{2s+(1/2)}}.$$

But

$$|c_n| = \frac{\left|\int_0^1 xf(x)J_p(\lambda_n x)\,dx\right|}{\left|\int_0^1 xJ_p^2(\lambda_n x)\,dx\right|},$$

and since

$$\int_0^1 xJ_p^2(\lambda_n x)\,dx \geqslant \frac{K}{\lambda_n} \qquad (K > 0),$$

according to equation (14.1) we finally obtain

$$|c_n| \leqslant \frac{L\sqrt{M}}{K}\frac{1}{\lambda_n^{2s-(1/2)}},$$

which proves the inequality (20.1).

The next result is a consequence of Theorem 1:

THEOREM 2. *If the hypotheses of Theorem* 1 *are met for* $s \geqslant 1$, *then*

1) *For* $p \geqslant 0$

$$|c_n J_p(\lambda_n x)| \leqslant \frac{H}{\lambda_n^{2s-(1/2)}} \qquad (H = \text{const}) \tag{20.2}$$

for any x $(0 \leqslant x \leqslant 1)$;

2) *For* $p \geqslant -\frac{1}{2}$

$$|c_n J_p(\lambda_n x)| \leqslant \frac{L}{\sqrt{x}\,\lambda_n^{2s}} \qquad (L = \text{const}) \tag{20.3}$$

for all x $(0 < x \leqslant 1)$ *uniformly*;

3) *For* $p > -1$, *the inequality* (20.3) *holds for each* x $(0 < x \leqslant 1)$ *if* $n > n(x)$. (*In this case, there is no uniformity in* x.)

Proof. In Sec. 18 (see the proof of Theorem 1), we saw that the function $J_p(x)$ is bounded for $p \geqslant 0$. Hence, the inequality (20.2) is an immediate consequence of (20.1). For $p \geqslant -\frac{1}{2}$, since $J_p(\lambda_n x)$ satisfies the inequality (18.6), we need only apply (20.1) to get (20.3). Finally, for $p > -1$, it follows from the asymptotic formula (9.16) that

$$|J_p(\lambda_n x)| \leqslant \frac{L}{\sqrt{\lambda_n x}} \qquad (L = \text{const}), \tag{20.4}$$

for every x $(0 < x \leqslant 1)$ and $n > n(x)$. Again using (20.1), we obtain (20.3).

*21. Term by Term Differentiation of Fourier-Bessel Series

Given a Fourier-Bessel series

$$f(x) = \sum_{n=1}^{\infty} c_n J_p(\lambda_n x), \tag{21.1}$$

we now find sufficient conditions for the validity of the equality

$$f'(x) = \sum_{n=1}^{\infty} (c_n J_p(\lambda_n x))' = \sum_{n=1}^{\infty} c_n \lambda_n J_p'(\lambda_n x). \tag{21.2}$$

It follows from formula (7.8) that

$$\begin{aligned} |\lambda_n x J_p'(\lambda_n x)| &= |pJ_p(\lambda_n x) - \lambda_n x J_{p+1}(\lambda_n x)| \\ &\leqslant |pJ_p(\lambda_n x)| + |\lambda_n x J_{p+1}(\lambda_n x)|. \end{aligned} \tag{21.3}$$

We assume that $p > -1$, so that $p + 1 > 0$. Therefore, the quantity

$$|\sqrt{\lambda_n x}\, J_{p+1}(\lambda_n x)|$$

is bounded, by the asymptotic formula (9.16), and hence

$$|\lambda_n x J_p'(\lambda_n x)| \leqslant |pJ_p(\lambda_n x)| + \sqrt{\lambda_n}\, H \qquad (H = \text{const}), \tag{21.4}$$

where we consider only values of x such that $0 \leqslant x \leqslant 1$. If

$$|c_n| \leqslant \frac{C}{\lambda_n^{(3/2)+\varepsilon}}, \tag{21.5}$$

where $\varepsilon > 0$ and C are constants, then for $x > 0$

$$|c_n \lambda_n J_p'(\lambda_n x)| \leqslant \frac{Cp}{\lambda_n^{(3/2)+\varepsilon}} \left| \frac{J_p(\lambda_n x)}{x} \right| + \frac{CH}{\lambda_n^{1+\varepsilon} x},$$

or by (20.4)

$$|c_n \lambda_n J_p'(\lambda_n x)| \leqslant \frac{CLp}{\lambda_n^{2+\varepsilon} x \sqrt{x}} + \frac{CH}{\lambda_n^{1+\varepsilon} x}.$$

This implies at once [see (18.3)] that the series (21.2) converges for $0 < x \leqslant 1$ and converges uniformly on every subinterval $[\delta, 1]$ where $0 < \delta < 1$. The last fact implies the validity of (21.2) for $0 < x \leqslant 1$.

As for the validity of (21.2) for $x = 0$, if $p < 1$, $p \neq 0$, it is easy to see that all the functions $J_p'(\lambda_n x)$ become infinite for $x = 0$, since $J_p(x) = x^p \varphi(x)$, where $\varphi(x)$ is differentiable and $\varphi(0) \neq 0$ [see (4.3)], and therefore in this case (18.2) becomes meaningless. If $p \geqslant 1$, then (7.9) implies that

$$J_p'(\lambda_n x) = \tfrac{1}{2}[J_{p-1}(\lambda_n x) - J_{p+1}(\lambda_n x)],$$

where the functions in the right-hand side have nonnegative indices and hence are bounded. Then

$$|c_n\lambda_n J_p'(\lambda_n x)| \leqslant |c_n\lambda_n|\, H \qquad (H = \text{const}),$$

so that if

$$|c_n| \leqslant \frac{C}{\lambda_n^{2+\varepsilon}}, \tag{21.6}$$

the series (21.2) is uniformly convergent [cf. (18.3)] on [0, 1], i.e., (21.2) holds *everywhere* on the interval [0, 1]. Finally, if $p = 0$ we have

$$|\lambda_n J_0'(\lambda_n x)| = |\lambda_n J_1(\lambda_n x)|$$

instead of (21.3). Since the function J_1 is bounded, then

$$|c_n\lambda_n J_0'(x)| \leqslant \frac{CH}{\lambda_n^{1+\varepsilon}}$$

if (21.6) holds, and the series in (21.2) is again uniformly convergent, so that (21.2) is valid for all x in the interval [0, 1].

Thus, finally, we have proved the following theorem:

THEOREM 1. *If* $p > -1$ *and if* c_n *satisfies the condition* (21.5), *then the series* (21.1) *can be differentiated term by term for* $0 < x \leqslant 1$. *If* $p = 0$ *or* $p \geqslant 1$ *and* c_n *satisfies the condition* (21.6), *then the series* (21.1) *can be differentiated term by term everywhere on* [0, 1].[3]

We now find a sufficient condition for differentiating the series (21.1) *twice* term by term, i.e., for the validity of the relation

$$f''(x) = \sum_{n=1}^{\infty} (c_n J_p(\lambda_n x))'' = \sum_{n=1}^{\infty} c_n\lambda_n^2 J_p''(\lambda_n x). \tag{21.7}$$

Since $J_p(x)$ is a solution of Bessel's equation, we have

$$\lambda_n^2 x^2 J_p''(\lambda_n x) + \lambda_n x J_p'(\lambda_n x) + (\lambda_n^2 x^2 - p^2) J_p(\lambda_n x) = 0.$$

Thus

$$\begin{aligned}|\lambda_n^2 x^2 J_p''(\lambda_n x)| &= |-\lambda_n x J_p'(\lambda_n x) - \lambda_n^2 x^2 J_p(\lambda_n x) + p^2 J_p(\lambda_n x)| \\ &\leqslant |\lambda_n x J_p'(\lambda_n x)| + |\lambda_n^2 x^2 J_p(\lambda_n x)| + |p^2 J_p(\lambda_n x)|,\end{aligned}$$

so that by (21.4)

$$|\lambda_n^2 x^2 J_p''(\lambda_n x)| \leqslant |pJ_p(\lambda_n x)| + \sqrt{\lambda_n}\, H + |\lambda_n^2 x^2 J_p(\lambda_n x)| + |p^2 J_p(\lambda_n x)|.$$

[3] Note that the convergence of the series (21.1) itself follows from (21.5) and (21.6) and the theorems of Sec. 18.

Therefore, we have

$$|c_n\lambda_n^2 J_p''(\lambda_n x)| \leqslant |c_n| \frac{H\sqrt{\lambda_n}}{x^2} + \left(\frac{|p| + p^2}{x^2} + \lambda_n^2\right) |c_n J_p(\lambda_n x)|.$$

If

$$|c_n| \leqslant \frac{C}{\lambda_n^{(5/2)+\varepsilon}}, \tag{21.8}$$

where $\varepsilon > 0$ and C are constants, then by (20.4), we obtain

$$|c_n\lambda_n^2 J_p''(\lambda_n x)| \leqslant \frac{CH}{\lambda_n^{2+\varepsilon} x^2} + \frac{CL(|p| + p^2)}{\lambda_n^{3+\varepsilon} x^2\sqrt{x}} + \frac{CL}{\lambda_n^{1+\varepsilon}\sqrt{x}}.$$

This together with (18.3) implies the convergence of the series in (21.7) on every interval $[\delta, 1]$, where $0 < \delta < 1$. Since the condition (21.8) implies the convergence of the series (21.2) for $0 < x \leqslant 1$, the uniform convergence of the series in (21.7) on every interval $[\delta, 1]$ implies that (21.7) is valid for $0 < x \leqslant 1$.

Next we observe that for $-1 < p < 2$, $p \neq 0$, $p \neq 1$, all the functions $J_p''(\lambda_n x)$ become infinite at $x = 0$, since $J_p(x) = x^p\varphi(x)$, where $\varphi(x)$ is the sum of a power series [$\varphi(0) \neq 0$] and hence is differentiable any number of times [see equation (4.3)]. Therefore, in this case the relation (21.7) becomes meaningless. Finally, we note that an argument similar to that given in connection with Theorem 1 can be used to show that for $p \geqslant 2$, $p = 0$ or $p = 1$ and for

$$|c_n| \leqslant \frac{C}{\lambda_n^{3+\varepsilon}}, \tag{21.9}$$

where $\varepsilon > 0$ and C are constants), the relation (21.7) holds *everywhere* on [0, 1]. Thus we have

THEOREM 2. *If* $p > -1$ *and if the coefficients* c_n *satisfy the inequality* (21.8), *then the series* (21.1) *can be differentiated twice term by term for* $0 < x \leqslant 1$.[4] *If* $p = 0, p = 1$ *or* $p \geqslant 2$, *and if the* c_n *satisfy the condition* (21.9), *then term by term differentiation is possible everywhere on* [0, 1].

As a corollary of Theorem 2, we obtain

THEOREM 3. *If the hypotheses of Theorem* 1 *of Sec.* 20 *are met for* $s = 2$, *then the Fourier-Bessel series of the function* $f(x)$ *can be differentiated twice term by term for* $0 < x \leqslant 1$ *if* $p > -1$ *and everywhere on* [0, 1] *if* $p = 0, p = 1$ *or* $p \geqslant 2$.

---

[4] The convergence of the series (21.1) itself follows from (21.8) or (21.9) and the theorems of Sec. 18.

In fact, in this case, according to (20.1)

$$|c_n| \leqslant \frac{C}{\lambda_n^{7/2}} \qquad (C = \text{const}),$$

and we need only apply Theorem 2.

22. Fourier-Bessel Series of the Second Type

We now let $\lambda_1, \lambda_2, \ldots, \lambda_n, \ldots$ be the roots of the equation

$$xJ_p'(x) - HJ_p(x) = 0 \qquad (H = \text{const}), \tag{22.1}$$

arranged in increasing order. For $H = 0$, this equation becomes

$$J_p'(x) = 0. \tag{22.2}$$

The existence of an infinite set of positive roots of equation (22.1) [and in particular of equation (22.2)] was proved in Sec. 10. Moreover, according to Sec. 12, for $p > -1$ the functions

$$J_p(\lambda_1 x),\ J_p(\lambda_2 x), \ldots,\ J_p(\lambda_n x), \ldots \tag{22.3}$$

form an orthogonal system on [0, 1], with weight x. The considerations given below pertain to the case where the λ_i are the roots of (22.1); all the results are valid for the special case where the system (22.3) corresponds to the roots of (22.2).

Let $f(x)$ be any function which is absolutely integrable on [0, 1]. Then we can form the Fourier series of $f(x)$ with respect to the system (22.3). We shall call such a series

$$f(x) \sim c_1 J_p(\lambda_1 x) + c_2 J_p(\lambda_2 x) + \cdots \tag{22.4}$$

a *Fourier-Bessel series of the second type*; here the constants

$$\begin{aligned} c_n &= \frac{\int_0^1 xf(x)J_p(\lambda_n x)\,dx}{\int_0^1 xJ_p^2(\lambda_n x)\,dx} \\ &= \frac{2\lambda_n^2}{\lambda_n^2 J_p'^2(\lambda_n) + (\lambda_n^2 - p^2)J_p^2(\lambda_n)} \int_0^1 xf(x)J_p(\lambda_n x)\,dx \end{aligned} \tag{22.5}$$

just like ordinary Fourier-Bessel coefficients (or for that matter, just like Fourier coefficients in general), can be obtained by the usual formal argument. We have the following theorem, which closely resembles Theorem 1 of Sec. 16:

THEOREM. *Let $f(x)$ be a piecewise smooth (continuous or discontinuous) function defined on [0, 1]. Then the Fourier-Bessel series of*

$f(x)$ of the second type ($p \geqslant -\frac{1}{2}, p > H$) converges for $0 < x < 1$. Moreover, its sum equals $f(x)$ at every point of continuity of $f(x)$ and

$$\tfrac{1}{2}[f(x + 0) + f(x - 0)]$$

at every point of discontinuity of $f(x)$.

For $x = 1$, the series converges to the value $f(1 - 0)$; if $f(x)$ is continuous at $x = 1$, the series converges to $f(1)$. If $p > 0$, the series converges to zero, since in this case all the functions of the system (22.3) vanish at zero. Since all the functions (22.3) become infinite at $x = 0$, it is meaningless to talk about the convergence of the series at $x = 0$.

We omit the proof of this theorem, noting only that the condition $p > H$ (which has no analog in the theorems of Sec. 18) is required because of complications that arise when $p \leqslant H$. It turns out that if $p \leqslant H$, then the theorem is true only if we add a new function to the system (22.3). For example, if $p = H$ the new function is x^p, and instead of (22.3) we have to consider the new orthogonal system

$$x^p,\ J_p(\lambda_1 x),\ J_p(\lambda_2 x), \ldots.$$

In this case, the orthogonality of the function x^p (with weight 1) to the other functions is a consequence of the following two facts:

1) $x = 0$ is a root of the equation

$$xJ_p'(x) - pJ_p(x) = 0$$

[see equation (7.8)];

2) x^p satisfies Bessel's equation with parameter $\lambda = 0$, i.e., the equation

$$x^2y'' + xy' - p^2y = 0,$$

as can be verified directly.

Therefore, we can apply to the functions x^p and $J_p(\lambda_n x)$ all the considerations of Sec. 12 (where the orthogonality of the solutions of the parametric form of Bessel's equation was discussed). For $p < H$, the "additional" function is much more complicated. Thus, for simplicity, we restrict ourselves to the case $p > H$. A remark like that made after the statement of Theorem 1 of Sec. 16 also applies to the present theorem, and moreover, there are also two theorems completely analogous to Theorems 2 and 3 of Sec. 16, with the supplementary condition $p > H$ in both cases and the elimination of the condition $f(1) = 0$ in the second case. (The remark made in connection with Theorems 2 and 3 of Sec. 16 remains in force.)

Example. *Make a series expansion of the function $f(x) = x^p$ $(0 \leqslant x \leqslant 1)$ with respect to the system*

$$J_p(\lambda_1 x),\ J_p(\lambda_2 x), \ldots,\ J_p(\lambda_n x), \ldots, \tag{22.6}$$

where the λ_n are the roots of the equation $J_p'(x) = 0$ $(p > 0)$.

By formula (22.5)

$$c_n = \frac{2\lambda_n^2}{(\lambda_n^2 - p^2)J_p^2(\lambda_n)} \int_0^1 x^{p+1} J_p(\lambda_n x)\, dx,$$

and by (16.1)

$$\int_0^1 x^{p+1} J_p(\lambda_n x)\, dx = \frac{1}{\lambda_n} J_{p+1}(\lambda_n).$$

Therefore, by the theorem given above, we can write

$$x^p = 2 \sum_{n=1}^{\infty} \frac{\lambda_n J_{p+1}(\lambda_n) J_p(\lambda_n x)}{(\lambda_n^2 - p^2)J_p^2(\lambda_n)}. \tag{22.7}$$

We now ask whether this expansion holds for $p = 0$. The answer is negative, since for $p = 0$

$$J_{p+1}(\lambda_n) = J_1(\lambda_n) = -J_0'(\lambda_n) = 0$$

[where we have used equation (7.8)]. Thus, the right-hand side of (22.7) is zero, while the left-hand side is $f(x) = x^0 = 1$. The reason why the expansion fails for $p = 0$ is the following: In our case $H = 0$, so that $p = H$. Then, as remarked above, the system (22.6) must be supplemented by the function $x^p = x^0 = 1$. If we denote the coefficient corresponding to this function by c_0, we obtain

$$c_0 = \frac{\int_0^1 x f(x) \cdot 1\, dx}{\int_0^1 x \cdot 1^2\, dx} = 1,$$

since $f(x) = 1$. Moreover, $c_n = 0$ if $n = 1, 2, \ldots$. Therefore, instead of (22.7) we obtain the expansion

$$1 = 1 + 0 + 0 + \cdots,$$

which is obviously valid!

*23. Extension of the Results of Secs. 17–21 to Fourier-Bessel Series of the Second Type

In the theorems of Secs. 17 and 18, we essentially made no use whatsoever of the assumption that the numbers λ_n are the roots of the equation $J_p(x) = 0$. Therefore, these theorems are equally true for Fourier-Bessel series of the second type. However, in the lemma of Sec. 19, we did use the condition $J_p(\lambda) = 0$ to prove the formula (19.3). Thus, we now need a new lemma, which we proceed to prove.

LEMMA. *Let $F(x)$ be a twice differentiable function defined on* $[0, 1]$, *such that* $F(0) = F'(0) = 0$, $F'(1) - (H + \frac{1}{2})F(1) = 0$, *and such that* $F''(x)$ *is bounded (the second derivative may not exist at certain points). Then, if* λ *is a root of the equation*

$$xJ_p'(x) - HJ_p(x) = 0,$$

where $p > -1$, *it follows that*

$$I = \left| \int_0^1 \sqrt{x}\, F(x) J_p(\lambda x)\, dx \right| \leqslant \frac{R}{\lambda^{5/2}} \qquad (R = \text{const}). \qquad (23.1)$$

Proof. We arrive at the equality

$$I = \frac{1}{\lambda^2} \int_0^1 \left(F \frac{m}{x^2} - F'' \right) z\, dx + [F' z - F z']_{x=0}^{x=1}$$

just as in the lemma of Sec. 19, and the whole problem reduces to showing that the last term vanishes. This last term is the difference

$$[F'(1)z(1) - F(1)z'(1)] - [F'(0)z(0) - F(0)z'(0)]. \qquad (23.2)$$

We begin by calculating the first term in brackets. Since $z = \sqrt{x} J_p(\lambda x)$, we have

$$z(1) = J_p(\lambda),$$

$$z'(1) = \left[\frac{J_p(\lambda x)}{2\sqrt{x}} + \lambda \sqrt{x}\, J_p'(\lambda x) \right]_{x=1}$$

$$= \frac{J_p(\lambda)}{2} + \lambda J_p'(\lambda) = (H + \tfrac{1}{2}) J_p(\lambda),$$

where we have used the condition $\lambda J_p'(\lambda) - HJ_p(\lambda) = 0$. But then the first term in brackets is just

$$[F'(1) - (H + \tfrac{1}{2})F(1)] J_p(\lambda),$$

which vanishes by the hypothesis of the lemma. The fact that the second term in brackets in (23.2) vanishes, and indeed the rest of the proof follows by just the same method as in the lemma of Sec. 19.

The lemma just proved leads to the following result:

THEOREM 1. *Let $f(x)$ be a twice differentiable function defined on the interval* $[0, 1]$, *such that* $f(0) = f'(0) = 0$, $f'(1) - Hf(1) = 0$,[5] *and such*

[5] The condition $f'(1) - Hf(1) = 0$ appears artificial at this point, but it arises quite naturally in the applications.

that $f''(x)$ is bounded (the second derivative may not exist at certain points). Then the inequality

$$|c_n| \leqslant \frac{C}{\lambda_n^{3/2}} \qquad (C = \text{const}), \tag{23.3}$$

is obeyed by the Fourier coefficients of $f(x)$ with respect to the system (22.3).

Proof. If $F(x) = \sqrt{x}\, f(x)$, then obviously $F(0) = F'(0) = 0$ and $F''(x)$ is bounded. Moreover

$$F'(x) = \frac{f(x)}{2\sqrt{x}} + \sqrt{x}\, f'(x),$$

so that

$$F'(1) - (H + \tfrac{1}{2})F(1)$$
$$= \tfrac{1}{2}f(1) + f'(1) - (H + \tfrac{1}{2})f(1) = f'(1) - Hf(1) = 0.$$

Thus, the lemma can be applied to the function $F(x)$; the result is

$$\left|\int_0^1 xf(x)J(\lambda_n x)\,dx\right| = \left|\int_0^1 \sqrt{x}\,F(x)J(\lambda_n x)\,dx\right| \leqslant \frac{R}{\lambda_n^{5/2}} \quad (R = \text{const}).$$

Since by formula (14.1)

$$\int_0^1 xJ_p^2(\lambda_n x)\,dx \geqslant \frac{K}{\lambda_n} \qquad (K > 0),$$

we obtain (23.3), taking account of (22.5).

Theorem 1 and the results of Sec. 18 imply

THEOREM 2. *Let $f(x)$ be a twice differentiable function on the interval* $[0, 1]$, *such that* $f(0) = f'(0) = 0$, $f'(1) - Hf(1) = 0$ *and such that* $f''(x)$ *is bounded (the second derivative may not exist at certain points). Then the Fourier-Bessel series of $f(x)$ of the second type converges absolutely and uniformly on every subinterval* $[\delta, 1]$, *where* $0 < \delta < 1$, *if* $p > -1$ *and on the whole interval* $[0, 1]$ *if* $p \geqslant 0$.[6]

Finally Theorems 1 and 2 of Sec. 21 apply equally well to Fourier-Bessel series of the second type.

24. Fourier-Bessel Expansions of Functions Defined on the Interval $[0, l]$

Let $f(x)$ be an absolutely integrable function defined on the interval $[0, l]$, and set $x = lt$ or $t = x/l$. Then the function $\varphi(t) = f(lt)$ is defined on

[6] If the condition $p > H$ appearing in the theorem of Sec. 22 is not met, then the series need not have $f(x)$ as its sum.

the interval [0, 1] of the t-axis, and we can write

$$\varphi(t) \sim c_1 J_p(\lambda_1 t) + c_2 J_p(\lambda_2 t) + \cdots + c_n J_p(\lambda_n t) + \cdots, \tag{24.1}$$

where

$$c_n = \frac{2}{J_{p+1}^2(\lambda_n)} \int_0^1 t\varphi(t) J_p(\lambda_n t)\, dt \qquad (n = 1, 2, \ldots)$$

in the case of a Fourier-Bessel series of the first type, or

$$c_n = \frac{2\lambda_n^2}{\lambda_n^2 J_n'^2(\lambda_n) + (\lambda_n^2 - p^2) J_p^2(\lambda_n)} \int_0^1 t\varphi(t) J_p(\lambda_n t)\, dt \qquad (n = 1, 2, \ldots)$$

in the case of a Fourier-Bessel series of the second type. Returning to the variable x, we obtain

$$f(x) \sim c_1 J_p\left(\frac{\lambda_1}{l} x\right) + c_2 J_p\left(\frac{\lambda_2}{l} x\right) + \cdots + c_n J_p\left(\frac{\lambda_n}{l} x\right) + \cdots, \tag{24.2}$$

where

$$c_n = \frac{2}{l^2 J_{p+1}^2(\lambda_n)} \int_0^l x f(x) J_p\left(\frac{\lambda_n}{l} x\right) dx \qquad (n = 1, 2, \ldots), \tag{24.3}$$

or

$$c_n = \frac{2\lambda_n^2}{l^2[\lambda_n^2 J_p'^2(\lambda_n) + (\lambda_n^2 - p^2) J_p^2(\lambda_n)]} \int_0^l x f(x) J_p\left(\frac{\lambda_n}{l} x\right) dx$$

$$(n = 1, 2, \ldots). \tag{24.4}$$

If the series (24.1) converges, then the series (24.2) converges, and conversely.

It should be noted that we can also obtain the expansion (24.2) directly, avoiding the auxiliary function $\varphi(t)$, if we observe that the system

$$J_p\left(\frac{\lambda_1}{l} x\right), \ldots, J_p\left(\frac{\lambda_n}{l} x\right), \ldots \qquad (p > -1) \tag{24.5}$$

is orthogonal with weight x on the interval $[0, l]$. In fact, we have

$$\int_0^l x J_p\left(\frac{\lambda_m}{l} x\right) J_p\left(\frac{\lambda_n}{l} x\right) dx = l^2 \int_0^1 t J(\lambda_m t) J(\lambda_n t)\, dt = 0 \qquad (m \neq n),$$

where we have changed the variable of integration by setting $x = lt$. Once we have established the orthogonality of the system (24.5), we calculate the Fourier coefficients in the usual way, obtaining (24.3) or (24.4), depending on whether the numbers λ_n are the roots of the equation $J_p(x) = 0$ or of the equation $xJ_p'(x) - HJ_p(x) = 0$. Just as in the case of the interval [0, 1], the series (24.2) with the coefficients (24.3) is called a *Fourier-Bessel series of the first type*, and the series (24.2) with the coefficients (24.4) is called a *Fourier-Bessel series of the second type*. Since we can go from the series (24.2) to the series (24.1) by making the substitution $x = lt$, all our previous results for the interval $[0, 1]$ also hold for the case of the interval $[0, l]$. In particular, the convergence criteria are valid, provided of course that they are formulated for the interval $[0, l]$ instead of for $[0, 1]$.

PROBLEMS

1. Write the general solution of the differential equation

$$y'' + \frac{5}{x}y' + y = 0.$$

2. Calculate

$$\Gamma\left(\frac{2n+1}{2}\right)$$

where n is a positive integer.

3. Find an asymptotic formula for $J'_p(x)$.
Hint. Use formulas (7.9) and (9.16).

4. Find an asymptotic formula for $J''_p(x)$.
Hint. Use equation (1.1), formula (9.16) and the preceding problem.

5. Show that the functions $J_p(x)$ $(p \geqslant 0)$ and $\sqrt{x}\,J_p(x)$ $(p \geqslant -\frac{1}{2})$ are bounded for $0 < x < \infty$.
Hint. Use formulas (4.3) and (9.16).

6. Show that

$$\int_0^x tJ_0(t)\,dt = xJ_1(x).$$

7. By multiplying the power series for $e^{xt/2}$ and $e^{-x/2t}$, show that

$$\exp\left[\frac{x}{2}\left(t - \frac{1}{t}\right)\right] = \sum_{n=-\infty}^{\infty} J_n(x)t^n.$$

8. By substituting $t = e^{i\theta}$ in the formula of the preceding problem, taking real and imaginary parts, and using the formula for trigonometric Fourier coefficients, show that

$$J_{2n}(x) = \frac{1}{\pi}\int_0^{\pi} \cos(x \sin\theta)\cos n\theta\,d\theta, \qquad (n = 0, 1, 2, \ldots)$$

$$J_{2n+1}(x) = \frac{1}{\pi}\int_0^{\pi} \sin(x \sin\theta)\sin n\theta\,d\theta.$$

9. Show that

a) $\left|\frac{d^k}{dx^k} J_n(x)\right| \leqslant 1$ for all x $\quad (k, n = 0, 1, 2, \ldots)$;

b) $J_{2n+1}(0) = 0 \quad (n = 0, 1, 2, \ldots);$

c) $\lim_{n\to\infty} J_n(x) = 0$ for all x.

10. Expand the function x^{-p} $(0 < x < 1)$ in Fourier-Bessel series of the first kind.

11. Expand the function x^p $(0 < x \leqslant 1)$ in Fourier-Bessel series with respect to the system

$$J_p(\lambda_1 x), J_p(\lambda_2 x), \ldots,$$

where the λ_n are the roots of the equation $xJ_p'(x) - HJ_p(x) = 0$.

Hint. Use the example in Sec. 22.

12. Expand the function x^3 $(0 \leqslant x < 2)$ in Fourier-Bessel series of the first kind, with $p = 3$.

Hint. Use formula (24.3) and the results of the example in Sec. 16.

13. Let $\lambda_1, \lambda_2, \ldots$ be the positive roots of the equation $J_0(x) = 0$. Show that

$$\tfrac{1}{8}(1 - x^2) = \sum_{n=1}^{\infty} \frac{J_0(\lambda_n x)}{\lambda_n^3 J_1(\lambda_n)} \qquad (0 \leqslant x \leqslant 1).$$

Hint. Use formulas (7.1), (7.2), and (15.3).

14. Let $\lambda_1, \lambda_2, \ldots$ be the positive roots of the equation $J_p(x) = 0$ $(p > -\tfrac{1}{2})$. Show that

a) $$x^{p+1} = 2^2(p + 1) \sum_{n=1}^{\infty} \frac{J_{p+1}(\lambda_n x)}{\lambda_n^2 J_{p+1}(\lambda_n)}; \qquad \text{(I)}$$

b) $$x^{p+2} = 2^3(p + 1)(p + 2) \sum_{n=1}^{\infty} \frac{J_{p+2}(\lambda_n x)}{\lambda_n^3 J_{p+1}(\lambda_n)};$$

c) $$\sum_{n=1}^{\infty} \frac{J_{p+2}(\lambda_n x)}{\lambda_n J_{p+1}(\lambda_n)} = 0.$$

(In each case, $0 \leqslant x \leqslant 1$.)

Hint. a) Multiply the last formula of Sec. 16 by x^{p+1}, and integrate from 0 to x, using (7.1); b) Multiply (I) by x^{p+2} and integrate; c) Multiply (I) by $x^{-(p+1)}$ and differentiate, using (7.2).

9

THE EIGENFUNCTION METHOD AND ITS APPLICATIONS TO MATHEMATICAL PHYSICS

Part I. THEORY

1. The Gist of the Method

Many problems of mathematical physics lead to linear partial differential equations. Examples of such equations are

$$P\frac{\partial^2 u}{\partial x^2} + R\frac{\partial u}{\partial x} + Qu = \frac{\partial^2 u}{\partial t^2}, \tag{1.1}$$

$$P\frac{\partial^2 u}{\partial x^2} + R\frac{\partial u}{\partial x} + Qu = \frac{\partial u}{\partial t}, \tag{1.2}$$

where P, R, Q are continuous functions of the variable x, and $u = u(x, t)$ is an unknown function of the variables x and t. The first equation arises in problems involving vibrations of strings and rods, while the second arises in problems of heat flow. We shall confine our attention to equation (1.1), since this will be quite sufficient to explain the gist of the method to be discussed here.

In every concrete problem leading to an equation of the form (1.1), one requires a solution u of (1.1) which satisfies certain conditions. For example, suppose that we have to find a solution $u = u(x, t)$, defined for $a \leqslant x \leqslant b$ and all $t \geqslant 0$, which satisfies the *boundary conditions*

$$\begin{aligned} \alpha u(a, t) + \beta\frac{\partial u(a, t)}{\partial x} &= 0, \\ \gamma u(b, t) + \delta\frac{\partial u(b, t)}{\partial x} &= 0 \end{aligned} \tag{1.3}$$

for any $t \geqslant 0$, where α, β, γ, and δ are constants, and which satisfies the *initial conditions*

$$u(x, 0) = f(x), \qquad \frac{\partial u(x, 0)}{\partial t} = g(x) \tag{1.4}$$

for $a \leqslant x \leqslant b$, where $f(x)$ and $g(x)$ are given continuous functions. Usually, x stands for length and t for time, which explains the terms *boundary* conditions and *initial* conditions. We shall assume that neither the pair α, β nor the pair γ, δ can vanish simultaneously [for otherwise, instead of the relations (1.3), we would have the vacuous identities $0 = 0$]. This assumption can be written as

$$\alpha^2 + \beta^2 \neq 0, \qquad \gamma^2 + \delta^2 \neq 0. \tag{1.5}$$

To solve the equation (1.1), we first look for particular solutions of the form[1]

$$u = \Phi(x)T(t), \tag{1.6}$$

which satisfy only the boundary conditions (1.3). (We are only interested in solutions which do not vanish identically.) With this in mind, we differentiate (1.6) and substitute the result in (1.1), obtaining

$$P\Phi''T + R\Phi'T + Q\Phi T = \Phi T'',$$

whence

$$\frac{P\Phi'' + R\Phi' + Q\Phi}{\Phi} = \frac{T''}{T}.$$

Since the left-hand side of this last equation is a function only of x, while the right-hand side is a function of t, the equality is possible only if both sides equal a constant:

$$\frac{P\Phi'' + R\Phi' + Q\Phi}{\Phi} = \frac{T''}{T} = -\lambda \qquad (\lambda = \text{const}).$$

This leads to the following two ordinary linear differential equations of the second order:

$$P\Phi'' + R\Phi' + Q\Phi = -\lambda\Phi, \tag{1.7}$$

$$T'' + \lambda T = 0. \tag{1.8}$$

Obviously, a function $u \not\equiv 0$ of the form (1.6) satisfies the boundary conditions (1.3) if and only if the function $\Phi(x)$ satisfies the boundary conditions

$$\begin{aligned} \alpha\Phi(a) + \beta\Phi'(a) &= 0, \\ \gamma\Phi(b) + \delta\Phi'(b) &= 0. \end{aligned} \tag{1.9}$$

[1] This is known as the method of *separation of variables*. (*Translator*)

The problem of finding a solution of the equation (1.7) satisfying the conditions (1.9) will be called the *boundary value problem* for the equation (1.7) with the conditions (1.9).

In general, the boundary value problem for a linear differential equation of the second order does not have a solution for every value of λ; in particular, this is true of the boundary value problem for the equation (1.7) with the conditions (1.9). Nevertheless, it can be shown that there exists an *infinite set* of values $\lambda_0, \lambda_1, \ldots, \lambda_n, \ldots$ for which the boundary value problem has a solution, provided only that $P \neq 0$. Every value λ for which the boundary value problem has a solution $\Phi \not\equiv 0$ is called an *eigenvalue*, and the solution Φ corresponding to λ is called an *eigenfunction*. It will be shown below that in our case only one eigenfunction corresponds to each eigenvalue (to within a constant factor). Thus, for our problem, there is an infinite set of eigenvalues $\lambda_0, \lambda_1, \ldots, \lambda_n, \ldots$, with corresponding eigenfunctions

$$\Phi_0(x), \Phi_1(x), \ldots, \Phi_n(x), \ldots. \tag{1.10}$$

As will be shown in Sec. 4, the functions (1.10) form an orthogonal system on $[a, b]$, with a certain weight.

Having finished with equation (1.7), we next solve equation (1.8) for each $\lambda = \lambda_n$ and find the corresponding function $T_n(t)$, which depends on two arbitrary constants A_n and B_n. Thus, if $\lambda_n > 0$ for $n = 0, 1, 2, \ldots$ (and this case occurs quite often in concrete problems), we obviously have

$$T_n(t) = A_n \cos \sqrt{\lambda_n}\, t + B_n \sin \sqrt{\lambda_n}\, t, \tag{1.11}$$

where A_n and B_n are arbitrary constants. Then, each function

$$u_n(x, t) = \Phi_n(x)T_n(t) \qquad (n = 0, 1, 2, \ldots)$$

will be a solution of equation (1.1) satisfying the boundary conditions (1.3). Because of the linearity and homogeneity (with respect to u and its derivatives) of equation (1.1), every finite sum of solutions of (1.1) is also a solution. The same is also true of the infinite series

$$u = \sum_{n=0}^{\infty} u_n(x, t) = \sum_{n=0}^{\infty} T_n(t)\Phi_n(x), \tag{1.12}$$

if it converges and if it can be differentiated term by term twice with respect to x and t. If this is the case, we have

$$\begin{aligned} P\frac{\partial^2 u}{\partial x^2} + R\frac{\partial u}{\partial x} + Qu - \frac{\partial^2 u}{\partial t^2} \\ &= P\sum_{n=0}^{\infty}\frac{\partial^2 u_n}{\partial x^2} + R\sum_{n=0}^{\infty}\frac{\partial u_n}{\partial x} + Q\sum_{n=0}^{\infty} u_n - \sum_{n=0}^{\infty}\frac{\partial^2 u_n}{\partial t^2} \\ &= \sum_{n=0}^{\infty}\left(P\frac{\partial^2 u_n}{\partial x^2} + R\frac{\partial u_n}{\partial x} + Qu_n - \frac{\partial^2 u_n}{\partial t^2}\right). \end{aligned}$$

Since each term in parentheses in the last sum vanishes [because u_n is a solution of (1.1)], the entire sum vanishes, which means that the function (1.12) is a solution of (1.1). Moreover, since each term of the series (1.12) satisfies the boundary conditions (1.3), the sum of the series, i.e., the function u, also satisfies these conditions.

We must now satisfy the initial conditions (1.4). This can be achieved by suitably choosing the values of the constants A_n and B_n appearing in the expressions for the functions $u_n(x, t)$. With this in mind, we require that the relations

$$\begin{aligned} u(x, 0) &= f(x) = \sum_{n=0}^{\infty} \Phi_n(x)T_n(0), \\ \frac{\partial u(x, 0)}{\partial t} &= g(x) = \sum_{n=0}^{\infty} \Phi_n(x)T_n'(0) \end{aligned} \tag{1.13}$$

hold, which is equivalent to requiring that the functions $f(x)$ and $g(x)$ can be expanded in series with respect to the eigenfunctions (1.10). The possibility of making such expansions can be proved under rather broad conditions on the coefficients in equation (1.1) and on the functions which are to be expanded. Thus let

$$\begin{aligned} f(x) &= \sum_{n=0}^{\infty} C_n\Phi_n(x), \\ g(x) &= \sum_{n=0}^{\infty} c_n\Phi_n(x). \end{aligned} \tag{1.14}$$

Then we need only set

$$\begin{aligned} T_n(0) &= C_n, \\ T_n'(0) &= c_n, \end{aligned} \qquad (n = 0, 1, 2, \ldots), \tag{1.15}$$

in order to find A_n and B_n. Hence, if (1.11) holds, we have

$$u(x, 0) = f(x) = \sum_{n=0}^{\infty} A_n\Phi_n(x),$$

$$\frac{\partial u(x, 0)}{\partial t} = g(x) = \sum_{n=0}^{\infty} B_n\sqrt{\lambda_n}\,\Phi_n(x),$$

so that

$$A_n = C_n, \quad B_n = \frac{c_n}{\sqrt{\lambda_n}} \qquad (n = 0, 1, 2, \ldots). \tag{1.16}$$

Our results are based on the supposition that the series (1.12) converges and can be differentiated term by term twice with respect to x and t. Therefore, the coefficients A_n and B_n just found must be such as to guarantee that this is the case. However, in actual problems, the coefficients A_n and B_n often do not have this property. Whether or not the series (1.12) will converge in such a case will be discussed in Sec. 7. In the meantime, we make the following remarks:

We know that series often define discontinuous functions. Hence, in order to avoid confusion, a few words concerning boundary conditions and initial conditions are in order. By the conditions (1.3) and (1.4), we mean more precisely

$$\alpha \lim_{x\to a} u(x, t) + \beta \lim_{x\to a} \frac{\partial u(x, t)}{\partial x} = 0,$$

$$\gamma \lim_{x\to b} u(x, t) + \delta \lim_{x\to b} \frac{\partial u(x, t)}{\partial x} = 0$$

instead of (1.3), and

$$\lim_{t\to 0} u(x, t) = f(x), \quad \lim_{t\to 0} \frac{\partial u(x, t)}{\partial t} = g(x)$$

instead of (1.4). In other words, in (1.3) and (1.4), the values of $u(a, t)$, $\partial u(a, t)/\partial x$, etc., are to be interpreted as the limits to which $u(x, t)$, $\partial u(x, t)/\partial x$, etc., converge as the point (x, t) lying in the region $a < x < b$, $t > 0$ converges to the corresponding boundary point. It is quite clear that only such an interpretation of the boundary and initial conditions can correspond to the physical content of the problem. In the same way, when we say that the function $u(x, t)$ is *continuous* in the region $a \leqslant x \leqslant b$, $t \geqslant 0$, we mean that $u(x, t)$ is continuous in the region $a < x < b$, $t > 0$ and that the limit

$$\lim_{\substack{x\to x_0\\ t\to t_0}} u(x, t) \qquad (a < x < b,\ t > 0)$$

has a finite value for every point (x_0, t_0) on the boundary of the region.[2] Then, it is easy to show that the boundary conditions vary continuously as the point (x_0, t_0) is moved along the boundary.

Subsequently, by the solution of equation (1.1), or of some similar equation, we shall always mean a solution which is continuous in the sense just indicated. It is easy to see that if such a solution is to exist, then the boundary conditions and the initial conditions must "agree" at the points $(a, 0)$ and $(b, 0)$ in such a way that the boundary values do not undergo a discontinuity. Thus, for example, if the conditions (1.3) have the form $u(0, t) = 0$, $u(1, t) = 0$, and the conditions (1.4) have the form $u(x, 0) = x + 1$, $\partial u(x, 0)/\partial t = x^2$, then it is clear that the boundary values undergo a discontinuity at the points $(0, 0)$ and $(1, 0)$, so that the problem certainly cannot have a continuous solution.

---

[2] The analytic expression for $u(x, t)$ may not be a continuous function, i.e., may undergo a jump on the boundary.

2. The Usual Statement of the Boundary Value Problem

We shall assume that the function P in equation (1.1) *does not vanish.* Obviously, equation (1.7) neither loses nor gains eigenvalues and eigenfunctions if we multiply all its terms by a nonvanishing function. We now show that by carrying out such a multiplication, we can transform (1.7) into the form

$$(p\Phi')' + q\Phi = -\lambda r\Phi, \tag{2.1}$$

where p, q, and r are continuous functions of x on $[a, b]$, p is a positive function with a continuous derivative, and r is a positive function.

Proof. First we assume that $P > 0$. (This can be done without loss of generality, since otherwise we need only multiply all the terms of (1.7) by -1 and replace $-\lambda$ by $\tilde{\lambda}$.) Then we solve the system

$$p = rP, \quad p' = rR, \tag{2.2}$$

obtaining

$$\frac{p'}{p} = \frac{R}{P}, \quad \ln p = \int_{x_0}^{x} \frac{R}{P}\,dx, \quad p = \exp\left\{\int_{x_0}^{x} \frac{R}{P}\,dx\right\}, \quad r = \frac{p}{P},$$

where x_0 is some point of the interval $[a, b]$. (We take the constant of integration to be zero.) Obviously, $p > 0$, p' is continuous and $r > 0$. Now, we need only consider

$$rP\Phi'' + rR\Phi' + rQ\Phi = -\lambda r\Phi$$

and set

$$q = rQ. \tag{2.3}$$

Then, according to (2.2),

$$p\Phi'' + p'\Phi' + q\Phi = -\lambda r\Phi,$$

which is just the desired equation (2.1).

The boundary value problem is usually posed for an equation of the form (2.1), with coefficients satisfying the requirements just stated. The boundary conditions are still given by the same formulas (1.9).

3. The Existence of Eigenvalues

We shall not give a complete proof of the existence of eigenvalues for the boundary value problem under consideration; instead we just describe the basic idea of such a proof. Thus, in equation (2.1) we give λ a fixed value (real or complex) and find a solution satisfying the conditions

$$[\Phi]_{x=a} = \beta, \quad [\Phi']_{x=a} = -\alpha.$$

We denote this solution by $\Phi(x, \lambda)$. Obviously

$$\alpha\Phi(a, \lambda) + \beta\Phi'(a, \lambda) = 0, \tag{3.1}$$

i.e., $\Phi(x, \lambda)$ satisfies the first of the boundary conditions (1.9). (The prime denotes differentiation with respect to x.) As λ changes, $\Phi(x, \lambda)$ also changes, but nevertheless continues to satisfy the condition (3.1). Thus, a *known* function of x and the parameter λ satisfies the first of the conditions (1.9) for *any* λ. (In the theory of differential equations, it is shown that $\Phi(x, \lambda)$ can be represented in the form of a power series in λ, and hence is an *analytic* function of λ for all values of λ.)

We now form the function

$$\gamma\Phi(x, \lambda) + \delta\Phi'(x, \lambda) \tag{3.2}$$

and set

$$D(\lambda) = \gamma\Phi(b, \lambda) + \delta\Phi'(b, \lambda).$$

$D(\lambda)$ is a *known* function of *one* variable λ, and every value of λ for which

$$D(\lambda) = \gamma\Phi(b, \lambda) + \delta\Phi'(b, \lambda) = 0 \tag{3.3}$$

is obviously an eigenvalue of our problem, since for such values of λ, (3.1) and (3.3) are satisfied *simultaneously*, i.e., *both* of the conditions (1.9) are met. Thus, the problem of the existence of eigenvalues reduces to a study of the roots of $D(\lambda)$. By using this fact, it can be shown that the problem has an infinite set of *real* eigenvalues (see Sec. 4), which can be written as a sequence of the form

$$\lambda_0 < \lambda_1 < \cdots < \lambda_n < \cdots,$$

where

$$\lim_{n\to\infty} \lambda_n = +\infty.$$

4. Eigenfunctions and Their Orthogonality

Let λ be an eigenvalue of our boundary value problem, and let $\Phi(x)$ be an eigenfunction corresponding to λ. Then, it is easy to see that every function of the form $C\Phi(x)$, where C is an arbitrary nonzero constant, is also an eigenfunction corresponding to λ. We shall not consider such *linearly dependent* eigenfunctions to be different, and in fact any of them can be regarded as a "representative" of the family of functions of the form $C\Phi(x)$, where $C \neq 0$.

We now ask whether two linearly independent eigenfunctions $\Phi(x)$ and $\Psi(x)$ can belong to the same eigenvalue.[3] With our hypotheses, the answer

[3] If $a\Phi(x) + b\Psi(x) \equiv 0$ implies $a = b = 0$, then $\Phi(x)$ and $\Psi(x)$ are said to be *linearly independent*. Cf. Ch. 2, Prob. 9. (*Translator*)

is negative. In fact, suppose that $\Phi(x)$ and $\Psi(x)$ belong to the same eigenvalue. Then, by a familiar property of linearly independent solutions of a differential equation, we would have

$$\begin{vmatrix} \Phi(x) & \Phi'(x) \\ \Psi(x) & \Psi'(x) \end{vmatrix} \neq 0 \tag{4.1}$$

everywhere on $[a, b]$, and in particular at $x = a$.[4] But according to the first of the conditions (1.9), we have

$$\alpha\Phi(a) + \beta\Phi'(a) = 0,$$
$$\alpha\Psi(a) + \beta\Psi'(a) = 0,$$

which by (4.1) would imply that $\alpha = 0$, $\beta = 0$, thereby contradicting the hypothesis (1.5). Thus, *to within a constant factor, only one eigenfunction corresponds to each eigenvalue.*

LEMMA 1. *Let*

$$L(\Phi) = \frac{d}{dx}\left(p\frac{d\Phi}{dx}\right) + q\Phi, \tag{4.2}$$

where Φ is a function depending on x. (If Φ also depends on other variables, e.g., on t, we write the partial derivative with respect to x.) Then, the identity

$$\Phi L(\Psi) - \Psi L(\Phi) = \frac{d}{dx}[p(\Phi\Psi' - \Phi'\Psi)] \tag{4.3}$$

holds for any twice differentiable functions Φ and Ψ.

The proof is an immediate consequence of replacing $L(\Phi)$ and $L(\Psi)$ by their expressions as given by (4.2).

LEMMA 2. *If Φ and Ψ satisfy the boundary conditions* (1.9), *then*

$$[\Phi\Psi' - \Phi'\Psi]_{x=a} = [\Phi\Psi' - \Phi'\Psi]_{x=b} = 0. \tag{4.4}$$

Proof. The numbers α and β, which do not vanish simultaneously according to (1.5), satisfy the homogeneous system

$$\alpha\Phi(a) + \beta\Phi'(a) = 0,$$
$$\alpha\Psi(a) + \beta\Psi'(a) = 0.$$

This is possible only if the determinant of the system vanishes, i.e.,

$$\begin{vmatrix} \Phi(a) & \Phi'(a) \\ \Psi(a) & \Psi'(a) \end{vmatrix} = [\Phi\Psi' - \Phi'\Psi]_{x=0} = 0.$$

[4] The symbol $\begin{vmatrix} a & b \\ c & d \end{vmatrix}$ denotes the determinant $ad - bc$. The functional determinant (4.1) is usually called a *Wronskian*. (*Translator*)

The second part of equation (4.4) is proved in the same way.

We now show that any two eigenfunctions $\Phi(x)$ and $\Psi(x)$ corresponding to different eigenvalues λ and μ, respectively, are orthogonal on $[a, b]$, with weight r.

Proof. Let Φ and Ψ satisfy the equations

$$L(\Phi) = -\lambda r\Phi,$$
$$L(\Psi) = -\mu r\Psi,$$

with the same boundary conditions (1.9). We multiply the first equation by Ψ and the second by Φ and then subtract the first equation from the second. By Lemma 1, we obtain

$$[p(\Phi\Psi' - \Phi'\Psi)]' = (\lambda - \mu)r\Phi\Psi.$$

Therefore we have

$$[p(\Phi\Psi' - \Phi'\Psi)]_{x=a}^{x=b} = (\lambda - \mu)\int_a^b r\Phi\Psi\,dx. \tag{4.5}$$

By Lemma 2

$$[p(\Phi\Psi' - \Phi'\Psi]_{x=a}^{x=b} = 0,$$

so that (4.5) implies that

$$(\lambda - \mu)\int_a^b r\Phi\Psi\,dx = 0.$$

Since $\lambda \neq \mu$, it follows that

$$\int_a^b r\Phi\Psi\,dx = 0,$$

as was to be shown.

Remark. In Sec. 3, we said that the eigenvalues are real, but we did not give a proof. The reality of the eigenvalues can be deduced from the orthogonality of the eigenfunctions just proved. In fact, if $\lambda = \mu + i\nu$ $(\nu \neq 0)$ is an eigenvalue and if $\Phi(x) = \varphi(x) + i\psi(x)$ is the eigenfunction corresponding to λ, then substituting in (2.1), we obtain

$$[p(\varphi' + i\psi')]' + q(\varphi + i\psi) = -(\mu + i\nu)r(\varphi + i\psi).$$

But then we also have the equation

$$[p(\varphi' - i\psi')]' + q(\varphi - i\psi) = -(\mu - i\nu)r(\varphi - i\psi),$$

which means that $\bar{\lambda} = \mu - i\nu$ is also an eigenvalue and that the function

$\overline{\Phi}(x) = \varphi(x) - i\psi(x)$ is the eigenfunction corresponding to $\bar{\lambda}$. But this implies that

$$\int_a^b r\Phi\overline{\Phi}\,dx = \int_a^b r(\varphi^2 + \psi^2)\,dx \neq 0,$$

which is impossible, for as just proved, Φ and $\overline{\Phi}$ must be orthogonal since $\lambda \neq \bar{\lambda}$.

5. Sign of the Eigenvalues

The following theorem gives more precise information about the eigenvalues for the case where $q \leqslant 0$ for $a \leqslant x \leqslant b$, a situation which is encountered very often in the applications.

THEOREM. *If* $r > 0$, $q \leqslant 0$ *and if the boundary conditions imply that*

$$[p\Phi\Phi']_{x=a}^{x=b} \leqslant 0, \tag{5.1}$$

then all the eigenvalues of the boundary value problem for equation (2.1) *are nonnegative.*

Proof. Let λ be an eigenvalue and let $\Phi(x)$ be the corresponding eigenfunction. Multiplying (2.1) by $\Phi(x)$ and integrating, we obtain

$$\int_a^b (p\Phi')'\Phi\,dx + \int_a^b q\Phi^2\,dx = -\lambda\int_a^b r\Phi^2\,dx,$$

whence, integrating by parts:

$$[p\Phi\Phi']_{x=a}^{x=b} - \int_a^b p\Phi'^2\,dx + \int_a^b q\Phi^2\,dx = -\lambda\int_a^b r\Phi^2\,dx. \tag{5.2}$$

It follows from (5.1) and the condition $q \leqslant 0$ that the left-hand side of (5.2) is $\leqslant 0$. Therefore $\lambda \geqslant 0$, and moreover, $\lambda = 0$ is possible only if $q \equiv 0$, $\Phi' \equiv 0$, i.e., only if the equation (2.1) has the form

$$(p\Phi')' = -\lambda r\Phi$$

and the function $\Phi = \text{const}$ is an eigenfunction.

Remark. The condition (5.1) seems to be quite artificial, but actually it is not. In fact, it is satisfied for the very boundary conditions most often encountered in the applications, namely

1) $\Phi(a) = \Phi(b) = 0;$
2) $\Phi'(a) = \Phi'(b) = 0;$
3) $\Phi'(a) - h\Phi(a) = 0,\ \Phi'(b) + H\Phi(b) = 0,$

where h and H are nonnegative constants. This is obvious for the first and second cases. In the third case,

$$[p\Phi\Phi']_{x=a}^{x=b} = -Hp(b)\Phi^2(b) - hp(a)\Phi^2(a) \leqslant 0,$$

since $\Phi'(a) = h\Phi(a)$, $\Phi'(b) = -H\Phi(b)$.

6. Fourier Series with Respect to the Eigenfunctions

Let $\lambda_0, \lambda_1, \ldots, \lambda_n, \ldots$ be all the eigenvalues of our boundary value problem, arranged in increasing order, and let

$$\Phi_0(x), \; \Phi_1(x), \ldots, \; \Phi_n(x), \ldots \tag{6.1}$$

be the corresponding eigenfunctions, which for simplicity we regard as having the normalization

$$\int_a^b r\Phi_n^2(x)\, dx = 1 \qquad (n = 0, 1, 2, \ldots). \tag{6.2}$$

Then, for every function $f(x)$ which is absolutely integrable on $[a, b]$, we can form the Fourier series

$$f(x) \sim c_0\Phi_0(x) + c_1\Phi_1(x) + \cdots,$$

where

$$c_n = \int_a^b rf(x)\Phi_n(x)\, dx \qquad (n = 0, 1, 2, \ldots), \tag{6.3}$$

and the following propositions, which we cite without proof, are valid:

THEOREM 1. *If $f(x)$ is continuous on $[a, b]$, if $f(x)$ has a piecewise smooth (but possibly discontinuous) derivative, and if $f(x)$ satisfies the boundary conditions of our boundary value problem, i.e.,*

$$\alpha f(a) + \beta f'(a) = 0, \quad \gamma f(b) + \delta f'(b) = 0, \tag{6.4}$$

then the Fourier series of $f(x)$ with respect to the eigenfunctions converges to $f(x)$ absolutely and uniformly.

The conditions (6.4) may appear artificial, but if we recall how our problem arose [see the first formula in (1.4) and (1.14)] and regard $f(x)$ as the initial value of the function $u(x, t)$, writing $f(x) = u(x, 0)$, then for $t = 0$, the conditions (1.3) become just the conditions (6.4).

THEOREM 2. *If the function $f(x)$ is piecewise smooth on $[a, b]$ (but either continuous or discontinuous), then the Fourier series of $f(x)$ with* *respect to the eigenfunctions converges for $a < x < b$ to the value $f(x)$ at every point of continuity and to the value*

$$\tfrac{1}{2}[f(x + 0) + f(x - 0)]$$

at every point of discontinuity.

Instead of the system (6.1), which is orthogonal with weight r, we can consider the system

$$\sqrt{r}\,\Phi_0(x),\ \sqrt{r}\,\Phi_1(x), \ldots, \tag{6.5}$$

which is orthogonal in the ordinary sense, and for which

$$\|\sqrt{r}\,\Phi_n(x)\| = \sqrt{\int_a^b r\Phi_n^2(x)\,dx} = 1 \qquad (n = 0, 1, 2, \ldots).$$

Let $\sqrt{r}\,f(x)$ be a square integrable function. Then, with respect to the new system, its Fourier coefficients have the form

$$c_n = \int_a^b rf(x)\Phi_n(x)\,dx,$$

i.e., they agree with the Fourier coefficients of $f(x)$ with respect to the system (6.1).

Applied to the function $\sqrt{r}\,f(x)$, the completeness condition for the system (6.5) becomes [see equation (7.1) of Ch. 2]

$$\int_a^b rf^2(x)\,dx = \sum_{n=0}^{\infty} c_n^2\|\sqrt{r}\,\Phi_n(x)\|^2 = \sum_{n=0}^{\infty} c_n^2. \tag{6.6}$$

If this equation holds for any square integrable function $f(x)$, then instead of reducing the problem to the system (6.5), we simply say that the system (6.1) is *complete with weight r*. We now prove that this is actually the case.

It is clear from what has just been said that it is sufficient to prove that the ordinary orthogonal system (6.5) is complete. Now, any continuous function $\Phi(x)$ can be approximated in the mean to any degree of accuracy by a function $g(x)$ (with two continuous derivatives) satisfying the boundary conditions of our boundary value problem. [For example, we can choose for $g(x)$ a function such that $g(a) = g'(a) = g(b) = g'(b) = 0$.] We shall not give a detailed proof of this fact, which is quite clear from geometric considerations.

Thus, let

$$\int_a^b [\Phi(x) - g(x)]^2\,dx \leqslant \frac{\varepsilon}{4}, \tag{6.7}$$

where $\varepsilon > 0$ is arbitrarily small. According to Theorem 1, the Fourier series of $g(x)$ with respect to the system (6.1) converges uniformly to $g(x)$.

This means that there exists a linear combination

$$\sigma_n(x) = \gamma_0\Phi_0(x) + \gamma_1\Phi_1(x) + \cdots + \gamma_n\Phi_n(x) \tag{6.8}$$

such that

$$|g(x) - \sigma_n(x)| \leqslant \sqrt{\frac{\varepsilon}{4(b-a)}} \qquad (a \leqslant x \leqslant b).$$

This implies that

$$\int_a^b [g(x) - \sigma_n(x)]^2\,dx \leqslant \frac{\varepsilon}{4}. \tag{6.9}$$

By the elementary inequality

$$(A + B)^2 \leqslant 2(A^2 + B^2),$$

it follows from (6.7) and (6.9) that

$$\int_a^b [\Phi(x) - \sigma_n(x)]^2\,dx = \int_a^b [(\Phi(x) - g(x)) + (g(x) - \sigma_n(x))]^2\,dx$$
$$\leqslant 2\int_a^b [\Phi(x) - g(x)]^2\,dx + 2\int_a^b [g(x) - \sigma_n(x)]^2\,dx \leqslant \varepsilon.$$

This proves that any continuous function can be approximated in the mean to any degree of accuracy by an expression of the form (6.8).

Now let $F(x)$ be any continuous function. Then the function $F(x)/\sqrt{r(x)}$ is also continuous. If we write

$$h = \max_{a \leqslant x \leqslant b} r(x),$$

then by what has just been proved, there exists a linear combination $\sigma_n(x)$ of the functions (6.1) for which

$$\int_a^b \left[\frac{F(x)}{\sqrt{r}} - \sigma_n(x)\right]^2 dx \leqslant \frac{\varepsilon}{h}.$$

Therefore, we have

$$\int_a^b [F(x) - \sqrt{r}\,\sigma_n(x)]^2\,dx = \int_a^b r\left[\frac{F(x)}{\sqrt{r}} - \sigma_n(x)\right]^2 dx \leqslant \varepsilon.$$

But then the function $\sqrt{r}\,\sigma_n(x)$ is a linear combination of functions of the system (6.5), and hence this system satisfies the completeness criterion for ordinary orthogonal systems (see Ch. 2, Sec. 9), i.e., the system (6.5) is complete, as was to be shown. Thus, we have proved

THEOREM 3. *The system* (6.1) *is complete with weight* r, *i.e., the relation* (6.6) *holds for every square integrable function* $f(x)$.

The following result is an easy consequence of Theorem 3:

THEOREM 4. *Let $f(x)$ be any square integrable function. Then*

$$\lim_{n\to\infty} \int_a^b r \left[f(x) - \sum_{k=0}^{\infty} c_k \Phi_k(x) \right]^2 dx = 0,$$

where the c_n are the Fourier coefficients of the function $f(x)$ with respect to the system (6.1).

In other words, the Fourier series of $f(x)$ always converges in the mean to $f(x)$, with weight r. To prove this, we need only apply equation (7.3) of Ch. 2 to the function $\sqrt{r}\, f(x)$ and to the system (6.5).

THEOREM 5. *Any continuous function $f(x)$ which is orthogonal with weight r to all the function of the system* (6.1) *must be identically zero.*

In fact, if $f(x)$ is orthogonal to all the functions of the system (6.1), then all its Fourier coefficients are zero. But then by (6.6)

$$\int_a^b r f^2(x)\, dx = 0,$$

whence $f(x) \equiv 0$.

The theorem on the expansion of a function in series with respect to the eigenfunctions of a boundary value problem (under quite broad hypotheses), as well as the related completeness theorem and its consequences, were first proved by the prominent mathematician V. A. Steklov.[5]

7. Does the Eigenfunction Method Always Lead to a Solution of the Problem?

The eigenfunction method will certainly lead to a solution of the problem posed in Sec. 1 if first of all, the functions $f(x)$ and $g(x)$ defined by (1.4) have series expansions (1.14), with respect to the eigenfunctions $\Phi_n(x)$, which converge to $f(x)$ and $g(x)$, and if secondly, the coefficients A_n and B_n defined by (1.16) are such as to guarantee the convergence of the series (1.12) and justify differentiating it twice term by term. However, *whether or not these conditions are met, every time the problem has a solution, the solution can be found in the form of a series* (1.12) *by the method indicated in Sec.* 1. This implies that the solution is *unique*, a fact which can often be deduced by an argument based on the physical content of the problem. This physical content also allows us to decide whether the problem has a solution in the

[5] The theory given in this chapter is usually associated with the names of the mathematicians J. C. F. Sturm and J. Liouville. (*Translator*)

first place. In view of what has just been said, this explains why the physicist or engineer, by using the eigenfunction method and manipulating the series as if term by term differentiation and other operations were justified, even when in fact this is not the case, nevertheless arrives at correct results.

A more precise formulation of these remarks goes as follows:

THEOREM. *Let the function* $u(x, t)$ *be a continuous solution of the equation* (1.1) *in the region* $a \leqslant x \leqslant b$, $t \geqslant 0$, *satisfying the boundary conditions* (1.3) *and the initial conditions* (1.4). *Then*

$$u(x, t) = \sum_{n=0}^{\infty} T_n(t)\Phi_n(x), \tag{7.1}$$

where the $\Phi_n(x)$ *are the eigenfunctions associated with the boundary value problem.*[6] *The functions* $T_n(t)$ *can be found from the equation*

$$T_n'' + \lambda_n T_n = 0 \qquad (n = 0, 1, 2, \ldots), \tag{7.2}$$

subject to the initial conditions

$$T_n(0) = C_n, \; T_n'(0) = c_n \qquad (n = 0, 1, 2, \ldots),$$

where the quantities C_n *and* c_n *are the Fourier coefficients of* $f(x)$ *and* $g(x)$ [*cf.* *the initial conditions* (1.4)] *with respect to the system of eigenfunctions* $\Phi_n(x)$. (*It is assumed that the derivatives* $\partial u/\partial t$ *and* $\partial^2 u/\partial t^2$ *are continuous and bounded in every region of the form* $a < x < b$, $0 < t < t_0$.)

Proof. We multiply equation (1.1) by the function

$$r = \frac{1}{P}\exp\left\{\int_{x_0}^{x} \frac{R}{P}\,dx\right\} = \frac{p}{P}$$

(see Sec. 2). Then according to (2.2) and (2.3), we obtain

$$p\frac{\partial^2 u}{\partial x^2} + p'\frac{\partial u}{\partial x} + qu = r\frac{\partial^2 u}{\partial t^2}$$

or

$$\frac{\partial}{\partial x}\left(p\frac{\partial u}{\partial x}\right) + qu = r\frac{\partial^2 u}{\partial t^2}.$$

By (4.2), this can be written as

$$L(u) = r\frac{\partial^2 u}{\partial t^2}, \tag{7.3}$$

and instead of (2.1) we can write

$$L(\Phi) = -\lambda r\Phi.$$

---

[6] For simplicity, we assume that the eigenfunctions are normalized as in (6.2).

Therefore, the relation

$$L(\Phi_n) = -\lambda_n r\Phi_n \qquad (n = 0, 1, 2, \ldots) \tag{7.4}$$

holds for the eigenfunctions of the boundary value problem.

By the hypothesis of the theorem and by Theorem 2 of Sec. 6, for $a < x < b$ and every $t > 0$, the function $u(x, t)$ can be expanded in a series of the form (7.1), where

$$T_n(t) = \int_a^b ru(x, t)\Phi_n(x)\,dx \qquad (n = 0, 1, 2, \ldots). \tag{7.5}$$

It follows from (7.4) that

$$r\Phi_n = -\frac{1}{\lambda_n}L(\Phi_n),$$

so that

$$T_n(t) = -\frac{1}{\lambda_n}\int_a^b u(x, t)L(\Phi_n)\,dx,$$

or by (4.3)

$$T_n(t) = -\frac{1}{\lambda_n}\int_a^b \Phi_n(x)L(u)\,dx + \frac{1}{\lambda_n}\left[p\left(\Phi_n\frac{\partial u}{\partial x} - \Phi_n' u\right)\right]_{x=a}^{x=b}$$

The last term vanishes because of Lemma 2 of Sec. 4.

Thus we have

$$T_n(t) = -\frac{1}{\lambda_n}\int_a^b \Phi_n(x)L(u)\,dx, \tag{7.6}$$

whence, using (7.3), we obtain

$$T_n(t) = -\frac{1}{\lambda_n}\int_a^b r\frac{\partial^2 u}{\partial t^2}\Phi_n(x)\,dx. \tag{7.7}$$

On the other hand, differentiating (7.5) twice with respect to t gives

$$T_n''(t) = \int_a^b r\frac{\partial^2 u}{\partial t^2}\Phi_n(x)\,dx, \tag{7.8}$$

where the differentiation behind the integral sign is legitimate because of our hypotheses concerning $\partial u/\partial t$ and $\partial^2 u/\partial t^2$. Comparing (7.7) and (7.8), we obtain (7.2). Furthermore, since $u(x, t)$ is continuous in the region $a \leqslant x \leqslant b$, $t \geqslant 0$ and since $\lim_{t\to 0} u(x, t) = f(x)$, it follows from (7.5) that

$$\begin{aligned}\lim_{t\to 0} T_n(t) &= \lim_{t\to 0}\int_a^b ru(x, t)\Phi_n(x)\,dx \\ &= \int_a^b rf(x)\Phi_n(x)\,dx = C_n \qquad (n = 0, 1, 2, \ldots),\end{aligned} \tag{7.9}$$

where the C_n are the Fourier coefficients of the function $f(x)$. Since the function $T_n(t)$ is continuous, this is equivalent to the relations

$$T_n(0) = C_n \qquad (n = 0, 1, 2, \ldots).$$

Similarly, we show that

$$T_n'(0) = c_n \qquad (n = 0, 1, 2, \ldots),$$

where the c_n are the Fourier coefficients of the function $g(x)$. This completes the proof of the theorem.

Thus, if the problem in which we are interested has a solution at all, the solution can be found in the form of a series (1.12) by the method of Sec. 1. On the other hand, quite often this method leads to a function $u(x, t)$ which does not have a derivative everywhere. Such a $u(x, t)$ cannot be regarded as a *solution* of the problem in the *exact* sense of the word, since $u(x, t)$ certainly must satisfy the differential equation! However, because of the theorem just proved, in this case it is useless to look for an *exact* solution, since if such existed, it would have to coincide with $u(x, t)$. This compels us to be satisfied with the function $u(x, t)$ that we have already found; we shall call this function a *generalized* solution of the problem. It can be shown that with our hypotheses, the series (1.12) obtained by the method of Sec. 1 *always* defines a function to which it converges either in the ordinary sense or in the mean, and therefore the eigenfunction method always gives a generalized solution, if it fails to give an exact solution.

8. The Generalized Solution

What is the practical value of the generalized solution described above? Does it represent anything of use to the physicist or engineer, or is it of purely mathematical interest? The generalized solution is in fact valuable, as will emerge from the following theorem:

THEOREM. *Let*

$$u(x, t) \sim \sum_{n=0}^{\infty} T_n(t)\Phi_n(x)$$

be either the exact or the generalized solution of equation (1.1), *subject to the conditions* (1.3) *and* (1.4). *If*

$$\lim_{m\to\infty} \int_a^b r[f(x) - f_m(x)]^2\,dx = \lim_{m\to\infty} \int_a^b r[g(x) - g_m(x)]^2\,dx = 0, \qquad (8.1)$$

i.e., if $f_m(x)$ and $g_m(x)$ converge in the mean (with weight r) to $f(x)$ and $g(x)$, respectively, as $m \to \infty$,[7] *and if*

$$u_m(x, t) = \sum_{n=0}^{\infty} T_{mn}(t)\Phi_n(x)$$

is either the exact or the generalized solution of the equation (1.1), *subject to the boundary conditions* (1.3) *and the initial conditions*

$$u_m(x, 0) = f_m(x), \quad \frac{\partial u_m(x, 0)}{\partial t} = g_m(x),$$

then $u_m(x, t)$ converges to $u(x, t)$ in the mean as $m \to \infty$.

Proof. We recall that

$$\begin{aligned} T_n'' + \lambda_n T_n = 0, \ T_n(0) = C_n, \ T_n'(0) = c_n \\ (n = 0, 1, 2, \ldots), \\ T_{mn}'' + \lambda_n T_{mn} = 0, \ T_{mn}(0) = C_{mn}, \ T_{mn}'(0) = c_{mn} \\ (n = 0, 1, 2, \ldots), \end{aligned} \tag{8.2}$$

where C_n, c_n, C_{mn}, c_{mn} are the Fourier coefficients of the functions $f(x)$, $g(x)$, $f_m(x)$, $g_m(x)$, respectively. Since

$$\lambda_0 < \lambda_1 < \lambda_2 < \cdots < \lambda_n < \cdots$$

and

$$\lim_{n \to \infty} \lambda_n = +\infty$$

(see Sec. 3), only a finite number of the λ_n can be negative. Let $\lambda_n \leqslant 0$ for $n \leqslant N$ and $\lambda_n > 0$ for $n > N$. Then by (8.2), we have[8]

$$\begin{aligned} T_n = \frac{1}{2}\left(C_n + \frac{c_n}{\sqrt{-\lambda_n}}\right) e^{\sqrt{-\lambda_n}\,t} \\ + \frac{1}{2}\left(C_n - \frac{c_n}{\sqrt{-\lambda_n}}\right) e^{-\sqrt{-\lambda_n}\,t} \quad (n \leqslant N), \\ T_n = C_n \cos\sqrt{\lambda_n}\,t + \frac{c_n}{\sqrt{\lambda_n}} \sin\sqrt{\lambda_n}\,t \quad (n > N), \end{aligned} \tag{8.3}$$

[7] In particular, this is the case if $f_m(x) \to f(x)$ and $g_m(x) \to g(x)$ uniformly.

[8] It may happen that $\lambda_N = 0$, in which case we have

$$T_N = C_N + c_N t.$$

and similarly

$$T_{mn} = \frac{1}{2}\left(C_{mn} + \frac{c_{mn}}{\sqrt{-\lambda_n}}\right) e^{\sqrt{-\lambda_n}\, t} + \frac{1}{2}\left(C_{mn} - \frac{c_{mn}}{\sqrt{-\lambda_n}}\right) e^{-\sqrt{-\lambda_n}\, t} \qquad (n \leqslant N), \tag{8.4}$$

$$T_{mn} = C_{mn} \cos \sqrt{\lambda_n}\, t + \frac{c_{mn}}{\sqrt{\lambda_n}} \sin \sqrt{\lambda_n}\, t \qquad (n > N).$$

Now consider

$$\int_a^b r[f(x) - f_m(x)]^2\, dx = \sum_{n=0}^{\infty} (C_n - C_{mn})^2,$$

$$\int_a^b r[g(x) - g_m(x)]^2\, dx = \sum_{n=0}^{\infty} (c_n - c_{mn})^2$$

(see Theorem 3 of Sec. 6). In view of (8.1), these formulas imply that

$$\lim_{m\to\infty} \sum_{n=0}^{\infty} (C_n - C_{mn})^2 = 0, \qquad \lim_{m\to\infty} \sum_{n=0}^{\infty} (c_n - c_{mn})^2 = 0, \tag{8.5}$$

whence

$$\lim_{m\to\infty} C_{mn} = C_n, \qquad \lim_{m\to\infty} c_{mn} = c_n. \tag{8.6}$$

Then, by (8.3), (8.4) and (8.6)

$$\lim_{m\to\infty} [T_n - T_{mn}] = 0, \tag{8.7}$$

where

$$[T_n - T_{mn}]^2 = \left[(C_n - C_{mn}) \cos \sqrt{\lambda_n}\, t + \frac{c_n - c_{mn}}{\sqrt{\lambda_n}} \sin \sqrt{\lambda_n}\, t\right]^2 \leqslant 2\left[(C_n - C_{mn})^2 + \left(\frac{c_n - c_{mn}}{\sqrt{\lambda_n}}\right)^2\right] \tag{8.8}$$

for $n > N$. All that remains is to consider the relation

$$\int_a^b r[u(x, t) - u_m(x, t)]^2\, dx = \sum_{n=0}^{\infty} (T_n - T_{mn})^2 = \sum_{n=0}^{N} (T_n - T_{mn})^2 + \sum_{n=N+1}^{\infty} (T_n - T_{mn})^2$$

(see Theorem 3 of Sec. 6). By (8.5), (8.7), and (8.8)

$$\int_a^b r[u(x, t) - u_m(x, t)]^2 \, dx \to 0,$$

which means that the function $u_m(x, t)$ converges in the mean to $u(x, t)$ as $m \to \infty$. This proves the theorem.

What we have just proved can be summarized as follows: *If $f_m(x)$ is close to $f(x)$ and $g_m(x)$ is close to $g(x)$ (in the sense of uniform closeness or even closeness in the mean), then the function $u_m(x, t)$ is close to $u(x, t)$ in the mean.*

We now observe that in actual problems of physics and engineering, the functions $f(x)$ and $g(x)$ are in general not exact, but rather represent approximations to certain exact functions. Nevertheless, by the theorem just proved, even if the solution of equation (1.1), subject to the conditions (1.3) and (1.4), is not an exact solution but only a generalized solution, it will differ only slightly (in the sense of uniform closeness or closeness in the mean) from the true solution of the problem. This constitutes the practical value of generalized solutions.

Finally, we note one further consequence of the theorem proved above:

If the functions $f_m(x)$ and $g_m(x)$ are chosen in such a way that the functions $u_m(x, t)$ are exact solutions of the corresponding problems (such $f_m(x)$ and $g_m(x)$ can always be chosen![9]), then the exact or generalized solution of equation (1.1), *subject to the conditions* (1.3) *and* (1.4), *is the limit of the exact solutions $u_m(x, t)$ as $f_m(x) \to f(x)$ and $g_m(x) \to g(x)$ either uniformly or in the mean.*

It follows immediately that the generalized solution is unique.

9. The Inhomogeneous Problem

Instead of equation (1.1), consider the more general equation

$$P\frac{\partial^2 u}{\partial x^2} + R\frac{\partial u}{\partial x} + Qu = \frac{\partial^2 u}{\partial t^2} + F(x, t) \tag{9.1}$$

subject to the same boundary and initial conditions (1.3) and (1.4). In vibration problems, equation (9.1) corresponds to the case of *forced vibrations*, while equation (1.1) corresponds to the case of *free vibrations*. When multiplied by the function

$$r = \frac{1}{P}\exp\left\{\int_{x_0}^x \frac{R}{P}\,dx\right\} = \frac{p}{P}$$

[9] For example, we can choose the functions $f_m(x)$ and $g_m(x)$ to be the mth partial sums of the Fourier series of $f(x)$ and $g(x)$.

(see Sec. 2), (9.1) takes the form

$$L(u) = r\frac{\partial^2 u}{\partial t^2} + rF(x, t). \tag{9.2}$$

Suppose now that the boundary value problem corresponding to equation (9.1) has a solution and that $F(x, t)$ has a series expansion in terms of the eigenfunctions of the boundary value problem for the equation

$$L(\Phi) = -\lambda r\Phi$$

(see Sec. 2). Then, for $t > 0$, we write $u(x, t)$ as the series

$$u(x, t) = \sum_{n=0}^{\infty} T_n(t)\Phi_n(x), \tag{9.3}$$

where

$$T_n(t) = \int_a^b ru(x, t)\Phi_n(x)\,dx \qquad (n = 0, 1, 2, \ldots), \tag{9.4}$$

which is possible because of Theorem 2 of Sec. 6. Repeating the argument given in the proof of the theorem of Sec. 7, we obtain

$$T_n(t) = -\frac{1}{\lambda_n}\int_a^b \Phi_n(x)L(u)\,dx$$

[see (7.6)] or

$$T_n(t) = -\frac{1}{\lambda_n}\int_a^b r\frac{\partial^2 u}{\partial t^2}\Phi_n(x)\,dx - \frac{1}{\lambda_n}\int_a^b rF(x, t)\Phi_n(x)\,dx, \tag{9.5}$$

where we have used (9.2). Assuming that $\partial u/\partial t$ and $\partial^2 u/\partial t^2$ are bounded, we find after differentiating (9.4) twice with respect to t that

$$T_n'' = \int_a^b r\frac{\partial^2 u}{\partial t^2}\Phi_n(x)\,dx.$$

Finally, if we set

$$\begin{aligned} F(x, t) &= \sum_{n=0}^{\infty} F_n(t)\Phi_n(x)\,dx, \\ F_n(t) &= \int_a^b rF(x, t)\Phi_n(x)\,dx \qquad (n = 0, 1, 2, \ldots), \end{aligned} \tag{9.6}$$

then (9.5) gives

$$T_n = -\frac{1}{\lambda_n}T_n'' - \frac{1}{\lambda_n}F_n$$

or

$$T_n'' + \lambda_n T_n + F_n = 0 \qquad (n = 0, 1, 2, \ldots). \tag{9.7}$$

Then, proceeding as in the proof of the theorem of Sec. 7 [see (7.9) *et seq.*], we find

$$T_n(0) = C_n, \quad T_n'(0) = c_n \qquad (n = 0, 1, 2, \ldots), \tag{9.8}$$

where C_n and c_n are the Fourier coefficients of the functions $f(x)$ and $g(x)$, respectively. Thus, if a solution of the problem exists, it is given by the series (9.3), where T_n is determined from the equation (9.7), subject to the conditions (9.8).

It is remarkable that just as in the case of equation (1.1), we can arrive at the series (9.3) by carrying out formal operations with series, without regard to whether or not these operations are legitimate. In fact, if in equation (9.2) we substitute the series (9.3) and the series (9.6) for $F(x, t)$, differentiate (9.3) term by term, and then equate the coefficients of $\Phi_n(x)$ to zero, we obtain (9.7). Moreover, if we set $t = 0$ in (9.3) and require that

$$u(x, 0) = \sum_{n=0}^{\infty} T_n(0)\Phi_n(x) = f(x),$$

we obtain the first of the equations (9.8). If we differentiate (9.3) term by term and again set $t = 0$, we find that

$$\frac{\partial u(x, 0)}{\partial t} = \sum_{n=0}^{\infty} T_n'(0)\Phi_n(x) = g(x),$$

which is the second of the equations (9.8).

Just as in the case of equation (1.1) in Sec. 7, we can introduce the concept of a generalized solution for equation (9.1). Then it turns out that for any continuous $F(x, t)$, the series (9.3) defines a function $u(x, t)$ to which it converges either in the ordinary sense or in the mean, so that the problem always has either an exact or a generalized solution. Moreover, this solution is unique, since if U and V are two solutions of the boundary value problem for equation (9.1), the function $u = U - V$ is also a solution of the boundary value problem for equation (1.1) with zero initial conditions. But as observed in Secs. 7 and 8, the boundary value problem for equation (1.1) has a unique solution (exact or generalized). Since the function which is identically zero is a solution of the boundary value problem for equation (1.1) with zero initial conditions, it follows that $u \equiv 0$, i.e., that $U \equiv V$. As concerns the practical value of generalized solutions, we can repeat the considerations of Sec. 8.

10. Supplementary Remarks

We considered equation (1.1) only for the sake of being explicit. All our considerations are also applicable to equation (1.2), with the same boundary conditions and with the initial condition

$$u(x, 0) = f(x). \tag{10.1}$$

In this case, the method of Sec. 1 gives

$$u(x, t) = \sum_{n=0}^{\infty} T_n(t)\Phi_n(x), \tag{10.2}$$

where the quantities T_n are found from the *first order* differential equation

$$T_n' + \lambda_n T_n = 0, \quad T_n(0) = C_n \qquad (n = 0, 1, 2, \ldots), \tag{10.3}$$

where C_n $(n = 0, 1, 2, \ldots)$ are the Fourier coefficients of $f(x)$.

The theorem of Sec. 7 needs only a corresponding change in its statement, i.e., one has to consider (10.3) instead of (7.2) and require the continuity and boundedness of only $\partial u/\partial t$. We suggest that the reader prove the new version of the theorem as an exercise.

The concept of a generalized solution can also be introduced in the case of equation (1.2). But this case is essentially different from the case of equation (1.1), in that the generalized solution turns out to be the exact solution as well. This results from the fact that all but a finite number of the λ_n are positive, and for such λ_n, (10.3) gives

$$T_n = C_n e^{-\lambda_n t}.$$

Consequently, it turns out that the series (10.2) is convergent and can be differentiated term by term *any number of times*, i.e., (10.2) gives the exact solution.

The considerations of Sec. 9 are also applicable to the inhomogeneous equation

$$P\frac{\partial^2 u}{\partial x^2} + R\frac{\partial u}{\partial x} + Qu = \frac{\partial u}{\partial t} + F(x, t),$$

with the same boundary and initial conditions (1.3) and (10.1). The solution is given by the series

$$u(x, t) = \sum_{n=0}^{\infty} T_n(t)\Phi_n(x), \tag{10.2}$$

where T_n is now defined by the equations

$$T_n' + \lambda_n T + F_n = 0, \quad T_n(0) = C_n \qquad (n = 0, 1, 2, \ldots).$$

In view of the results of Secs. 7–10, we shall not worry about the convergence of series in the problems to be solved in Part II, since we now know that the series obtained as a result of solving such a problem *always* defines a solution (exact or generalized). It is true that strictly speaking, some of the problems in Part II do not fall into the category already studied, but it can be shown that the previous considerations apply to them as well.

Part II. APPLICATIONS

11. Equation of a Vibrating String

Consider a stretched homogeneous string, fastened at both ends, and suppose that the equilibrium position of the string is a straight line. Take this straight line to be the x-axis, and let the ends of the string be located at the points $x = 0$ and $x = l$, where l is the length of the string. If the string is displaced from its equilibrium position (or if certain velocities are imparted to the points of the string), and the string is then released, it begins to vibrate. We shall consider only the case of small vibrations; then, the length of the string can be regarded as unchanged. Moreover, we shall regard the vibrations as taking place in one plane, in such a way that each point of the string moves in the direction perpendicular to the x-axis.

Let $u(x, t)$ be the displacement at time t of the point of the string with abscissa x. Then, for every fixed value of t, the graph of the function $u(x, t)$ obviously represents the form of the string at the time t. The element AB of the string (see Fig. 47) is acted upon by the tension forces T_1

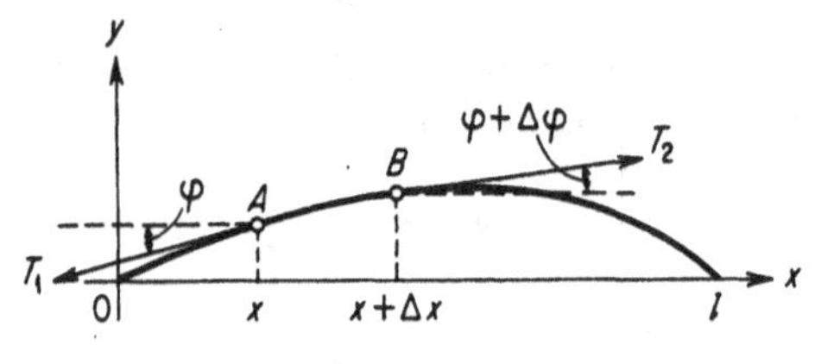

FIGURE 47

and T_2, which are directed along the tangent to the string. For the time being, we assume that no other forces act on the string. In the equilibrium position, the tension T is the same at all points of the string. To the extent that we can assume that the length of the string does not change (see above remark), we can also assume that the tension in the string does not change. Therefore, T_1 and T_2 have the same magnitude as T, although they have different directions, and because of the curvature of the element AB, one direction is not quite the negative of the other. Hence, the force acting on the element AB in the direction of the u-axis is

$$\begin{aligned} T[\sin(\varphi + \Delta\varphi) - \sin\varphi] \\ &\approx T[\tan(\varphi + \Delta\varphi) - \tan\varphi] = T\left[\frac{\partial u(x + \Delta x, t)}{\partial x} - \frac{\partial u(x, t)}{\partial x}\right] \\ &= T\frac{\partial^2 u(x + \theta\Delta x, t)}{\partial x^2}\Delta x \qquad (0 < \theta < 1), \end{aligned}$$

where $\approx$ denotes approximate equality. Regarding the element Δx as being very small, and using Newton's second law of motion, we can write

$$\rho\Delta x \frac{\partial^2 u}{\partial t^2} = T \frac{\partial^2 u}{\partial x^2} \Delta x, \tag{11.1}$$

where ρ is the linear density of the string (i.e., the mass per unit length). Setting $a^2 = T/\rho$ and dividing by Δx, we obtain

$$\frac{\partial^2 u}{\partial t^2} = a^2 \frac{\partial^2 u}{\partial x^2}, \tag{11.2}$$

which is the equation for *free vibrations* of the string.

Next, suppose that in addition to the tension T, the string is acted upon by a force of amount $F(x, t)$ per unit length of the string. Then, instead of equation (11.1), we obtain

$$\rho \Delta x \frac{\partial^2 u}{\partial t^2} = T \frac{\partial^2 u}{\partial x^2} \Delta x + F(x, t)\Delta x,$$

or

$$\frac{\partial^2 u}{\partial t^2} = a^2 \frac{\partial^2 u}{\partial x^2} + \frac{F(x, t)}{\rho}, \tag{11.3}$$

which is the equation for *forced vibrations* of the string.

We now study the following problem: Given the form of the string and the velocity of its points at the initial time $t = 0$, what is its form at the arbitrary time t? Mathematically, this problem reduces to solving equation (11.2) in the case of free vibrations, and equation (11.3) in the case of forced vibrations, subject to the boundary conditions

$$u(0, t) = u(l, t) = 0 \tag{11.4}$$

and the initial conditions

$$u(x, 0) = f(x), \quad \frac{\partial u(x, 0)}{\partial t} = g(x), \tag{11.5}$$

where $f(x)$ and $g(x)$ are given continuous functions, which vanish for $x = 0$ and $x = l$. Equations (11.2) and (11.3) are special cases of the equations studied in Part I.

12. Free Vibrations of a String

Instead of starting from the formulas already found in Part I, we shall once more go through the derivation given in Sec. 1. We are looking for a solution (different from $u \equiv 0$) of the form

$$u(x, t) = \Phi(x)T(t) \tag{12.1}$$

which satisfies the boundary conditions. Substituting (12.1) in (11.2) gives

$$\Phi T'' = a^2\Phi'' T,$$

whence

$$\frac{\Phi''}{\Phi} = \frac{T''}{a^2 T} = -\lambda = \text{const},$$

so that

$$\Phi'' = -\lambda\Phi, \tag{12.2}$$

$$T'' = -a^2\lambda T. \tag{12.3}$$

If a function of the form (12.1), which is not identically zero, is to satisfy the conditions (11.4), then obviously the condition

$$\Phi(0) = \Phi(l) = 0 \tag{12.4}$$

must be met. Thus, we obtain a boundary value problem for equation (12.2), subject to the condition (12.4). In view of the theorem of Sec. 5 (see also the remark made there), all the eigenvalues of our problem are positive.[10] Therefore, it is permissible to write λ^2 instead of λ. Then, equations (12.2) and (12.3) take the form

$$\Phi'' + \lambda^2\Phi = 0, \tag{12.5}$$

$$T'' + a^2\lambda^2 T = 0. \tag{12.6}$$

The solution of (12.5) is

$$\Phi = C_1 \cos \lambda x + C_2 \sin \lambda x \qquad (C_1 = \text{const},\ C_2 = \text{const}),$$

where for $x = 0$ and $x = l$ we must have

$$\Phi(0) = C_1 = 0,$$

$$\Phi(l) = C_2 \sin \lambda l = 0.$$

Assuming that $C_2 \neq 0$, since otherwise Φ would be identically zero, we find that $\lambda l = \pi n$, where n is an integer. Setting $C_2 = 1$ gives

$$\lambda_n = \frac{\pi n}{l} \qquad (n = 1, 2, \ldots),$$

and the corresponding eigenfunctions are

$$\Phi_n(x) = \sin \frac{\pi n x}{l} \qquad (n = 1, 2, \ldots).$$

---

[10] This can also be verified directly by examining the solution of equation (12.2) for $\lambda \leqslant 0$. By doing this, the reader can convince himself that the condition (12.4) cannot be satisfied.

We do not consider negative values of n, since they give the same eigenfunctions (up to a constant factor) as the corresponding positive values of n. Thus, in the sense indicated in Sec. 4, only one eigenfunction corresponds to each value of λ^2.

For $\lambda = \lambda_n$, equation (12.6) gives

$$T_n = A_n \cos a\lambda_n t + B_n \sin a\lambda_n t$$
$$= A_n \cos \frac{a\pi nt}{l} + B_n \sin \frac{a\pi nt}{l} \qquad (n = 1, 2, \ldots),$$

so that

$$u_n(x, t) = \left(A_n \cos \frac{a\pi nt}{l} + B_n \sin \frac{a\pi nt}{l}\right) \sin \frac{\pi nx}{l} \qquad (n = 1, 2, \ldots). \tag{12.7}$$

Thus, to solve our problem, we set

$$u(x, t) = \sum_{n=1}^{\infty} u_n(x, t)$$
$$= \sum_{n=1}^{\infty} \left(A_n \cos \frac{a\pi nt}{l} + B_n \sin \frac{a\pi nt}{l}\right) \sin \frac{\pi nx}{l}, \tag{12.8}$$

and require that[11]

$$u(x, 0) = \sum_{n=1}^{\infty} A_n \sin \frac{\pi nx}{l} = f(x),$$

$$\frac{\partial u(x, 0)}{\partial t} = \left[\sum_{n=1}^{\infty} \left(-A_n \frac{a\pi n}{l} \sin \frac{a\pi nt}{l} + B_n \frac{a\pi n}{l} \cos \frac{a\pi nt}{l}\right) \sin \frac{\pi nx}{l}\right]_{t=0}$$
$$= \sum_{n=1}^{\infty} B_n \frac{a\pi n}{l} \sin \frac{\pi nx}{l} = g(x).$$

Therefore, we have to expand $f(x)$ and $g(x)$ in Fourier series with respect to the system $\{\sin (\pi nx/l)\}$. The formulas for the Fourier coefficients give

$$A_n = \frac{2}{l} \int_0^l f(x) \sin \frac{\pi nx}{l}\, dx \qquad (n = 1, 2, \ldots), \tag{12.9}$$

$$B_n \frac{a\pi n}{l} = \frac{2}{l} \int_0^l g(x) \sin \frac{\pi nx}{l}\, dx,$$

or

$$B_n = \frac{2}{a\pi n} \int_0^l g(x) \sin \frac{\pi nx}{l}\, dx \qquad (n = 1, 2, \ldots), \tag{12.10}$$

i.e., the solution of our problem is given by the series (12.8), where A_n and B_n are determined from formulas (12.9) and (12.10).

---

[11] As in the method of Sec. 1, we differentiate the series term by term.

Thus, we see that the vibrational motion of the string is a superposition of separate harmonic vibrations of the form (12.7), or of the equivalent form

$$u_n = H_n \sin\left(\frac{a\pi nt}{l} + \alpha_n\right) \sin\frac{\pi nx}{l},$$

where

$$H_n = \sqrt{A_n^2 + B_n^2}, \quad \sin\alpha_n = \frac{A_n}{H_n}, \quad \cos\alpha_n = \frac{B_n}{H_n}.$$

The amplitude of the vibration of the point with coordinate x is

$$H_n \left|\sin\frac{\pi nx}{l}\right|,$$

and is independent of t. The points for which $x = 0, l/n, 2l/n, \ldots, (n-1)l/n, l$, remain fixed during the motion, and are known as *nodes*. Hence, a string whose vibrations are described by formula (12.7) is divided into n segments, the end points of which do not vibrate. Moreover, in adjacent segments, the displacements of the string have opposite signs, and the midpoints of the segments, the so-called *antinodes*, vibrate with the largest amplitudes. This whole phenomenon is known as a *standing wave*.

Figure 48 shows consecutive positions of a string whose vibrations are described by formula (12.7), where $n = 1, 2, 3, 4$. In the general case, where

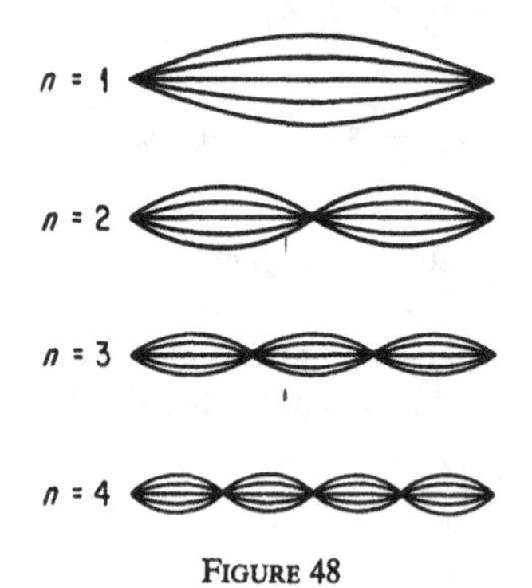

FIGURE 48

the vibrations of the string are described by formula (12.8), the *fundamental* (*mode*) corresponds to the component u_1 with frequency

$$\omega_1 = \frac{a\pi}{l} = \frac{\pi}{l}\sqrt{\frac{T}{\rho}}$$

and period

$$\tau_1 = \frac{2\pi}{\omega_1} = 2l\sqrt{\frac{\rho}{T}}.$$

The other vibrational modes of the string, i.e., the *overtones*, with frequencies

$$\omega_n = \frac{a\pi n}{l} = \frac{\pi n}{l}\sqrt{\frac{T}{\rho}}$$

and periods

$$\tau_n = \frac{2\pi}{\omega_n} = \frac{2l}{n}\sqrt{\frac{\rho}{T}},$$

characterize the *timbre* or "color" of the sound. If the string is held fixed at its midpoint, then clearly the even overtones of the string, for which the midpoint of the string is a node, are preserved. However, the fundamental and the odd overtones are immediately extinguished, since by holding the midpoint of the string fixed we essentially go from a string of length l to a string of length $l/2$, and changing l to $l/2$ in (12.8) leads to a series containing only even components. Then the overtone with period $\tau_2 = 2\pi/\omega_2 = \tau_1/2$ plays the role of the fundamental.

13. Forced Vibrations of a String

Next we consider the case of a periodic perturbing force, i.e., we write

$$\frac{F(x, t)}{\rho} = A \sin \omega t.$$

Then

$$\frac{F(x, t)}{\rho} = A \sin \omega t = \sum_{n=1}^{\infty} F_n(t) \sin \frac{\pi n x}{l}, \tag{13.1}$$

where

$$\begin{aligned} F_n(t) &= \frac{2}{l}\int_0^l A \sin \omega t \sin \frac{\pi n x}{l}\,dx \\ &= \frac{2A}{\pi n}[1 - (-1)^n] \sin \omega t \qquad (n = 1, 2, \ldots). \end{aligned}$$

If we write

$$u(x, t) = \sum_{n=1}^{\infty} T_n(t) \sin \frac{\pi n x}{l}, \tag{13.2}$$

substitute (13.2) and (13.1) in (11.3), and carry out the required term by term differentiation, we obtain

$$\sum_{n=1}^{\infty}\left(T_n'' + \frac{a^2\pi^2n^2}{l^2}T_n - \frac{2A}{\pi n}[1 - (-1)^n] \sin \omega t\right) \sin \frac{\pi n x}{l} = 0,$$

whence

$$T_n'' + \frac{a^2\pi^2n^2}{l^2}T_n - \frac{2A}{\pi n}[1 - (-1)^n]\sin\omega t = 0. \tag{13.3}$$

Writing

$$\omega_n = \frac{a\pi n}{l} \qquad (n = 1, 2, \ldots)$$

for simplicity (these will be recognized as the frequencies of the free or *characteristic* vibrations of the string), we can rewrite equation (13.3) as

$$T_n'' + \omega_n^2 T_n = \frac{2A}{\pi n}[1 - (-1)^n]\sin\omega t. \tag{13.4}$$

Solving this equation, we obtain

$$T_n = A_n \cos\omega_n t + B_n \sin\omega_n t + \frac{2A[1 - (-1)^n]}{\pi n(\omega_n^2 - \omega^2)}\sin\omega t, \tag{13.5}$$

if $\omega_n \neq \omega$. To satisfy the conditions (11.4) and (11.5), we require that

$$u(x, 0) = \sum_{n=1}^{\infty} T_n(0)\sin\frac{\pi n x}{l} = f(x),$$

$$\frac{\partial u(x, 0)}{\partial t} = \sum_{n=1}^{\infty} T_n'(0)\sin\frac{\pi n x}{l} = g(x).$$

A calculation of the Fourier coefficients of $f(x)$ and $g(x)$ gives

$$\begin{aligned} T_n(0) &= A_n = \frac{2}{l}\int_0^l f(x)\sin\frac{\pi n x}{l}\,dx, \\ T_n'(0) &= \omega_n B_n + \frac{2A\omega[1 - (-1)^n]}{\pi n(\omega_n^2 - \omega^2)} \\ &= \frac{2}{l}\int_0^l g(x)\sin\frac{\pi n x}{l}, \end{aligned} \tag{13.6}$$

or

$$B_n = \frac{2}{l\omega_n}\int_0^l g(x)\sin\frac{\pi n x}{l}\,dx - \frac{2A\omega[1 - (-1)^n]}{\pi n\omega_n(\omega_n^2 - \omega^2)} \tag{13.7}$$

where we have used (13.5). Substituting (13.6) and (13.7) in (13.5), and then substituting the resulting expression for T_n in (13.2), we obtain

$$\begin{aligned} u(x, t) = &\sum_{n=1}^{\infty} (A_n \cos\omega_n t + \bar{B}_n \sin\omega_n t)\sin\frac{\pi n x}{l} \\ &+ \frac{4A}{\pi}\sin\omega t \sum_{k=0}^{\infty} \frac{\sin[(2k+1)\pi x/l]}{(2k+1)(\omega_{2k+1}^2 - \omega^2)} \\ &- \frac{4A\omega}{\pi}\sum_{k=0}^{\infty} \frac{\sin\omega_{2k+1}t\,\sin[(2k+1)\pi x/l]}{(2k+1)\omega_{2k+1}(\omega_{2k+1}^2 - \omega^2)}, \end{aligned} \tag{13.8}$$

where we have written

$$\bar{B}_n = \frac{2}{l\omega_n} \int_0^l g(x) \sin \frac{\pi n x}{l}\, dx.$$

Recalling the expression for ω_n, the reader will easily recognize that the first sum in the right-hand side of (13.8) is the function giving the *free* vibrations of the string, subject to the conditions (11.4) and (11.5). [See (12.8), (12.9), and (12.10).] Therefore, the second and third sums give the "correction" caused by the presence of the perturbing force. The second term represents what is sometimes referred to as the "pure" forced vibrations, since they occur with the frequency of the perturbing force.

Equation (13.5) holds if $\omega_n \neq \omega$. We now examine the situation when $\omega_n = \omega$, i.e., when the frequency of the perturbing force is the same as one of the characteristic frequencies of the string. Then, equation (13.4) gives

$$T_n = A_n \cos \omega t + B_n \sin \omega t - \frac{At}{\pi n \omega} [1 - (-1)^n] \cos \omega t.$$

This shows that when n is odd, the amplitude of the nth vibration in the term $T_n(t) \sin (\pi n x/l)$ of the sum (13.2) is

$$H = \sqrt{\left(A_n - \frac{2At}{\pi n \omega}\right)^2 + B_n^2}\, \left|\sin \frac{\pi n x}{l}\right|$$

which becomes unbounded as t increases. In this case, we say that *resonance* occurs.

14. Equation of the Longitudinal Vibrations of a Rod

Consider a homogeneous rod of length l. If the rod is stretched or compressed along its longitudinal axis, and then released, it executes longitudinal vibrations. We choose the x-axis along the axis of the rod, and we assume that in the equilibrium position, the ends of the rod are located at the points $x = 0$ and $x = l$. Let x be the abscissa of a cross section of the rod at rest, and let $u(x, t)$ be the displacement of this cross section at the time t. Consider an element AB of the rod, whose ends are located at the points x and $x + \Delta x$ when the rod is at rest, and let the ends of this element have the new positions A' and B' at the time t, with abscissas $x + u(x, t)$ and $x + \Delta x + u(x + \Delta x, t)$, respectively (see Fig. 49). Thus, at the time t, the element AB has length $\Delta x + u(x + \Delta x, t) - u(x, t)$, i.e., its absolute increase in length is

$$u(x + \Delta x, t) - u(x, t) = \frac{\partial u(x + \theta \Delta x, t)}{\partial x} \Delta x \qquad (0 < \theta < 1),$$

while its relative increase in length is

$$\frac{\partial u(x + \theta\Delta x, t)}{\partial x}.$$

In the limit as $\Delta x \to 0$, the relative increase in length of the cross section of the rod at x is

$$\frac{\partial u(x, t)}{\partial x}.$$

According to Hooke's law, the force T acting on the cross section at x is given by the formula

$$T = Es\frac{\partial u}{\partial x},$$

where E is the modulus of elasticity (Young's modulus) of the material from which the rod is made, and s is the cross-sectional area of the rod.

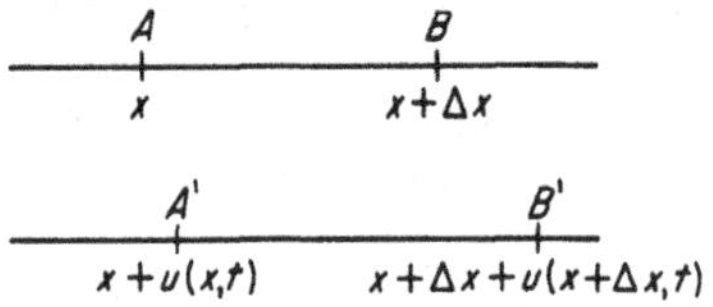

FIGURE 49

Returning to the element AB which has the new position $A'B'$ at time t, we note that AB is acted upon by forces T_1 and T_2, applied at the points A' and B' and directed along the x-axis. (For the time being, we consider no other forces.) The resultant of these forces is

$$\begin{aligned} T_2 - T_1 &= Es\left(\frac{\partial u(x + \Delta x, t)}{\partial x} - \frac{\partial u(x, t)}{\partial x}\right) \\ &= Es\frac{\partial^2 u(x + \theta\Delta x, t)}{\partial x^2}\Delta x, \end{aligned}$$

and is also directed along the x-axis. Regarding the element AB as being very small, we can write

$$\rho s\,\Delta x\frac{\partial^2 u}{\partial t^2} = Es\frac{\partial^2 u}{\partial x^2}\Delta x, \tag{14.1}$$

where ρ is the density of the material from which the rod is made. Setting $E/\rho = a^2$ and dividing by Δx and s, we obtain

$$\frac{\partial^2 u}{\partial t^2} = a^2\frac{\partial^2 u}{\partial x^2}, \tag{14.2}$$

which is the equation for *free* vibrations of the rod. This equation has the same form as the equation for free vibrations of a string.

If the rod is acted upon by an additional force of amount $F(x, t)$ per unit volume, then instead of (14.1), we have

$$\rho s \Delta x \frac{\partial^2 u}{\partial t^2} = Es \frac{\partial^2 u}{\partial x^2} \Delta x + F(x, t) s \Delta x,$$

so that

$$\frac{\partial^2 u}{\partial t^2} = a^2 \frac{\partial^2 u}{\partial x^2} + \frac{F(x, t)}{\rho}. \tag{14.3}$$

This is the equation for *forced* oscillations of the rod [cf. (11.3)].

We now pose the problem of finding the displacement of the cross sections of the rod at any time t, with specified initial and boundary conditions. The boundary conditions can take various forms:

Case 1) The rod is fastened at both ends:

$$u(0, t) = u(l, t) = 0;$$

Case 2) One end of the rod is fastened, and the other end is free:

$$u(0, t) = 0, \quad \frac{\partial u(l, t)}{\partial x} = 0 \tag{14.4}$$

(the force on the free end of the rod is zero and hence $\partial u/\partial x = 0$);

Case 3) Both ends are free.

We shall devote our attention to Case 2, where the boundary conditions (14.4) are met. The initial conditions have the familiar form

$$u(x, 0) = f(x), \quad \frac{\partial u(x, 0)}{\partial t} = g(x), \tag{14.5}$$

i.e., the initial displacements and initial velocities of the cross sections of the rod are specified.

15. Free Vibrations of a Rod

As in the case of the vibrating string (see Sec. 12), we look for particular solutions of the form

$$u(x, t) = \Phi(x)T(t),$$

and arrive at the equations

$$\Phi'' + \lambda^2\Phi = 0, \tag{15.1}$$

$$T'' + a^2\lambda^2 T = 0, \tag{15.2}$$

with the boundary conditions

$$\Phi(0) = \Phi'(l) = 0. \tag{15.3}$$

Equation (15.1) has the solution

$$\Phi(x) = C_1 \cos \lambda x + C_2 \sin \lambda x \qquad (C_1 = \text{const},\ C_2 = \text{const}).$$

According to (15.3), for $x = 0$ and $x = l$ we must have

$$\Phi(0) = C_1 = 0, \quad \Phi'(l) = C_2 \lambda \cos \lambda l = 0.$$

Assuming that $C_2 \neq 0$, since otherwise $\Phi(x)$ would be identically zero, we find that $\lambda l = (2n + 1)\pi/2$, where n is an integer. We write

$$\begin{aligned} \lambda_n &= \frac{(2n+1)\pi}{2l} \qquad (n = 0, 1, 2, \ldots), \\ \Phi_n(x) &= \sin \lambda_n x = \sin \frac{(2n+1)\pi x}{2l} \qquad (n = 0, 1, 2, \ldots). \end{aligned} \tag{15.4}$$

(The negative values of n give no new eigenfunctions.) For $\lambda = \lambda_n$, equation (15.2) leads to

$$T_n(t) = A_n \cos a\lambda_n t + B_n \sin a\lambda_n t \qquad (n = 0, 1, 2, \ldots),$$

and therefore

$$u_n(x, t) = (A_n \cos a\lambda_n t + B_n \sin a\lambda_n t) \sin \lambda_n x \qquad (n = 0, 1, 2, \ldots).$$

To solve our problem, we form the series

$$\frac{u(x, t)}{\partial t} = \sum_{n=0}^{\infty} u_n(x, t) = \sum_{n=0}^{\infty} (A_n \cos a\lambda_n t + B_n \sin a\lambda_n t) \sin \lambda_n x, \tag{15.5}$$

and require that

$$u(x, 0) = \sum_{n=0}^{\infty} A_n \sin \lambda_n x = f(x),$$

$$\begin{aligned} \frac{\partial u(x, 0)}{\partial t} &= \left[\sum_{n=0}^{\infty} (-A_n a\lambda_n \sin a\lambda_n t + B_n a\lambda_n \cos a\lambda_n t) \sin \lambda x \right]_{t=0} \\ &= \sum_{n=0}^{\infty} B_n a\lambda_n \sin \lambda_n x = g(x). \end{aligned}$$

A calculation of the Fourier coefficients of $f(x)$ and $g(x)$ with respect to the system $\{\sin \lambda_n x\}$ gives

$$A_n = \frac{\int_0^l f(x) \sin \lambda_n x \, dx}{\int_0^l \sin^2 \lambda_n x \, dx} \qquad (n = 0, 1, 2, \ldots),$$

$$B_n a\lambda_n = \frac{\int_0^l g(x) \sin \lambda_n x \, dx}{\int_0^l \sin^2 \lambda_n x \, dx} \qquad (n = 0, 1, 2, \ldots).$$

But

$$\int_0^l \sin^2 \lambda_n x \, dx = \frac{1}{2}\int_0^l (1 - \cos 2\lambda_n x) \, dx = \frac{l}{2},$$

and therefore

$$A_n = \frac{2}{l}\int_0^l f(x) \sin \lambda_n x \, dx \qquad (n = 0, 1, 2, \ldots),$$

$$B_n = \frac{2}{a\lambda_n l}\int_0^l g(x) \sin \lambda_n x \, dx \tag{15.6}$$

$$= \frac{4}{(2n + 1)a\pi}\int_0^l g(x) \sin \lambda_n x \, dx \qquad (n = 0, 1, 2, \ldots)$$

[see (15.4)].

Thus, the solution of our problem is given by formula (15.5), where A_n and B_n are determined from (15.6), i.e., the vibrational motion of the rod is a superposition of separate harmonic vibrations

$$u_n = (A_n \cos a\lambda_n t + B_n \sin a\lambda_n t) \sin \lambda_n x, \tag{15.7}$$

or

$$u_n = H_n \sin (a\lambda_n t + \alpha_n) \sin \lambda_n x,$$

where

$$H_n = \sqrt{A_n^2 + B_n^2}, \qquad \sin \alpha_n = \frac{A_n}{H_n}, \qquad \cos \alpha_n = \frac{B_n}{H_n}.$$

The amplitude of the vibrational motion described by (15.7) is

$$H_n|\sin \lambda_n x| = H_n\left|\sin \frac{(2n + 1)\pi x}{2l}\right|,$$

which depends only on the position x of the cross section, and not on the time t. As for the frequency, it is given by

$$\omega_n = a\lambda_n = \frac{(2n + 1)a\pi}{2l} = \frac{(2n + 1)\pi}{2l}\sqrt{\frac{E}{\rho}},$$

and hence the period is

$$\tau_n = \frac{2\pi}{\omega_n} = \frac{4l}{(2n + 1)a} = \frac{4l}{2n + 1}\sqrt{\frac{\rho}{E}}.$$

In the case of a vibrating rod, the fundamental mode is obtained for $n = 0$; it has amplitude

$$\left|A_0 \sin \frac{\pi x}{2l}\right|,$$

frequency

$$\omega_0 = \frac{\pi}{2l}\sqrt{\frac{E}{\rho}}$$

and period

$$\tau_0 = 4l\sqrt{\frac{\rho}{E}}.$$

Therefore, the fundamental mode corresponds to a node at the fixed end ($x = 0$), and an antinode at the free end ($x = l$), as shown in Fig. 50.

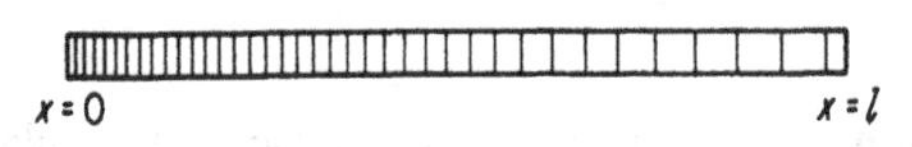

FIGURE 50

16. Forced Vibrations of a Rod

Next, we consider the case where the rod is suspended from the end $x = 0$ and the perturbing force is the force of gravity, i.e.,

$$F(x, t) = \rho g,$$

where $F(x, t)$ is the force per unit volume, ρ is the density of the rod, and g is the acceleration due to gravity. In this case, the equation for the vibrations takes the form

$$\frac{\partial^2 u}{\partial t^2} = a^2 \frac{\partial^2 u}{\partial x^2} + g \tag{16.1}$$

[see (14.3)], again subject to the boundary conditions (14.4) and the initial conditions (14.5). We set

$$\begin{aligned} u(x, t) &= \sum_{n=0}^{\infty} T_n(t) \sin \lambda_n x, \\ \frac{F(x, t)}{\rho} &= g = \sum_{n=0}^{\infty} F_n \sin \lambda_n x, \end{aligned} \tag{16.2}$$

where

$$F_n = \frac{\int_0^l g \sin \lambda_n x \, dx}{\int_0^l \sin^2 \lambda_n x \, dx} = \frac{2g}{l\lambda_n} \qquad (n = 0, 1, 2, \ldots).$$

Substituting the series (16.2) in (16.1) and transposing all the terms to the left, we obtain

$$\sum_{n=0}^{\infty}\left(T_n'' + a^2\lambda_n^2 T_n - \frac{2g}{l\lambda_n}\right)\sin\lambda_n x = 0,$$

whence

$$T_n'' + a^2\lambda_n^2 T_n - \frac{2g}{l\lambda_n} = 0 \qquad (n = 0, 1, 2, \ldots).$$

The solution of this equation has the form

$$T_n = A_n \cos a\lambda_n t + B_n \sin a\lambda_n t + \frac{2g}{la^2\lambda_n^3} \qquad (n = 0, 1, 2, \ldots).$$

To satisfy the conditions (14.5), we require that

$$u(x, 0) = \sum_{n=0}^{\infty} T_n(0)\sin\lambda_n x = f(x),$$

$$\frac{\partial u(x, 0)}{\partial t} = \sum_{n=0}^{\infty} T_n'(0)\sin\lambda_n x = g(x).$$

A calculation of the Fourier coefficients of $f(x)$ and $g(x)$ with respect to the system $\{\sin\lambda_n x\}$ gives

$$T_n(0) = A_n + \frac{2g}{la^2\lambda_n^3} = \frac{2}{l}\int_0^l f(x)\sin\lambda_n x\,dx,$$

$$T_n'(0) = B_n a\lambda_n = \frac{2}{l}\int_0^l g(x)\sin\lambda_n x\,dx$$

[cf. (15.6)], so that

$$A_n = \frac{2}{l}\int_0^l f(x)\sin\lambda_n x\,dx - \frac{2}{la^2\lambda_n^3} = \bar{A}_n - \frac{2g}{la^2\lambda_n^3},$$

$$B_n = \frac{2}{al\lambda_n}\int_0^l g(x)\sin\lambda_n x\,dx \qquad (n = 0, 1, 2, \ldots).$$

Therefore, we have

$$u(x, t) = \sum_{n=0}^{\infty}(\bar{A}_n \cos a\lambda_n t + B_n \sin a\lambda_n t)\sin\lambda_n x$$

$$- \frac{2g}{la^2}\sum_{n=0}^{\infty}\frac{\cos a\lambda_n t \sin\lambda_n x}{\lambda_n^3} + \frac{2g}{la^2}\sum_{n=0}^{\infty}\frac{\sin\lambda_n x}{\lambda_n^3}.$$

The reader will recognize at once that the first sum on the right is the function giving the solution of the problem of the free vibrations of a rod with the same conditions (14.5). Hence, the second and third terms give the corrections due to the force of gravity.

17. Vibrations of a Rectangular Membrane

By a *membrane*, we mean an elastic film supported by a closed plane curve. When the membrane is at rest, all its points lie in one plane, which we take to be the xy-plane. If the membrane is displaced from its equilibrium position and then released, it begins to vibrate. We consider only small vibrations of the membrane, assuming that the area of the membrane does not change and that each of its points vibrates in a direction perpendicular to the xy-plane. Let $u(x, y, t)$ denote the displacement at time t of the point (x, y) of the membrane from its equilibrium position. Then, by a derivation similar to that made in the case of the string, one finds that the equation for the free vibrations of the membrane has the form

$$\frac{\partial^2 u}{\partial t^2} = c^2\left(\frac{\partial^2 u}{\partial x^2} + \frac{\partial^2 u}{\partial y^2}\right), \tag{17.1}$$

while the equation for forced vibrations of the membrane is

$$\frac{\partial^2 u}{\partial t^2} = c^2\left(\frac{\partial^2 u}{\partial x^2} + \frac{\partial^2 u}{\partial y^2}\right) + \frac{F(x, y, t)}{\rho}, \tag{17.2}$$

where $c^2 = T/\rho$, T is the tension in the membrane, ρ is its surface density, and $F(x, y, t)$ is the force per unit area acting on the membrane.

The problem of the vibrating membrane can be posed as follows: Find the solution of equation (17.1) or (17.2), i.e., find the displacement of the points of the membrane at any time t, subject to the condition

$$u = 0 \tag{17.3}$$

met on the (fixed) boundary of the membrane, and the initial conditions

$$u(x, y, 0) = f(x, y) \tag{17.4}$$

(specifying the initial displacement of the membrane) and

$$\frac{\partial u(x, y, 0)}{\partial t} = g(x, y) \tag{17.5}$$

(specifying the initial velocities of the points of the membrane).

We now study the case of the *free* vibrations of a membrane in the shape of a rectangle R: $0 \leqslant x \leqslant a$, $0 \leqslant y \leqslant b$. The problem differs from the problem considered in Part I in that the function u depends on three rather than two variables. Nevertheless, we again apply the eigenfunction method and begin by looking for particular solutions of the form

$$u = \Phi(x, y)T(t), \tag{17.6}$$

which are not identically zero and which satisfy the condition (17.3) on the boundary of the rectangle R. Differentiating (17.6) and substituting the result in (17.1), we obtain

$$\Phi T'' = c^2 \left(\frac{\partial^2 \Phi}{\partial x^2} + \frac{\partial^2 \Phi}{\partial y^2} \right) T,$$

so that[12]

$$\frac{\dfrac{\partial^2 \Phi}{\partial x^2} + \dfrac{\partial^2 \Phi}{\partial y^2}}{\Phi} = \frac{T''}{c^2 T} = -\lambda^2 = \text{const.}$$

Hence

$$\frac{\partial^2 \Phi}{\partial x^2} + \frac{\partial^2 \Phi}{\partial y^2} + \lambda^2 \Phi = 0, \qquad (17.7)$$

$$T'' + c^2 \lambda^2 T = 0, \qquad (17.8)$$

where, as is easily seen, the function Φ satisfies the condition

$$\Phi = 0 \qquad (17.9)$$

on the boundary of the rectangle.

Thus, we now consider equation (17.7), with the boundary condition (17.9). We fix λ and look for particular solutions of (17.7) of the form

$$\Phi(x, y) = \varphi(x)\psi(y), \qquad (17.10)$$

which satisfy the condition (17.9) on the boundary. Substituting (17.10) in (17.7) gives

$$\varphi''\psi + \varphi\psi'' + \lambda^2 \varphi\psi = 0,$$

or[13]

$$\frac{\varphi''}{\varphi} = -\frac{\psi'' + \lambda^2 \psi}{\psi} = -k^2 = \text{const,}$$

whence

$$\varphi'' + k^2 \varphi = 0, \quad \psi'' + l^2 \psi = 0, \qquad (17.11)$$

where we have written

$$l^2 = \lambda^2 - k^2. \qquad (17.12)$$

---

[12] The fact that this constant should not be chosen to be positive is clear just from the fact that otherwise the coefficient of T in equation (17.8) would be negative, so that the solution of (17.8) would not be periodic, i.e., contrary to experience, we would not have vibrations.

[13] If the constant were positive, we could not satisfy the boundary condition.

For the solutions of (17.11), we find

$$\varphi(x) = C_1 \cos kx + C_2 \sin kx, \quad \psi(x) = C_3 \cos ly + C_4 \sin ly,$$

where C_1, C_2, C_3, and C_4 are constants. The boundary condition implies that

$$\varphi(0) = \varphi(a) = 0, \qquad \psi(0) = \psi(b) = 0,$$

and hence

$$C_1 = C_3 = 0, \qquad C_2 \sin ka = C_4 \sin lb = 0,$$

so that $ka = m\pi$, $lb = n\pi$, where m and n are integers. Setting $C_2 = C_4 = 1$ and writing

$$k_m = \frac{\pi m}{a}, \quad l_m = \frac{\pi n}{b} \qquad (m = 1, 2, \ldots;\ n = 1, 2, \ldots),$$

we find

$$\begin{aligned} \varphi_m(x) &= \sin k_m x = \sin \frac{\pi m x}{a}, \\ \psi_n(y) &= \sin l_n x = \sin \frac{\pi n y}{b}. \end{aligned} \qquad (m = 1, 2, \ldots;\ n = 1, 2, \ldots)$$

(We do not consider negative values of m and n, since they give the same functions φ_m and ψ_n to within a constant factor.) According to (17.10) and (17.12), if

$$\lambda = \lambda_{mn} = \sqrt{k_m^2 + l_n^2} = \pi \sqrt{(m^2/a^2) + (n^2/b^2)}, \tag{17.13}$$

we obtain the following particular solutions of equation (17.7) satisfying the boundary condition:

$$\Phi_{mn}(x, y) = \varphi_m(x)\psi_n(y) = \sin \frac{\pi m x}{a} \sin \frac{\pi n y}{b}.$$

We now solve equation (17.8) for every $\lambda = \lambda_{mn}$; the result is

$$T_{mn}(t) = A_{mn} \cos c\lambda_{mn}t + B_{mn} \sin c\lambda_{mn}t.$$

Therefore, the functions

$$u_{mn}(x, y, t) = (A_{mn} \cos c\lambda_{mn}t + B_{mn} \sin c\lambda_{mn}t) \sin \frac{\pi m x}{a} \sin \frac{\pi n y}{b}$$
$$(m = 1, 2, \ldots;\ n = 1, 2, \ldots) \tag{17.14}$$

are particular solutions of (17.1) satisfying the boundary condition (17.3).

To obtain the solution of equation (17.1) satisfying the *initial* conditions, we set

$$u(x, y, t) = \sum_{m,n=1}^{\infty} u_{mn}(x, y, t)$$

$$= \sum_{m,n=1}^{\infty} (A_{mn} \cos c\lambda_{mn}t + B_{mn} \sin c\lambda_{mn}t) \sin \frac{\pi mx}{a} \sin \frac{\pi ny}{b}, \quad (17.15)$$

and require that

$$u(x, y, 0) = \sum_{m,n=1}^{\infty} A_{mn} \sin \frac{\pi mx}{a} \sin \frac{\pi ny}{b} = f(x, y),$$

$$\frac{\partial u(x, y, 0)}{\partial t}$$

$$= \left[\sum_{m,n=1}^{\infty} (-A_{mn}c\lambda_{mn} \sin c\lambda_{mn}t + B_{mn}c\lambda_{mn} \cos c\lambda_{mn}t) \sin \frac{\pi mx}{a} \sin \frac{\pi ny}{b} \right]_{t=0}$$

$$= \sum_{m,n=1}^{\infty} B_{mn}c\lambda_{mn} \sin \frac{\pi mx}{a} \sin \frac{\pi ny}{b}.$$

Assuming that $f(x, y)$ and $g(x, y)$ can be expanded in double Fourier series with respect to the system $\{\sin (\pi mx/a) \sin (\pi ny/b)\}$ and using the results of Ch. 7, Sec. 4, we obtain

$$A_{mn} = \frac{4}{ab} \iint_R f(x, y) \sin \frac{\pi mx}{a} \sin \frac{\pi ny}{b} \, dx \, dy \quad (17.16)$$

and

$$B_{mn}c\lambda_{mn} = \frac{4}{ab} \iint_R g(x, y) \sin \frac{\pi mx}{a} \sin \frac{\pi ny}{b} \, dx \, dy,$$

or[14]

$$B_{mn} = \frac{4}{abc\lambda_{mn}} \iint_R g(x, y) \sin \frac{\pi mx}{a} \sin \frac{\pi ny}{b} \, dx \, dy. \quad (17.17)$$

Thus, the solution of our problem is given by the series (17.15), where A_{mn} and B_{mn} are calculated by using formulas (17.16) and (17.17).

---

[14] In Ch. 7, Sec. 4, we expanded functions in double Fourier series in a rectangle of the form $-l \leqslant x \leqslant l$, $-h \leqslant y \leqslant h$. If instead, we want to expand a function $f(x, y)$ in a rectangle of the form $0 \leqslant x \leqslant a$, $0 \leqslant y \leqslant b$, with respect to the system $\{\sin (\pi mx/a) \sin (\pi ny/b)\}$, this can be done by making the odd extension of $f(x, y)$, first in the variable x and then in the variable y. Then, the formulas for the Fourier coefficients take the form (17.16).

The frequencies of the characteristic vibrations of the rectangular membrane have the form

$$\omega_{mn} = c\lambda_{mn} = \pi c\sqrt{(m^2/a^2) + (n^2/b^2)} \qquad (m = 1, 2, \ldots;\ n = 1, 2, \ldots),$$

[see (17.13) and (17.14)], with the corresponding periods

$$\tau_{mn} = \frac{2\pi}{\omega_{mn}} = \frac{2}{c\sqrt{(m^2/a^2) + (n^2/b^2)}} \qquad (m = 1, 2, \ldots;\ n = 1, 2, \ldots).$$

There is an important difference between the case of the vibrating membrane and that of the vibrating string: In the case of a string, every characteristic frequency corresponds to its own configuration of nodes, whereas in the case of a membrane, the same characteristic frequency can correspond to several configurations of *nodal lines* (i.e., lines or curves which remain fixed during the course of the motion).

We illustrate this situation, using the particularly simple case of a square membrane, where we set $a = b = 1$. Then, the frequencies of vibration are

$$\omega_{mn} = c\lambda_{mn} = \pi c\sqrt{m^2 + n^2}$$

and instead of (17.14), we can write

$$u_{mn} = H_{mn} \sin(\omega_{mn}t + \alpha_{mn}) \sin \pi mx \sin \pi ny,$$

where

$$H_{mn} = \sqrt{A_{mn}^2 + B_{mn}^2},$$

$$\sin \alpha_{mn} = \frac{A_{mn}}{H_{mn}}, \qquad \cos \alpha_{mn} = \frac{B_{mn}}{H_{mn}}.$$

For the fundamental mode, i.e., for $m = n = 1$, we have

$$\omega_{11} = \pi c\sqrt{2},$$

$$u_{11} = H_{11} \sin(\omega_{11}t + \alpha_{11}) \sin \pi x \sin \pi y,$$

from which it is clear that there are no nodal lines at all. Next, we set $m = 1, n = 2$ or $m = 2, n = 1$. Then, the same frequency

$$\omega = \omega_{12} = \omega_{21} = \pi c\sqrt{5}$$

corresponds to the two modes

$$u_{12} = H_{12} \sin(\omega t + \alpha_{12}) \sin \pi x \sin 2\pi y,$$

$$u_{21} = H_{21} \sin(\omega t + \alpha_{21}) \sin 2\pi x \sin \pi y,$$

the first of which has the nodal line $y = \frac{1}{2}$ and the second the nodal line $x = \frac{1}{2}$. Moreover, the frequency ω also corresponds to the "compound" mode $u_{12} + u_{21}$, which in general leads to a new nodal line. In fact, writing $\alpha_{12} = \alpha_{21} = 0$ for simplicity, we obtain

$$u_{12} + u_{21} = \sin \omega t(H_{12} \sin \pi x \sin 2\pi y + H_{21} \sin 2\pi x \sin \pi y),$$

whence it follows that the points satisfying the equation

$$H_{12} \sin \pi x \sin 2\pi y + H_{21} \sin 2\pi x \sin \pi y = 0 \tag{17.18}$$

form a nodal line. In particular, for $H_{12} = H_{21}$ we obtain

$$\sin \pi x \sin 2\pi y + \sin 2\pi x \sin \pi y = 2 \sin \pi x \sin \pi y (\cos \pi x + \cos \pi y) = 0,$$

which gives the nodal line

$$x + y = 1.$$

Similarly, for $H_{12} = -H_{21}$, we find the nodal line

$$x - y = 0.$$

Thus, if we vary the coefficients H_{12} and H_{21} in the "compound" tone $u_{12} + u_{21}$, equation (17.18) leads to new nodal lines, i.e., an infinite number of nodal lines can correspond to a given frequency. In Fig. 51(a), we show the simple nodal lines just found for the frequency $\omega = \pi c\sqrt{5}$.

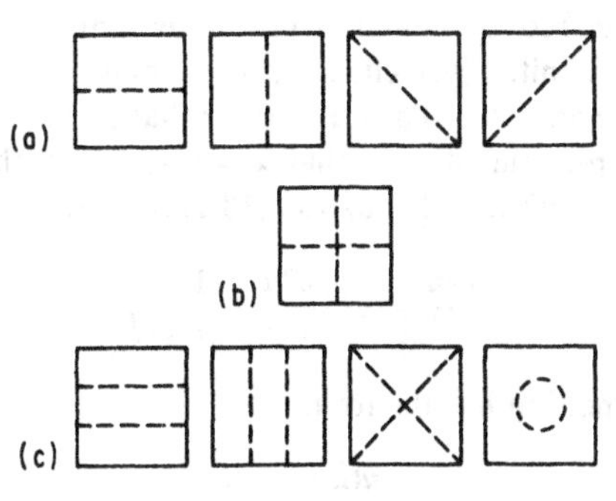

FIGURE 51

If we set $m = 2$, $n = 2$, we obtain the frequency

$$\omega_{22} = \pi c\sqrt{8},$$

corresponding to the unique mode

$$u_{22} = H_{22} \sin (\omega_{22}t + \alpha_{22}) \sin 2\pi x \sin 2\pi y.$$

In this case, the nodal line is the locus of the points in the plane for which either $x = \frac{1}{2}$ or $y = \frac{1}{2}$ [see Fig. 51(b)]. Finally, in Fig. 51(c), we show the simplest nodal lines corresponding to the frequency

$$\omega = \omega_{13} = \omega_{31} = \pi c\sqrt{10}.$$

18. Radial Vibrations of a Circular Membrane

Next, we consider a circular membrane of radius l. For convenience, we introduce polar coordinates in the xy-plane, choosing the center of the membrane as the origin. The change of variables

$$x = r\cos\theta, \quad y = r\sin\theta$$

transforms equation (17.1) into

$$\frac{\partial^2 u}{\partial t^2} = c^2\left(\frac{\partial^2 u}{\partial r^2} + \frac{1}{r}\frac{\partial u}{\partial r} + \frac{1}{r^2}\frac{\partial^2 u}{\partial\theta^2}\right), \tag{18.1}$$

and equation (17.2) into

$$\frac{\partial^2 u}{\partial t^2} = c^2\left(\frac{\partial^2 u}{\partial r^2} + \frac{1}{r}\frac{\partial u}{\partial r} + \frac{1}{r^2}\frac{\partial^2 u}{\partial\theta^2}\right) + \frac{F(r, \theta, t)}{\rho}. \tag{18.2}$$

We shall discuss only the case of free vibrations of a circular membrane, i.e., the case described by equation (18.1), assuming first that the initial displacements and initial velocities of the points of the membrane are independent of the angle θ. Then, it is clear that at any time t, the displacement is also independent of θ, so that $u = u(r, t)$. In this case, the vibrations are said to be *radial*, and equation (18.1) reduces to

$$\frac{\partial^2 u}{\partial t^2} = c^2\left(\frac{\partial^2 u}{\partial r^2} + \frac{1}{r}\frac{\partial u}{\partial r}\right), \tag{18.3}$$

The boundary condition has the form

$$u(l, t) = 0, \tag{18.4}$$

and the initial conditions have the form

$$u(r, 0) = f(r),$$
$$\frac{\partial u(r, 0)}{\partial t} = g(r). \tag{18.5}$$

Following our previous method, we look for solutions of equation (18.3) of the form

$$u(r, t) = R(r)T(t),$$

which satisfy the condition (18.4). Differentiating this expression and substituting the result in (18.3), we obtain

$$RT'' = c^2\left(R''T + \frac{1}{r}R'T\right),$$

whence

$$\frac{R'' + (1/r)R'}{R} = \frac{T''}{c^2T} = -\lambda^2 = \text{const},$$

so that

$$R'' + \frac{1}{r}R' + \lambda^2 R = 0, \tag{18.6}$$

$$T'' + c^2\lambda^2 T = 0. \tag{18.7}$$

Equation (18.6) is the parametric form of Bessel's equation, with index $p = 0$ (see Ch. 8, Sec. 11), and has the general solution

$$R(r) = C_1 J_0(\lambda r) + C_2 Y_0(\lambda r)$$

(see Ch. 8, Sec. 6). Since $Y_0(\lambda r)$ is unbounded at $r = 0$, we must set $C_2 = 0$. Moreover, to obtain a solution different from $R \equiv 0$, we have to assume that $C_1 \neq 0$. Then, it follows from the boundary condition (18.4) that

$$J_0(\lambda l) = 0,$$

i.e., $\mu = \lambda l$ must be a zero of the function $J_0(\mu)$. Setting $C_1 = 1$, we obtain

$$\lambda_n = \frac{\mu_n}{l},$$

$$R_n(r) = J_0(\lambda_n r) = J_0\left(\frac{\mu_n r}{l}\right) \qquad (n = 1, 2, \ldots), \tag{18.8}$$

where $\mu_n = \lambda_n l$ is the nth zero of the function $J_0(\mu)$.

For $\lambda = \lambda_n$, equation (18.7) has the solution

$$T_n(t) = A_n \cos c\lambda_n t + B_n \sin c\lambda_n t \qquad (n = 1, 2, \ldots).$$

Thus, we have finally found particular solutions of (18.3) of the form

$$u_n(r, t) = (A_n \cos c\lambda_n t + B_n \sin c\lambda_n t)J_0(\lambda_n r) \qquad (n = 1, 2, \ldots), \tag{18.9}$$

satisfying the boundary condition (18.4).

To find a solution of (18.3) satisfying both the boundary condition (18.4) and the initial conditions (18.5), we write

$$u(r, t) = \sum_{n=1}^{\infty} (A_n \cos c\lambda_n t + B_n \sin c\lambda_n t)J_0(\lambda_n r), \tag{18.10}$$

and require that

$$u(r, 0) = \sum_{n=1}^{\infty} A_n J_0(\lambda_n r) = f(r),$$

$$\frac{\partial u(r, 0)}{\partial t} = \left[\sum_{n=1}^{\infty} (-A_n c\lambda_n \sin c\lambda_n t + B_n c\lambda_n \cos c\lambda_n t) J_0(\lambda_n r) \right]_{t=0}$$

$$= \sum_{n=1}^{\infty} B_n c\lambda_n J_0(\lambda_n r) = g(r).$$

A calculation of the Fourier coefficients of $f(r)$ and $g(r)$ with respect to the system $\{J_0(\lambda_n r)\}$ gives (see Ch. 8, Sec. 24)

$$A_n = \frac{2}{l^2 J_1^2(\mu_n)} \int_0^l r f(r) J_0(\lambda_n r)\, dr,$$
$$B_n c\lambda_n = \frac{2}{l^2 J_1^2(\mu_n)} \int_0^l r g(r) J_0(\lambda_n r)\, dr, \tag{18.11}$$

or

$$B_n = \frac{2}{c l^2 \lambda_n J_1^2(\mu_n)} \int_0^l r g(r) J_0(\lambda_n r)\, dr. \tag{18.12}$$

Thus, the solution of our problem is given by the series (18.10), where the coefficients A_n and B_n are determined from the formulas (18.11) and (18.12).

The separate harmonic vibrations (18.9) which make up the complicated vibrational motion of the membrane, can be represented in the form

$$u_n(r, t) = H_n \sin(c\lambda_n t + \alpha_n) J_0(\lambda_n r),$$

where

$$H_n = \sqrt{A_n^2 + B_n^2}, \quad \sin \alpha_n = \frac{A_n}{H_n}, \quad \cos \alpha_n = \frac{B_n}{H_n},$$

and the characteristic frequencies of the membrane are

$$\omega_n = c\lambda_n = c\frac{\mu_n}{l}.$$

The amplitude of vibration of the nth mode is

$$H_n |J_0(\lambda_n r)|,$$

and depends only on r. The nodal lines are obtained from the equation

$$J_0(\lambda_n r) = J\left(\frac{\mu_n r}{l}\right) = 0 \qquad (0 \leqslant r < l)$$

[see (18.8)]. If $n = 1$, there are no nodal lines. If $n = 2$, a nodal line is obtained for $r = \mu_1 l/\mu_2$, if $n = 3$, nodal lines are obtained for $r = \mu_1 l/\mu_3$, $r = \mu_2 l/\mu_3$, etc. In Fig. 52, we show the nodal lines and corresponding

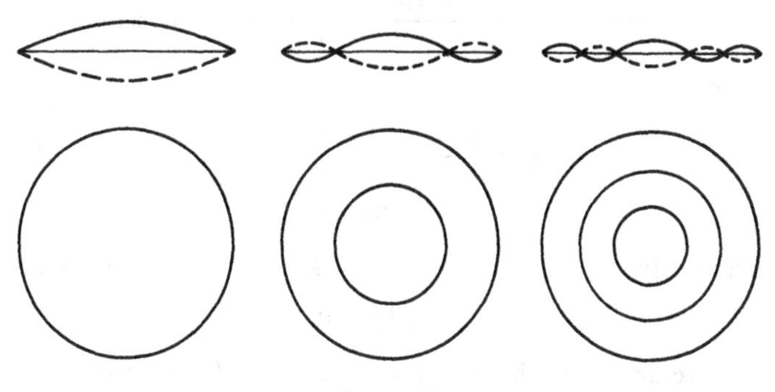

FIGURE 52

sections of a membrane whose vibrations are described by (18.9), for $n = 1, 2, 3$.

19. Vibrations of a Circular Membrane (General Case)

In the general case, the problem of the free vibrations of a circular membrane of radius l reduces to solving the equation

$$\frac{\partial^2 u}{\partial t^2} = c^2 \left(\frac{\partial^2 u}{\partial r^2} + \frac{1}{r}\frac{\partial u}{\partial r} + \frac{1}{r^2}\frac{\partial^2 u}{\partial \theta^2} \right) \tag{19.1}$$

(see beginning of Sec. 18), with the boundary condition

$$u(l, \theta, t) = 0, \tag{19.2}$$

and with the initial conditions

$$u(r, \theta, 0) = f(r, \theta),$$

$$\frac{\partial u(r, \theta, 0)}{\partial t} = g(r, \theta). \tag{19.3}$$

We look for particular solutions in the form of a product

$$u(r, \theta, t) = \Phi(r, \theta)T(t),$$

which are not identically zero and which satisfy the boundary condition (19.2). Substituting this expression in (19.1), we obtain

$$\Phi T'' = c^2 \left(\frac{\partial^2 \Phi}{\partial r^2} + \frac{1}{r}\frac{\partial \Phi}{\partial r} + \frac{1}{r^2}\frac{\partial^2 \Phi}{\partial \theta^2} \right) T,$$

so that[15]

$$\frac{\dfrac{\partial^2\Phi}{\partial r^2} + \dfrac{1}{r}\dfrac{\partial\Phi}{\partial r} + \dfrac{1}{r^2}\dfrac{\partial^2\Phi}{\partial\theta^2}}{\Phi} = \frac{T''}{c^2T} = -\lambda^2 = \text{const.}^{16}$$

Therefore

$$\frac{\partial^2\Phi}{\partial r^2} + \frac{1}{r}\frac{\partial\Phi}{\partial r} + \frac{1}{r^2}\frac{\partial^2\Phi}{\partial\theta^2} = -\lambda^2\Phi, \tag{19.4}$$

$$T'' + c^2\lambda^2 T = 0, \tag{19.5}$$

where to satisfy the boundary condition (19.2), we must require that

$$\Phi(l, \theta) = 0. \tag{19.6}$$

Thus, we arrive at a boundary value problem for equation (19.4). To solve this problem, we look for particular solutions of (19.4) of the form

$$\Phi(r, \theta) = R(r)F(\theta), \tag{19.7}$$

which are not identically zero and which satisfy the condition (19.6). Substituting (19.7) in (19.4) gives

$$R''F + \frac{1}{r}R'F + \frac{1}{r^2}RF'' + \lambda^2 RF = 0,$$

$$-\frac{R'' + (1/r)R' + \lambda^2 R}{(1/r^2)R} = \frac{F''}{F} = -\nu^2 = \text{const.}^{16}$$

Therefore, we have

$$r^2R'' + rR' + (\lambda^2r^2 - \nu^2)R = 0, \tag{19.8}$$

$$F'' + \nu^2 F = 0. \tag{19.9}$$

Equation (19.9) has solutions of the form $\cos \nu\theta$ and $\sin \nu\theta$. Since if we change θ by 2π, we come back to the same point of the membrane, the function u, and hence Φ and F, must have period 2π. Therefore, the number ν must be an integer, i.e.,

$$\nu = n, \qquad (n = 0, 1, 2, \ldots),$$

and the solutions of (19.9) are[17]

$$\cos n\theta,\ \sin n\theta, \qquad (n = 0, 1, 2, \ldots). \tag{19.10}$$

---

[15] We take the constant to be negative, since otherwise the function T would not turn out to be periodic, and we would not have vibrations, contrary to what actually happens.

[16] The constant is taken to be negative, since in a problem like this, the function $F(\theta)$ must be periodic [see (19.9) *et seq.*].

[17] For $\nu = 0$, equation (19.9) has another solution of the form $C\theta$, which can be discarded, since it is not periodic.

(The negative values of n give the same functions, to within a constant factor.)

Thus, equation (19.8) takes the form

$$r^2R'' + rR' + (\lambda^2r^2 - n^2)R = 0.$$

This equation is the parametric form of Bessel's equation, with integral index n, and has the general solution

$$R(r) = C_1J_n(\lambda r) + C_2Y_n(\lambda r).$$

Since the solution must be bounded, we are compelled to set $C_2 = 0$, because $Y_n(\lambda r) \to \infty$ as $r \to 0$ (see Ch. 8, Sec. 6). If we put $C_1 = 1$, then by the boundary condition (19.6)

$$R(l) = J_n(\lambda l) = 0,$$

i.e., $\lambda l = \mu$ must be a zero of the function $J_n(\mu)$.

We now write

$$\lambda_{nm} = \frac{\mu_{nm}}{l},$$

$$R_{nm} = J_n(\lambda_{nm}r) = J_n\left(\frac{\mu_{nm}r}{l}\right) \tag{19.11}$$

$$(m = 1, 2, \ldots;\ n = 0, 1, 2, \ldots),$$

where μ_{nm} is the mth positive zero (in order of increasing size) of the function $J_n(\mu)$. Then, the boundary value problem for equation (19.4), subject to the boundary condition (19.6), has the eigenvalues λ_{nm} [see (19.11)], and the eigenfunctions [see (19.7), (19.10), (19.11)]

$$\begin{aligned} \Phi_{nm}(r, \theta) &= J_n(\lambda_{nm}r) \cos n\theta, \\ \Phi^*_{nm}(r, \theta) &= J_n(\lambda_{nm}r) \sin n\theta \end{aligned} \qquad (n = 1, 2, \ldots;\ m = 1, 2, \ldots).$$

For $\lambda = \lambda_{nm}$, equation (19.5) gives

$$T_{nm} = A_{nm} \cos c\lambda_{nm}t + B_{nm} \sin c\lambda_{nm}t.$$

Therefore, we have the following particular solutions of (19.1), satisfying the boundary conditions (19.2):

$$u_{nm}(r, \theta, t) = (A_{nm} \cos c\lambda_{nm}t + B_{nm} \sin c\lambda_{nm}t)J_n(\lambda_{nm}r) \cos n\theta$$
$$(m = 1, 2, \ldots;\ n = 0, 1, 2, \ldots),$$

$$u^*_{nm}(r, \theta, t) = (A^*_{nm} \cos c\lambda_{nm}t + B^*_{nm} \sin c\lambda_{nm}t)J_n(\lambda_{nm}r) \sin n\theta$$
$$(m = 1, 2, \ldots;\ n = 1, 2, \ldots).$$

To obtain a solution of equation (19.1) which also satisfies the boundary conditions (19.3), we form the series

$$u(r, \theta, t) = \sum_{n=0}^{\infty} \sum_{m=1}^{\infty} [(A_{nm} \cos c\lambda_{nm}t + B_{nm} \sin c\lambda_{nm}t) \cos n\theta + (A_{nm}^* \cos c\lambda_{nm}t + B_{nm}^* \sin c\lambda_{nm}t) \sin n\theta] J_n(\lambda_{nm}r) \tag{19.12}$$

and require that

$$u(r, \theta, 0) = \sum_{n=0}^{\infty} \sum_{m=1}^{\infty} (A_{nm} \cos n\theta + A_{nm}^* \sin n\theta) J_n(\lambda_{nm}r) = f(r, \theta),$$

$$\begin{aligned} \frac{\partial u(r, \theta, 0)}{\partial t} &= \Big\{ \sum_{n=0}^{\infty} \sum_{m=1}^{\infty} [(-A_{nm}c\lambda_{nm} \sin c\lambda_{nm}t + B_{nm}c\lambda_{nm} \cos c\lambda_{nm}t) \cos n\theta \\ &\qquad + (-A_{nm}^*c\lambda_{nm} \sin c\lambda_{nm}t \\ &\qquad + B_{nm}^*c\lambda_{nm} \cos c\lambda_{nm}t) \sin n\theta] J_n(\lambda_{nm}r) \Big\}_{t=0} \\ &= \sum_{n=0}^{\infty} \sum_{m=1}^{\infty} (B_{nm}c\lambda_{nm} \cos n\theta \\ &\qquad + B_{nm}^*c\lambda_{nm} \sin n\theta) J_n(\lambda_{nm}r) = g(r, \theta). \end{aligned} \tag{19.13}$$

To find the coefficients in these expansions, we now argue as follows: Let

$$f(r, \theta) = \sum_{n=0}^{\infty} [f_n(r) \cos n\theta + f_n^*(r) \sin n\theta], \tag{19.14}$$

where

$$\begin{aligned} f_0(r) &= \frac{1}{2\pi} \int_{-\pi}^{\pi} f(r, \theta)\, d\theta, \\ f_n(r) &= \frac{1}{\pi} \int_{-\pi}^{\pi} f(r, \theta) \cos n\theta\, d\theta, \\ & \qquad\qquad (n = 1, 2, \ldots) \\ f_n^*(r) &= \frac{1}{\pi} \int_{-\pi}^{\pi} f(r, \theta) \sin n\theta\, d\theta \end{aligned} \tag{19.15}$$

i.e., we expand the function $f(r, \theta)$ in a trigonometric Fourier series with respect to the variable θ. We then expand each of the functions $f_n(r)$ and $f_n^*(r)$ in Fourier series with respect to the system $\{J_n(\lambda_{nm}r)\}$. The result is

$$f_n(r) = \sum_{m=1}^{\infty} C_{nm} J_n(\lambda_{nm}r), \quad f_n^*(r) = \sum_{m=1}^{\infty} C_{nm}^* J_n(\lambda_{nm}r),$$

where

$$C_{nm} = \frac{2}{l^2 J_{n+1}^2(\mu_{nm})} \int_0^l r f_n(r) J_n(\lambda_{nm} r)\, dr,$$
$$C_{nm}^* = \frac{2}{l^2 J_{n+1}^2(\mu_{nm})} \int_0^l r f_n^*(r) J_n(\lambda_{nm} r)\, dr. \tag{19.16}$$

Substituting (19.16) in (19.14) gives

$$f(r, \theta) = \sum_{n=0}^{\infty} \sum_{m=1}^{\infty} (C_{nm} \cos n\theta + C_{nm}^* \sin n\theta) J_n(\lambda_{nm} r).$$

Comparing this with the first of the formulas (19.13), and using (19.15) and (19.16), we obtain

$$A_{0m} = C_{0m} = \frac{1}{\pi l^2 J_1^2(\mu_{0m})} \int_0^l dr \int_{-\pi}^{\pi} r f(r, \theta) J_0(\lambda_{0m} r)\, d\theta,$$
$$A_{nm} = C_{nm}$$
$$= \frac{2}{\pi l^2 J_{n+1}^2(\mu_{nm})} \int_0^l dr \int_{-\pi}^{\pi} r f(r, \theta) \cos n\theta J_n(\lambda_{nm} r)\, d\theta$$
$$(n = 1, 2, \ldots;\ m = 1, 2, \ldots), \tag{19.17}$$
$$A_{nm}^* = C_{nm}^*$$
$$= \frac{2}{\pi l^2 J_{n+1}^2(\mu_{nm})} \int_0^l dr \int_{-\pi}^{\pi} r f(r, \theta) \sin n\theta J_n(\lambda_{nm} r)\, d\theta$$
$$(n = 1, 2, \ldots;\ m = 1, 2, \ldots).$$

In just the same way, we find that

$$B_{0m} c\lambda_{0m} = \frac{1}{\pi l^2 J_1^2(\mu_{0m})} \int_0^l dr \int_{-\pi}^{\pi} r g(r, \theta) J_0(\lambda_{0m} r)\, d\theta,$$
$$B_{nm} c\lambda_{nm} = \frac{2}{\pi l^2 J_{n+1}^2(\mu_{nm})} \int_0^l dr \int_{-\pi}^{\pi} r g(r, \theta) \cos n\theta J_n(\lambda_{nm} r)\, d\theta$$
$$(n = 1, 2, \ldots;\ m = 1, 2, \ldots), \tag{19.18}$$
$$B_{nm}^* c\lambda_{nm} = \frac{2}{\pi l^2 J_{n+1}^2(\mu_{nm})} \int_0^l dr \int_{-\pi}^{\pi} r g(r, \theta) \sin n\theta J_n(\lambda_{nm} r)\, d\theta$$
$$(n = 1, 2, \ldots;\ m = 1, 2, \ldots).$$

Thus, the solution of our problem is given by the series (19.12), where the coefficients are determined from formulas (19.17) and (19.18).

There are a great variety of nodal lines corresponding to the simple harmonic vibrations u_{nm} and u_{nm}^*, which make up the complex vibration (19.12). For example, for u_{01}, u_{02}, and u_{03}, one finds the nodal lines shown in Fig. 52, while for u_{12}, u_{22}, and u_{32}, one finds the nodal lines shown in Fig. 53.

As in the case of a rectangular membrane, an infinite set of different configurations of nodal lines can correspond to the same frequency (depending on the coefficients A_{nm}, B_{nm}, A^*_{nm}, B^*_{nm}).

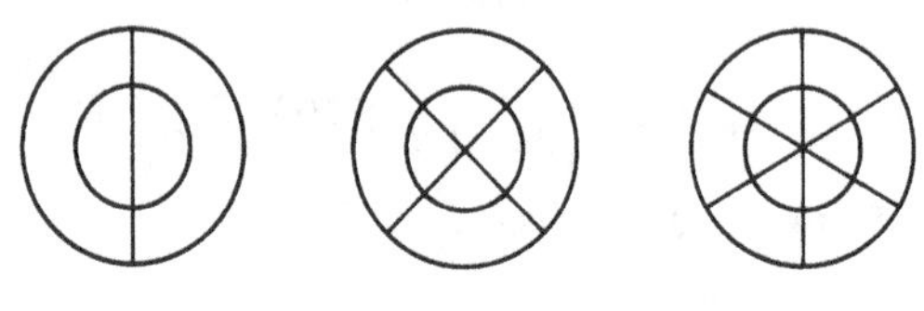

FIGURE 53

20. Equation of Heat Flow in a Rod

Consider a homogeneous cylindrical rod, whose lateral surface is insulated from the surrounding medium. Choose the x-axis along the axis of the rod, and let $u(x, t)$ denote the temperature at time t of the cross section of the rod with abscissa x. Let AB be the element of the rod lying between x and $x + \Delta x$ (see Fig. 49). Let Δt be a time interval which is so small that we can assume that the temperature of the cross sections at x and $x + \Delta x$ does not change (in time). It has been established experimentally that the amount of heat q flowing through a rod whose ends are held at two constant temperatures is proportional to the difference between the temperatures, to the cross-sectional area of the rod and to the time interval Δt, while q is inversely proportional to the length of the rod. Therefore, the amount of heat flowing through the element AB is

$$q = \frac{K[u(x + \Delta x, t) - u(x, t)]s\,\Delta t}{\Delta x} = \frac{K\,\partial u(x + \theta\Delta x, t)}{\partial x}\,s\,\Delta t \qquad (0 < \theta < 1),$$

where K is a constant of proportionality called the *thermal conductivity* of the material from which the rod is made, and s is the cross-sectional area of the rod. In the limit as $\Delta x \to 0$, we obtain the amount of heat $Q(x)$ flowing through the cross section at x in the time Δt:

$$Q(x) = K\frac{\partial u}{\partial t}\,s\,\Delta t. \tag{20.1}$$

Now, it can easily be seen that the amount of heat ΔQ which the element AB receives in the time Δt is

$$\begin{aligned} \Delta Q &= Q(x + \Delta x) - Q(x) \\ &= K\,s\,\Delta t\left[\frac{\partial u(x + \Delta x, t)}{\partial x} - \frac{\partial u(x, t)}{\partial x}\right] \\ &= K\,s\,\Delta t\,\Delta x\,\frac{\partial^2 u(x + \theta_1\Delta x, t)}{\partial x^2} \qquad (0 < \theta_1 < 1). \end{aligned} \tag{20.2}$$

(It should be kept in mind that heat flows in the direction opposite to the direction in which the temperature increases.) The quantity ΔQ can also be calculated by another method: Suppose that the element AB is so small that at any given time the temperature of all its cross sections can be regarded as being the same. Then

$$\Delta Q = c\rho s\,\Delta x\,[u(x, t + \Delta t) - u(x, t)]$$
$$= c\rho s\,\Delta x\,\Delta t\,\frac{\partial u(x, t + \theta_2\,\Delta t)}{\partial t} \qquad (0 < \theta_2 < 1), \tag{20.3}$$

where c is the heat capacity and ρ is the density (per unit length) of the material from which the rod is made. (Thus, $\rho s\,\Delta x$ is the mass of the element AB.)

A comparison of (20.2) and (20.3) shows that

$$c\rho\,\frac{\partial u(x, t + \theta_2\,\Delta t)}{\partial t} = K\,\frac{\partial^2 u(x + \theta_1\,\Delta x, t)}{\partial x^2},$$

and if we pass to the limit as $\Delta t \to 0$, $\Delta x \to 0$, then

$$\frac{\partial u}{\partial t} = a^2\,\frac{\partial^2 u}{\partial x^2}, \tag{20.4}$$

where $a^2 = K/c\rho$. In this way, we obtain the equation for heat flow (or heat conduction) in a rod.

We now pose a variety of problems, corresponding to different conditions imposed on the ends of the rod.

21. Heat Flow in a Rod With Ends Held at Zero Temperature

This problem consists in finding the solution of equation (10.4), with the boundary conditions

$$u(0, t) = u(l, t) = 0 \tag{21.1}$$

(the ends of the rod are at $x = 0$ and $x = l$), and with the initial condition

$$u(x, 0) = f(x), \tag{21.2}$$

where $f(x)$ is a given function. Equation (20.4) is a special case of equation (1.2) of Part I, and hence all the considerations given there apply to the present problem.

Thus, we look for particular solutions of the form

$$u(x, t) = \Phi(x)T(t)$$

which do not vanish identically and which satisfy the boundary conditions (21.1). Substituting this expression in (20.4) gives

$$\Phi T' = a^2\Phi'' T,$$

whence[18]

$$\frac{\Phi''}{\Phi} = \frac{T'}{a^2 T} = -\lambda^2 = \text{const.}$$

Therefore, we have

$$\Phi'' + \lambda^2\Phi = 0, \tag{21.3}$$

$$T' + a^2\lambda^2 T = 0. \tag{21.4}$$

The solution of (21.3) is

$$\Phi(x) = C_1 \cos \lambda x + C_2 \sin \lambda x,$$

and because of the condition (21.1), we must require that

$$\Phi(0) = C_1 = 0,$$

$$\Phi(l) = C_2 \sin \lambda l = 0.$$

Therefore, assuming that $C_2 \neq 0$, we obtain $\lambda l = \pi n$, where n is an integer. If we set $C_2 = 1$, then

$$\lambda_n = \frac{\pi n}{l},$$

$$\Phi_n(x) = \sin \lambda_n x = \sin \frac{\pi n x}{l} \qquad (n = 1, 2, \ldots).$$

For $\lambda = \lambda_n$, equation (21.4) gives

$$T_n(t) = A_n e^{-a^2\lambda_n^2 t} = A_n e^{-\frac{a^2\pi^2 n^2}{l^2} t}, \quad A_n = \text{const} \qquad (n = 1, 2, \ldots).$$

Hence, the functions

$$u_n(x, t) = A_n \sin \frac{\pi n x}{l} e^{-\frac{a^2\pi^2 n^2}{l^2} t} \qquad (n = 1, 2, \ldots)$$

represent particular solutions of (20.4) which satisfy the boundary conditions.

To satisfy the initial condition, we form the series

$$u(x, t) = \sum_{n=1}^{\infty} A_n \sin \frac{\pi n x}{l} e^{-\frac{a^2\pi^2 n^2}{l^2} t} \tag{21.5}$$

---

[18] We leave it to the reader to decide why the constant is chosen to be negative here (see Sec. 12).

and require that

$$u(x, 0) = \sum_{n=1}^{\infty} A_n \sin \frac{\pi n x}{l} = f(x).$$

Thus, we have to expand $f(x)$ with respect to the system $\{\sin (\pi n x/l)\}$. A calculation of the Fourier coefficients gives

$$A_n = \frac{2}{l}\int_0^l f(x) \sin \frac{\pi n x}{l}\, dx \qquad (n = 1, 2, \ldots), \tag{21.6}$$

i.e., the solution of our problem is given by the series (21.5), where the coefficients A_n are determined from the formulas (21.6). Because of the presence of the factors

$$e^{-\frac{a^2\pi^2 n^2}{l^2}t}$$

it is easy to see that the series (21.5) is uniformly convergent for $t \geqslant t_0 > 0$, for any $t_0 > 0$. The same is true for the series obtained by term by term differentiation of (21.5) with respect to x and t (any number of times). Therefore, the sum of the series (21.5) is continuous, and the term by term differentiation is legitimate (cf. Sec. 10).

22. Heat Flow in a Rod with Ends Held at Constant Temperatures

This problem consists in finding the solution of equation (20.4), with the boundary conditions[19]

$$u(0, t) = A = \text{const}, \quad u(l, t) = B = \text{const}, \tag{22.1}$$

and with the initial condition

$$u(x, 0) = f(x). \tag{22.2}$$

We look for a solution in the form of a series

$$u(x, t) = \sum_{n=1}^{\infty} T_n(t) \sin \frac{\pi n x}{l}, \tag{22.3}$$

where

$$T_n(t) = \frac{2}{l}\int_0^l u(x, t) \sin \frac{\pi n x}{l}\, dx. \tag{22.4}$$

[19] The boundary conditions in this problem, and also in the problem of Sec. 23, have a different form from those considered previously. We show below how to deal with cases like these.

Integrating by parts twice, we obtain

$$\frac{l}{2} T_n = \left[-\frac{lu(x, t)}{\pi n} \cos \frac{\pi n x}{l} \right]_{x=0}^{x=l} + \left[\frac{l^2}{\pi^2 n^2} \frac{\partial u(x, t)}{\partial x} \sin \frac{\pi n x}{l} \right]_{x=0}^{x=l} - \frac{l^2}{\pi^2 n^2} \int_0^l \frac{\partial^2 u}{\partial x^2} \sin \frac{\pi n x}{l} \, dx.$$

Since $u(x, t)$ satisfies equation (20.4) and the conditions (22.1), we have

$$\frac{1}{2} T_n = \frac{1}{\pi n} [A - (-1)^n B] - \frac{l}{a^2 \pi^2 n^2} \int_0^l \frac{\partial u}{\partial t} \sin \frac{\pi n x}{l} \, dx.$$

Differentiating (22.4) with respect to t, we obtain

$$T'_n = \frac{2}{l} \int_0^l \frac{\partial u}{\partial t} \sin \frac{\pi n x}{l} \, dx,$$

so that

$$\frac{1}{2} T_n = \frac{1}{\pi n} [A - (-1)^n B] - \frac{l^2}{2 a^2 \pi^2 n^2} T'_n,$$

or

$$T'_n + \frac{a^2 \pi^2 n^2}{l^2} T_n = \frac{2 a^2 \pi n}{l^2} [A - (-1)^n B]. \tag{22.5}$$

This equation has the solution

$$T_n = A_n e^{-\frac{a^2 \pi^2 n^2}{l^2} t} + 2 \frac{A - (-1)^n B}{\pi n}. \tag{22.6}$$

To satisfy the initial condition (22.2), we require that

$$u(x, 0) = \sum_{n=1}^{\infty} T_n(0) \sin \frac{\pi n x}{l} = f(x).$$

A calculation of the Fourier coefficients of $f(x)$ with respect to the system $\{\sin (\pi n x / l)\}$ gives

$$T_n(0) = A_n + 2 \frac{A - (-1)^n B}{\pi n} = \frac{2}{l} \int_0^l f(x) \sin \frac{\pi n x}{l} \, dx,$$

and hence

$$A_n = \frac{2}{l} \int_0^l f(x) \sin \frac{\pi n x}{l} \, dx - 2 \frac{A - (-1)^n B}{\pi n}. \tag{22.7}$$

Thus, the solution of our problem is given by the series (22.3), where the T_n are determined from the formulas (22.6) and (22.7).

23. Heat Flow in a Rod Whose Ends are at Specified Variable Temperatures

In this problem, it is required to find the solution of equation (20.4), with the boundary conditions

$$u(0, t) = \varphi(t), \qquad u(l, t) = \psi(t) \tag{23.1}$$

(where φ and ψ are given), and with the initial condition

$$u(x, 0) = f(x). \tag{23.2}$$

We again look for a solution in the form of a series (22.3). Repeating the argument of Sec. 22, we find for $T_n(t)$ the equation

$$T_n' + \frac{a^2\pi^2n^2}{l^2} T_n = \frac{2a^2\pi n}{l^2} [\varphi(t) - (-1)^n\psi(t)]$$

which is the same as formula (22.5), except that A and B have been replaced by φ and ψ. Solving this equation, we obtain

$$\begin{aligned} T_n = A_n e^{-\frac{a^2\pi^2n^2}{l^2}t} \\ + \frac{2a^2\pi n}{l^2} e^{-\frac{a^2\pi^2n^2}{l^2}t} \int_0^t e^{\frac{a^2\pi^2n^2}{l^2}t} [\varphi(t) - (-1)^n\psi(t)]\, dt. \end{aligned} \tag{23.3}$$

To satisfy the condition (23.2), we require that

$$u(x, 0) = \sum_{n=1}^{\infty} T_n(0) \sin\frac{\pi nx}{l} = f(x).$$

A calculation of the Fourier coefficients of $f(x)$ with respect to the system $\sin\{(\pi nx/l)\}$ gives

$$T_n(0) = A_n = \frac{2}{l}\int_0^l f(x) \sin\frac{\pi nx}{l}\, dx. \tag{23.4}$$

Therefore, the solution of our problem is given by the series (22.3), where T_n is determined from the formulas (23.3) and (23.4).

24. Heat Flow in a Rod Whose Ends Exchange Heat Freely with the Surrounding Medium

If the surface of a body exchanges heat with a surrounding gaseous medium, then the amount of heat flowing through an area s in the time Δt is given by the formula

$$Q = H(u - u_0)s\,\Delta t, \tag{24.1}$$

where u is the temperature of the body, u_0 is the temperature of the surrounding medium and H is a constant called the *emissivity*.

In the case of heat flow in a rod whose lateral surface is insulated, but whose ends exchange heat freely with the surrounding medium, a comparison of (20.1) and (24.1) leads to the boundary conditions

$$H(u - u_0) = K\frac{\partial u}{\partial x}$$

for $x = 0$, and

$$H(u - u_0) = -K\frac{\partial u}{\partial x}$$

for $x = l$. Setting $h = H/K$ ($h > 0$), these conditions become

$$\left[\frac{\partial u}{\partial x} - h(u - u_0)\right]_{x=0} = 0, \quad \left[\frac{\partial u}{\partial x} + h(u - u_0)\right]_{x=l} = 0. \tag{24.2}$$

First, let us assume that $u_0 = 0$. The boundary conditions then take the form

$$\left[\frac{\partial u}{\partial x} - hu\right]_{x=0} = 0, \quad \left[\frac{\partial u}{\partial x} + hu\right]_{x=l} = 0, \tag{24.3}$$

while the initial condition is

$$u(x, 0) = f(x), \tag{24.4}$$

as before. Following our usual method, we look for particular solutions of equation (20.4) of the form

$$u(x, t) = \Phi(x)T(t),$$

satisfying the conditions (24.3). Substituting this expression in (20.4) gives

$$\Phi T' = a^2\Phi'' T,$$

whence

$$\frac{\Phi''}{\Phi} = \frac{T'}{a^2T} = -\lambda = \text{const}, \tag{24.5}$$

so that

$$\Phi'' = -\lambda\Phi, \quad T' = -a^2\lambda T. \tag{24.6}$$

To satisfy the boundary conditions (24.3), we must obviously require that

$$\begin{aligned} \Phi'(0) - h\Phi(0) &= 0, \\ \Phi'(l) + h\Phi(l) &= 0. \end{aligned} \tag{24.7}$$

Thus, we have arrived at a boundary value problem for equation (24.6), with the boundary conditions (24.7). According to Sec. 5, all the eigenvalues of this boundary value problem are positive. Therefore, we can write λ^2 instead of λ; thus, instead of the equations (24.6), we obtain

$$\Phi'' + \lambda^2\Phi = 0, \tag{24.8}$$

$$T' + a^2\lambda^2 T = 0. \tag{24.9}$$

The solution of (24.8) is

$$\Phi(x) = C_1 \cos \lambda x + C_2 \sin \lambda x,$$

and by (24.7), we must have

$$C_2\lambda - hC_1 = 0,$$

$$\lambda(-C_1 \sin \lambda l + C_2 \cos \lambda l) + h(C_1 \cos \lambda l + C_2 \sin \lambda l) = 0.$$

Therefore

$$\frac{C_1}{\lambda} = \frac{C_2}{h}, \tag{24.10}$$

so that

$$\tan \lambda l = \frac{2\lambda h}{\lambda^2 - h^2}. \tag{24.11}$$

The positive roots of this equation give the eigenvalues of our problem. It is worth pointing out that these roots are the abscissas of the points of intersection (in the $\mu\lambda$-plane) of the function $\mu = 2/\tan \lambda l$ and the hyperbola $\mu = (\lambda^2 - h^2)/\lambda h$.

Now, let λ_n be the nth positive root of equation (24.11). According to (24.10), we can set $C_1 = \lambda_n$, $C_2 = h$. Then, the eigenfunctions become

$$\Phi_n(x) = \lambda_n \cos \lambda_n x + h \sin \lambda_n x \qquad (n = 1, 2, \ldots).$$

For $\lambda = \lambda_n$, the solution of equation (24.9) is

$$T_n(t) = A_n e^{-a^2\lambda_n^2 t} \qquad (n = 1, 2, \ldots).$$

This leads to the following particular solutions of (20.4):

$$u_n(x, t) = A_n(\lambda_n \cos \lambda_n x + h \sin \lambda_n x)e^{-a^2\lambda_n^2 t} \qquad (n = 1, 2, \ldots).$$

To satisfy the initial condition (24.4), we form the series

$$u(x, t) = \sum_{n=1}^{\infty} A_n(\lambda_n \cos \lambda_n x + h \sin \lambda_n x)e^{-a^2\lambda_n^2 t}, \tag{24.12}$$

and require that

$$u(x, 0) = \sum_{n=1}^{\infty} A_n(\lambda_n \cos \lambda_n x + h \sin \lambda_n x) = f(x).$$

A calculation of the Fourier coefficients of the function $f(x)$ with respect to the system[20]

$$\{\Phi_n(x)\} = \{\lambda_n \cos \lambda_n x + h \sin \lambda_n x\}$$

gives

$$A_n = \frac{\int_0^l f(x)\Phi_n(x)\, dx}{\int_0^l \Phi_n^2(x)\, dx} \qquad (n = 1, 2, \ldots). \tag{24.13}$$

Thus, the solution of our problem is given by the series (24.12), where the coefficients are determined from (24.13).

The integral in the denominator of (24.13) can be calculated as follows: The equation

$$\Phi_n'' + \lambda_n^2\Phi_n = 0$$

implies that

$$\lambda_n^2\Phi_n^2 = -\Phi_n\Phi_n''.$$

Therefore, we have

$$\lambda_n^2 \int_0^l \Phi_n^2\, dx = -[\Phi_n\Phi_n']_{x=0}^{x=l} + \int_0^l \Phi_n'^2\, dx. \tag{24.14}$$

But

$$\Phi_n = \lambda_n \cos \lambda_n x + h \sin \lambda_n x,$$
$$\Phi_n' = -\lambda_n^2 \sin \lambda_n x + h\lambda_n \cos \lambda_n x,$$

which means that

$$\lambda_n^2\Phi_n^2 + \Phi_n'^2 = \lambda_n^4 + h^2\lambda_n^2, \tag{24.15}$$

and hence

$$\lambda_n^2 \int_0^l \Phi_n^2\, dx + \int_0^l \Phi_n'^2\, dx = (\lambda_n^4 + h^2\lambda_n^2)l.$$

[20] According to Sec. 4, this system is orthogonal.

This, together with (24.14), implies

$$2\lambda_n^2 \int_0^l \Phi_n^2\, dx = (\lambda_n^4 + h^2\lambda_n^2)l - [\Phi_n\Phi_n']_{x=0}^{x=l}. \tag{24.16}$$

On the other hand, it follows from the boundary conditions and (24.15) that

$$\lambda_n^2\Phi_n^2 + h^2\Phi_n^2 = \lambda_n^4 + h^2\lambda_n^2,$$

or

$$\Phi_n^2 = \lambda_n^2, \tag{24.17}$$

at both $x = 0$ and $x = l$. Therefore, writing the boundary conditions in the form

$$[\Phi_n\Phi_n' - h\Phi_n^2]_{x=0} = 0,$$
$$[\Phi_n\Phi_n' + h\Phi_n^2]_{x=l} = 0,$$

and using (24.17), we find that

$$[\Phi_n\Phi_n']_{x=0}^{x=l} = -2h\lambda_n^2.$$

Substituting this expression in (24.16) gives

$$\int_0^l \Phi_n^2\, dx = \frac{(\lambda_n^2 + h^2)l + 2h}{2}.$$

Thus, instead of (24.13), we can write

$$A_n = \frac{2\int_0^l f(x)(\lambda_n \cos \lambda_n x + h \sin \lambda_n x)\, dx}{(\lambda_n^2 + h^2)l + 2h} \qquad (n = 1, 2, \ldots). \tag{24.18}$$

Next, we consider the case where the end of the rod at $x = 0$ exchanges heat with a medium at the temperature u_0, while the end of the rod at $x = l$ exchanges heat with a medium at the temperature u_1. This problem can be reduced to the problem just solved by making the substitution

$$u = v + w,$$

where the function $v = v(x)$ depends only on x and satisfies the equation

$$v'' = 0, \tag{24.19}$$

with the boundary conditions

$$[v' - h(v - u_0)]_{x=0} = 0, \quad [v' + h(v - u_1)]_{x=l} = 0, \tag{24.20}$$

while w satisfies the equation

$$\frac{\partial w}{\partial t} = a^2 \frac{\partial^2 w}{\partial x^2},$$

with the boundary conditions

$$\left[\frac{\partial w}{\partial x} - hw\right]_{x=0} = 0,$$

$$\left[\frac{\partial w}{\partial x} + hw\right]_{x=l} = 0$$

and the initial condition

$$w(x, 0) = f(x) - v(x).$$

According to (24.19), the function $v = v(x)$ has the form

$$v = Ax + B,$$

where the constants A and B are determined from the conditions (24.20). This leads to the system of equations

$$A - h(B - u_0) = 0,$$
$$A + h(Al + B - u_1) = 0,$$

which can be solved very easily. Then, the boundary value problem for w is of the type discussed above.

25. Heat Flow in an Infinite Rod

In the case of an infinite rod, there are no boundary conditions, and the problem reduces to finding a solution of equation (20.4) which is defined for all x and $t > 0$, and satisfies the initial condition

$$u(x, 0) = f(x) \qquad (-\infty < x < \infty). \tag{25.1}$$

As usual, we look for particular solutions of the form

$$u(x, t) = \Phi(x)T(t).$$

Substituting this expression in (20.4) gives

$$\Phi T' = a^2\Phi''T,$$

whence

$$\frac{\Phi''}{\Phi} = \frac{T'}{a^2T} = -\lambda^2 = \text{const}, \tag{25.2}$$

so that

$$\Phi'' + \lambda^2\Phi = 0, \tag{25.3}$$

$$T' + a^2\lambda^2 T = 0. \tag{25.4}$$

These equations have the solutions

$$\Phi(x) = C_1 \cos \lambda x + C_2 \sin \lambda x,$$
$$T(t) = C_3 e^{-a^2\lambda^2 t}.$$

Therefore, the required particular solutions can be written in the form[21]

$$u(x, t; \lambda) = (A \cos \lambda x + B \sin \lambda x)e^{-a^2\lambda^2 t}.$$

The constants A and B, being arbitrary, can be regarded as values of functions $A = A(\lambda)$ and $B = B(\lambda)$. It will be recalled that for discrete values of λ (as in the case of a finite rod), we formed an infinite series of particular solutions, and we then chose the coefficients of the series in such a way as to obtain a solution satisfying the initial condition. In the present case, λ varies continuously, and we set

$$\begin{aligned} u(x, t) &= \int_0^\infty u(x, t; \lambda)\, d\lambda \\ &= \int_0^\infty (A(\lambda) \cos \lambda x + B(\lambda) \sin \lambda x)e^{-a^2\lambda^2 t}\, d\lambda. \end{aligned} \tag{25.5}$$

If we can differentiate this function behind the integral sign (once with respect to t and twice with respect to x), then the function $u(x, t)$ will be a solution of equation (20.4). In fact, we then have

$$\begin{aligned} \frac{\partial u}{\partial t} - a^2 \frac{\partial^2 u}{\partial x^2} &= \int_0^\infty \frac{\partial u(x, t; \lambda)}{\partial t}\, d\lambda - a^2 \int_0^\infty \frac{\partial^2 u(x, t; \lambda)}{\partial x^2}\, d\lambda \\ &= \int_0^\infty \left(\frac{\partial u(x, t; \lambda)}{\partial t} - a^2 \frac{\partial^2 u(x, t; \lambda)}{\partial x^2}\right) d\lambda = 0. \end{aligned}$$

To satisfy the initial condition, we require that

$$u(x, 0) = \int_0^\infty (A(\lambda) \cos \lambda x + B(\lambda) \sin \lambda x)\, d\lambda = f(x).$$

This will be the case if we require that $f(x)$ can be represented as a Fourier integral (see Ch. 7, Sec. 5), a sufficient condition for which is that $f(x)$ be a piecewise smooth and absolutely integrable function on the whole x-axis. With these assumptions

$$\begin{aligned} A(\lambda) &= \frac{1}{\pi} \int_{-\infty}^\infty f(v) \cos \lambda v\, dv, \\ B(\lambda) &= \frac{1}{\pi} \int_{-\infty}^\infty f(v) \sin \lambda v\, dv \end{aligned} \tag{25.6}$$

[see equations (5.5) to (5.7) of Ch. 7].

[21] If we chose the constant to be positive in (25.2), then the exponential factor which falls off as t increases would be replaced by an exponential factor which increases without limit as t increases, contrary to the physical meaning of the problem.

For such $A(\lambda)$ and $B(\lambda)$, the integral (25.5) can be differentiated behind the integral sign any number of times with respect to x and t. In fact, because of the presence of the factor $e^{-a^2\lambda^2 t}$ in the integrand of (25.5) and because of the inequalities

$$|A(\lambda)| < \frac{1}{\pi}\int_{-\infty}^{\infty} |f(v)|\, dv = C,$$

$$|B(\lambda)| < \frac{1}{\pi}\int_{-\infty}^{\infty} |f(v)|\, dv = C \qquad (C = \text{const}),$$

the integral (25.5) and the integrals obtained from it by differentiating the integrand any number of times with respect to x and t are uniformly convergent for $t \geqslant t_0 > 0$ (where $t_0 > 0$ is arbitrary). This follows from the inequalities

$$|(A(\lambda)\cos\lambda x + B(\lambda)\sin\lambda x)|e^{-a^2\lambda^2 t} \leqslant 2Ce^{-a^2\lambda^2 t} \leqslant 2Ce^{-a^2\lambda^2 t_0},$$

$$\left|\frac{\partial^n}{\partial x^n}[(A(\lambda)\cos\lambda x + B(\lambda)\sin\lambda x)e^{-a^2\lambda^2 t}]\right| \leqslant 2C\lambda^n e^{-a^2\lambda^2 t} \leqslant 2C\lambda^n e^{-a^2\lambda^2 t_0},$$

$$\left|\frac{\partial^m}{\partial t^m}[(A(\lambda)\cos\lambda x + B(\lambda)\sin\lambda x)e^{-a^2\lambda^2 t}]\right| \leqslant 2Ca^{2m}\lambda^{2m} e^{-a^2\lambda^2 t} \leqslant 2Ca^{2m}\lambda^{2m}e^{-a^2\lambda^2 t_0},$$

and the fact that the majorizing functions on the right are integrable in λ from 0 to ∞. Thus, we need only apply Theorems 4 and 3 of Ch. 7, Sec. 6. It should be noted that although our argument shows that $u(x, t)$ is a solution of equation (20.4), it does not show that

$$\lim_{t\to 0} u(x, t) = f(x).$$

However, this relation is true and can easily be proved.

Using (25.6), we can rewrite the solution of our problem as

$$u(x, t) = \frac{1}{\pi}\int_0^{\infty} d\lambda \int_{-\infty}^{\infty} f(v)\cos\lambda(x - v)e^{-a^2\lambda^2 t}\, dv. \tag{25.7}$$

To further transform this expression, we begin by proving that it is possible to change the order of integration. To show this, we note first that for every $\varepsilon > 0$

$$\left|\int_l^{\infty} \cos\lambda(x - v)e^{-a^2\lambda^2 t}\, d\lambda\right| \leqslant \int_l^{\infty} e^{-a^2\lambda^2 t}\, d\lambda \leqslant \varepsilon$$

for sufficiently large l (where $t > 0$ is fixed). Therefore

$$\left|\frac{1}{\pi}\int_{-\infty}^{\infty} dv \int_l^{\infty} f(v)\cos\lambda(x - v)e^{-a^2\lambda^2 t}\, d\lambda\right| \leqslant \frac{\varepsilon}{\pi}\int_{-\infty}^{\infty} |f(v)|\, dv,$$

from which it follows that the integral in the left-hand side converges to zero as $l \to \infty$. But then

$$\frac{1}{\pi}\int_{-\infty}^{\infty} dv \int_0^{\infty} f(v) \cos \lambda(x - v)e^{-a^2\lambda^2 t}\, d\lambda$$

$$= \lim_{l\to\infty} \frac{1}{\pi}\int_{-\infty}^{\infty} dv \int_0^{l} f(v) \cos \lambda(x - v)e^{-a^2\lambda^2 t}\, d\lambda$$

$$= \lim_{l\to\infty} \frac{1}{\pi}\int_0^{l} d\lambda \int_{-\infty}^{\infty} f(v) \cos \lambda(x - v)e^{-a^2\lambda^2 t}\, dv = u(x, t)$$

[see (25.7)]. Here the change in order of integration is legitimate, because the integral

$$\int_{-\infty}^{\infty} f(v) \cos \lambda(x - v)e^{-a^2\lambda^2 t}\, dv$$

is uniformly convergent in λ for $0 \leqslant \lambda \leqslant l$. This follows from the fact that the integrand is dominated by $|f(v)|$ (see Theorems 4 and 2 of Ch. 7, Sec. 6).

Thus, we can write

$$u(x, t) = \frac{1}{\pi}\int_{-\infty}^{\infty} f(v)\, dv \int_0^{\infty} \cos \lambda(x - v)e^{-a^2\lambda^2 t}\, d\lambda, \tag{25.8}$$

instead of (25.7). It turns out that the inner integral can be evaluated. In fact, set

$$a\lambda\sqrt{t} = z, \qquad \lambda(x - v) = \mu z$$

so that

$$d\lambda = \frac{dz}{a\sqrt{t}}, \qquad \mu = \frac{x - v}{a\sqrt{t}}.$$

Then we have

$$\int_0^{\infty} \cos \lambda(x - v)e^{-a^2\lambda^2 t}\, d\lambda = \frac{1}{a\sqrt{t}}\int_0^{\infty} e^{-z^2} \cos \mu z\, dz = \frac{1}{a\sqrt{t}} I(\mu). \tag{25.9}$$

Differentiation with respect to μ behind the integral sign gives

$$I'(\mu) = -\int_0^{\infty} e^{-z^2} z \sin \mu z\, dz;$$

this differentiation is legitimate because the resulting integral is uniformly convergent in μ.

We now integrate by parts, obtaining

$$I'(\mu) = \frac{1}{2}\,[e^{-z^2} \sin \mu z]_{z=0}^{z=\infty} - \frac{\mu}{2}\int_0^{\infty} e^{-z^2} \cos \mu z\, dz = -\frac{\mu}{2} I(\mu).$$

It follows that

$$I(\mu) = Ce^{-\mu^2/4}.$$

To find C, we set $\mu = 0$. This gives

$$C = I(0) = \int_0^\infty e^{-z^2}\,dz,$$

an integral which, as we know, equals $\frac{1}{2}\sqrt{\pi}$ (see Ch. 8, Sec. 8). Therefore

$$I(\mu) = \tfrac{1}{2}\sqrt{\pi}\,e^{-\mu^2/4},$$

and by (25.9)

$$\int_0^\infty \cos\lambda(x - v)e^{-a^2\lambda^2 t}\,d\lambda = \frac{1}{2a}\sqrt{\pi/t}\,e^{-\frac{(x-v)^2}{4a^2 t}}.$$

Substituting this expression in (25.8), we finally find

$$u(x, t) = \frac{1}{2a\sqrt{\pi t}}\int_{-\infty}^{\infty} f(v)e^{-\frac{(x-v)^2}{4a^2 t}}\,dv. \qquad (25.10)$$

On the one hand, formula (25.10) shows that as t increases, $u(x, t) \to 0$, i.e., the heat "spreads out" along the rod. On the other hand, (25.10) shows that the heat is "transmitted" instantaneously along the rod. In fact, let the initial temperature be positive for $x_0 \leqslant v \leqslant x_1$ and zero outside this interval. Then the subsequent distribution of temperature is given by the formula

$$u(x, t) = \frac{1}{2a\sqrt{\pi t}}\int_{x_0}^{x_1} f(v)e^{-\frac{(x-v)^2}{4a^2 t}}\,dv,$$

from which it is clear that $u(x, t) > 0$ for arbitrarily small $t > 0$ and arbitrarily large x.

26. Heat Flow in a Circular Cylinder Whose Surface is Insulated

Let the axis of a circular cylinder of radius l be directed along the z-axis, and let its ends be insulated (or let the length of the cylinder be infinite). Suppose that the initial temperature distribution and the boundary conditions are independent of z. Then, it can be shown that the equation for heat flow is

$$\frac{\partial u}{\partial t} = a^2\left(\frac{\partial^2 u}{\partial x^2} + \frac{\partial^2 u}{\partial y^2}\right), \qquad (26.1)$$

where $a^2 = K/c\rho$, K is the conductivity of the material from which the rod is made, c is its heat capacity, and ρ its density. Thus, the temperature is independent of z (a fact which is of course a consequence of the assumptions just made), and we are essentially dealing with a problem in the plane. If we go over to polar coordinates by setting $x = r\cos\theta$, $y = r\sin\theta$, then, instead of equation (26.1), we obtain

$$\frac{\partial u}{\partial t} = a^2\left(\frac{\partial^2 u}{\partial r^2} + \frac{1}{r}\frac{\partial u}{\partial r} + \frac{1}{r^2}\frac{\partial^2 u}{\partial \theta^2}\right).$$

We now assume further that the initial and boundary conditions are independent of θ. Then, obviously u is a function only of r and t, and the heat flow equation takes the form

$$\frac{\partial u}{\partial t} = a^2\left(\frac{\partial^2 u}{\partial r^2} + \frac{1}{r}\frac{\partial u}{\partial r}\right). \tag{26.2}$$

We also assume that the surface of the cylinder is insulated from the surrounding medium, i.e.,

$$\frac{\partial u(l, t)}{\partial r} = 0 \tag{26.3}$$

(absence of heat flow), and that the initial temperature distribution is given by the condition

$$u(r, 0) = f(r). \tag{26.4}$$

We look for particular solutions of the form

$$u(r, t) = R(r)T(t).$$

Substituting this expression in (16.2) gives

$$RT' = a^2\left(R''T + \frac{1}{r}R'T\right),$$

whence

$$\frac{R'' + (1/r)R'}{R} = \frac{T'}{a^2T} = -\lambda^2 = \text{const},$$

so that

$$R'' + \frac{1}{r}R' + \lambda^2 R = 0, \tag{26.5}$$

$$T' + a^2\lambda^2 T = 0. \tag{26.6}$$

Equation (26.5) is the parametric form of Bessel's equation, with index $p = 0$ (see Ch. 8, Sec. 11). Its general solution is

$$R(r) = C_1 J_0(\lambda r) + C_2 Y_0(\lambda r).$$

Since $Y_0(\lambda r) \to \infty$ as $r \to 0$, we have to set $C_2 = 0$. Taking $C_1 = 1$, we find from the boundary condition (26.3) that

$$J_0'(\lambda l) = 0.$$

Therefore, $\mu = \lambda l$ is the nth positive root of the equation $J_0'(\mu) = 0$. We write

$$\lambda_n = \frac{\mu_n}{l},$$

$$R_n(r) = J_0(\lambda_n r) = J_0\left(\frac{\mu_n r}{l}\right) \qquad (n = 1, 2, \ldots),$$

where $\mu_n = \lambda_n l$ is the nth positive zero of the function $J_0'(\mu)$. Setting $\lambda = \lambda_n$ in equation (26.6), we find

$$T_n(t) = A_n e^{-a^2\lambda_n^2 t} \qquad (n = 1, 2, \ldots). \tag{26.7}$$

Thus, for equation (26.2), subject to the condition (26.3), we have found particular solutions of the form

$$u_n(r, t) = A_n J_0(\lambda_n r) e^{-a^2\lambda_n^2 t} \qquad (n = 1, 2, \ldots). \tag{26.8}$$

We now form the series

$$u(r, t) = \sum_{n=1}^{\infty} A_n J_0(\lambda_n r) e^{-a^2\lambda_n^2 t}, \tag{26.9}$$

and to satisfy the initial condition (26.4), we require that

$$u(r, 0) = \sum_{n=1}^{\infty} A_n J_0(\lambda_n r) = f(r). \tag{26.10}$$

A calculation of the Fourier coefficients of $f(r)$ with respect to the system $\{J_0(\lambda_n r)\}$ gives

$$A_n = \frac{2}{l^2 J_0^2(\mu_n)} \int_0^l r f(r) J_0(\lambda_n r)\, dr \qquad (n = 1, 2, \ldots) \tag{26.11}$$

(see Ch. 8, Sec. 24). Therefore, the solution of our problem is given by the series (26.9), where the coefficients A_n are determined from the formula (26.11).

27. Heat Flow in a Circular Cylinder Whose Surface Exchanges Heat with the Surrounding Medium

This problem reduces to solving equation (26.2) with the boundary condition

$$\frac{\partial u(l, t)}{\partial r} + hu(l, t) = 0 \tag{27.1}$$

and with the previous initial condition

$$u(r, 0) = f(r). \tag{27.2}$$

Repeating the argument of Sec. 26, we again obtain equations (26.5) and (26.6), and we again find that

$$R(r) = J_0(\lambda r).$$

The condition (27.1) gives

$$\lambda J_0'(\lambda l) + hJ_0(\lambda l) = 0$$

or

$$\lambda l J_0'(\lambda l) + hlJ_0(\lambda l) = 0.$$

Therefore, the number $\mu = \lambda l$ must be a root of the equation

$$\mu J_0'(\mu) + hlJ_0(\mu) = 0. \tag{27.3}$$

We now write

$$\lambda_n = \frac{\mu_n}{l},$$

$$R_n(r) = J_0(\lambda_n r) = J_0\left(\frac{\mu_n r}{l}\right) \qquad (n = 1, 2, \ldots),$$

where μ_n is the nth positive root of equation (27.3). For $\lambda = \lambda_n$ $(n = 1, 2, \ldots)$, the solution of equation (26.6) is given by (26.7), and (26.8) gives particular solutions of (26.2) satisfying the condition (27.1). We again form the series (26.9), and require that the relation (26.10) be satisfied. A calculation of the Fourier coefficients of $f(r)$ with respect to the system $\{J_0(\lambda_n r)\}$ leads to the formula

$$A_n = \frac{2}{l^2[J_0'^2(\mu_n) + J_0^2(\mu_n)]} \int_0^l rf(r)J_0(\lambda_n r)\, dr \tag{27.4}$$

(see Ch. 8, Sec. 24). Thus, the solution of equation (26.2), subject to the conditions (27.1) and (27.2), is given by the series (26.9), where the coefficients are determined from (27.4), and the numbers μ_n are the roots of equation (27.3).

28. Steady-State Heat Flow in a Circular Cylinder

We now assume that a constant temperature is maintained on the surface of the cylinder, and that the distribution of temperature is independent of z.

Then after a sufficiently long interval of time, a definite temperature is established at every point of the cylinder, i.e., the function u ceases to depend on t. Thus, instead of equation (26.1), we have

$$\frac{\partial^2 u}{\partial x^2} + \frac{\partial^2 u}{\partial y^2} = 0,$$

or in polar coordinates

$$\frac{\partial^2 u}{\partial r^2} + \frac{1}{r}\frac{\partial u}{\partial r} + \frac{1}{r^2}\frac{\partial^2 u}{\partial \theta^2} = 0. \tag{28.1}$$

Let the temperature on the boundary be specified by the condition

$$u(l, \theta) = f(\theta), \tag{28.2}$$

and look for particular solutions of the form

$$u(r, \theta) = R(r)\Phi(\theta).$$

Substituting this expression in (28.1) gives

$$R''\Phi + \frac{1}{r}R'\Phi + \frac{1}{r^2}R\Phi'' = 0,$$

whence

$$-\frac{R'' + (1/r)R'}{(1/r^2)R} = \frac{\Phi''}{\Phi} = -\lambda^2 = \text{const}, \tag{28.3}$$

so that

$$r^2R'' + rR' - \lambda^2 R = 0, \tag{28.4}$$

$$\Phi'' + \lambda^2\Phi = 0. \tag{28.5}$$

The solution of (28.5) is

$$\Phi(\theta) = A\cos\lambda\theta + B\sin\lambda\theta.$$

It follows from the physical meaning of the problem that the function $\Phi(\theta)$ must have period 2π, and hence λ must be an integer. (Incidentally, we note that $\Phi(\theta)$ would not have been periodic if we had taken the constant in (28.3) to be positive.) Thus, we write

$$\Phi_n(\theta) = A_n\cos n\theta + B_n\sin n\theta \qquad (n = 0, 1, 2, \ldots). \tag{28.6}$$

For $\lambda = n$, equation (28.4) takes the form

$$r^2R'' + rR' - n^2R = 0, \tag{28.7}$$

which is a second order linear differential equation. It can be verified by direct substitution that the functions r^n and r^{-n} satisfy this equation. Therefore, for $n > 0$, the general solution of (28.7) is

$$R_n = C_n r^n + D_n r^{-n}. \tag{28.8}$$

Since $r^{-n} \to \infty$ as $r \to 0$, we have to set $D_n = 0$. For $n = 0$, we easily find that

$$R_0 = C_0 + D_0 \ln r, \tag{28.9}$$

and hence we must again take $D_0 = 0$. Using (28.6), (28.8), (28.9), and the conditions $D_n = 0$ $(n = 0, 1, 2, \ldots)$, we can write the particular solutions of (28.1) as

$$u_n(r, \theta) = (\alpha_n \cos n\theta + \beta_n \sin n\theta)r^n \qquad (n = 1, 2, \ldots),$$

$$u_0 = \frac{\alpha_0}{2}.$$

We now form the series

$$u(r, \theta) = \frac{\alpha_0}{2} + \sum_{n=1}^{\infty} (\alpha_n \cos n\theta + \beta_n \sin n\theta)r^n,$$

and to satisfy the boundary condition (28.2), we require that

$$u(l, \theta) = \frac{\alpha_0}{2} + \sum_{n=1}^{\infty} (\alpha_n \cos n\theta + \beta_n \sin n\theta)l^n = f(\theta).$$

A calculation of the Fourier coefficients of $f(\theta)$ gives

$$\alpha_n l^n = \frac{1}{\pi}\int_{-\pi}^{\pi} f(\theta) \cos n\theta \, d\theta = a_n \qquad (n = 0, 1, 2, \ldots),$$

$$\beta_n l^n = \frac{1}{\pi}\int_{-\pi}^{\pi} f(\theta) \sin n\theta \, d\theta = b_n \qquad (n = 1, 2, \ldots),$$

so that

$$\alpha_n = \frac{a_n}{l^n}, \qquad \beta_n = \frac{b_n}{l^n}.$$

Therefore

$$u(r, \theta) = \frac{a_0}{2} + \sum_{n=1}^{\infty} (a_n \cos n\theta + b_n \sin n\theta)\left(\frac{r}{l}\right)^n. \tag{28.10}$$

For $r < l$, this series can be differentiated term by term any number of times with respect to r and θ, since each resulting series is uniformly convergent for $0 \leqslant r \leqslant r_0$, where $r_0 < l$ is arbitrary. It follows that (28.10) actually gives the solution of equation (28.1).

This solution can be given a more compact form if we use the Poisson integral (see Ch. 6, Sec. 7). Then we have

$$u(r, \theta) = \frac{1}{2\pi}\int_{-\pi}^{\pi} f(t) \frac{1 - (r/l)^2}{1 - 2(r/l)\cos(t - \theta) + (r/l)^2}\, dt$$

or

$$u(r, \theta) = \frac{1}{2\pi}\int_{-\pi}^{\pi} f(t) \frac{l^2 - r^2}{l^2 - 2lr\cos(t - \theta) + r^2}\, dt.$$

Moreover

$$\lim_{r \to l} u(r, \theta) = f(\theta)$$

wherever $f(\theta)$ is continuous, i.e., the solution just written satisfies the boundary condition (28.2).

PROBLEMS

1. Find the eigenvalues and normalized eigenfunctions of the differential equation $\Phi''(x) + \lambda\Phi(x) = 0$ on the interval $[0, 1]$, subject to the following boundary conditions:

a) $\Phi(0) = \Phi(1) = 0$;
b) $\Phi(0) = \Phi'(1) = 0$;
c) $\Phi(0) = \Phi(1) + h\Phi'(1) = 0$.

2. Consider a string of length l, tension T and linear density ρ, fastened at the points $x = 0$ and $x = l$. Suppose that at time $t = 0$, the point $x = c$ $(0 < c < l)$ is displaced by an amount h and then released (the "plucked string"). Write the initial conditions and find the subsequent motion of the string.

3. Consider a string of length $2l$, tension T and linear density ρ, fastened at the points $x = \pm l$. Let the initial position of the string be a parabola which is symmetric with respect to the center of the string, with maximum initial displacement equal to h, and let the initial velocity of the string be zero. Write the initial conditions and find the subsequent motion of the string.

4. Consider a string of length $2l$, tension T and linear density ρ, fastened at the points $x = \pm l$. Suppose that at time $t = 0$, the string receives an impulse of magnitude P at its midpoint. Find the subsequent motion of the string.

Hint. Solve the problem with the initial conditions

$$u(x, 0) = 0,$$

$$\frac{\partial u(x, 0)}{\partial t} = \begin{cases} \dfrac{P}{2\rho\varepsilon} & \text{for } |x| \leqslant \varepsilon, \\ 0 & \text{for } \varepsilon < |x| \leqslant l, \end{cases}$$

and then pass to the limit $\varepsilon \to 0$.

5. Consider a string of length l, tension T and density ρ, fastened at the points $x = 0$ and $x = l$. Let the string be initially at rest in its equilibrium position. Suppose the section of the string between x_1 and x_2 $(0 \leqslant x_1 < x_2 \leqslant l)$ is acted upon by a periodic perturbing force $F(x, t) = \rho A \sin \omega t$ (see Sec. 13). Find the subsequent motion of the string. Verify that when $x_1 = 0$, $x_2 = l$, the answer reduces to formula (13.8).

6. Let a concentrated periodic perturbing force $F = A \sin \omega t$ act upon the point $x = c$ of the string of Prob. 5. Find the subsequent motion of the string.

Hint. Pass to the limit $x_1 \to c$, $x_2 \to c$ in Prob. 5.

7. Consider a rod of length l, Young's modulus E, cross section s and density ρ, fastened at the end $x = 0$. Suppose that the rod is stretched by a force F acting on the end $x = l$, and then is suddenly released at time $t = 0$. Write the initial conditions and find the subsequent longitudinal vibrations of the rod.

Hint. The amount by which the rod is initially stretched is Fl/Es.

8. Let the free end of the rod of Prob. 7 receive a sudden impulse P at time $t = 0$. Find the subsequent longitudinal vibrations of the rod.

Hint. Solve the problem with the initial conditions

$$u(x, 0) = 0,$$

$$\frac{\partial u(x, 0)}{\partial t} = \begin{cases} 0 & \text{for } 0 \leqslant x < l - \varepsilon, \\ \dfrac{P}{\varepsilon \rho s} & \text{for } l - \varepsilon \leqslant x \leqslant l, \end{cases}$$

and then pass to the limit $\varepsilon \to 0$.

9. Suppose the free end of the rod of Prob. 7 is acted upon by a periodic perturbing force $A \sin \omega t$. Find the subsequent longitudinal vibrations of the rod, assuming that the rod is initially at rest.

10. Find the linear combinations of the modes u_{13} and u_{31} of a rectangular membrane corresponding to the nodal lines of Fig. 51(c) of Sec. 17. Write the equation of the "nodal ring" indicated by the fourth sketch in Fig. 51(c). Is this curve actually a circle?

11. Consider a circular membrane of radius l, tension T and density ρ per unit area. Find the radial vibrations of the membrane if it receives a sudden impulse P at time $t = 0$ distributed over the circle $r \leqslant \varepsilon$, assuming that the membrane is originally at rest in its equilibrium position.

Hint. Solve the problem with the initial conditions

$$u(r, 0) = 0,$$

$$\frac{\partial u(r, 0)}{\partial t} = \begin{cases} \dfrac{P}{\pi \varepsilon^2 \rho} & \text{for } 0 \leqslant r \leqslant \varepsilon, \\ 0 & \text{for } \varepsilon < r \leqslant l. \end{cases}$$

12. Find the radial vibrations of the membrane of Prob. 11 if it is acted upon by a periodic perturbing force $F = A \sin \omega t$ per unit area, uniformly distributed over the entire membrane, assuming that at time $t = 0$ the membrane is at rest in its equilibrium position.

13. Consider an infinite slab of width $2l$ bounded by the planes $x = \pm l$, made of material with thermal conductivity K, specific heat c and density ρ. The slab is first heated to temperature T_0 and then at time $t = 0$, its faces are held at zero temperature. Find the subsequent temperature distribution in the slab.

14. Let the slab of Prob. 13 have an initial temperature distribution $u(x, 0) = f(x)$. Find the subsequent temperature distribution if starting from time $t = 0$, the slab exchanges heat freely with the surrounding medium which is at zero temperature. (Let the emissivity of the slab be H (see Sec. 24.))

15. Find the steady-state temperature distribution in an infinite rectangular prism bounded by the planes $x = 0$, $x = a$, $y = 0$, $y = b$, if the faces formed by the planes $x = 0$, $x = a$, $y = 0$ are at zero temperature, while the face formed by the plane $y = b$ has the temperature distribution $u(x, b) = f(x)$, where $0 \leqslant x \leqslant a$. (Draw a figure.) Specialize the answer to the case $f(x) = T_0$.

ANSWERS TO PROBLEMS

CHAPTER 1

1. a) $$e^{ax} = \frac{e^{a\pi} - e^{-a\pi}}{\pi}\left[\frac{1}{2a} + \sum_{n=1}^{\infty}\frac{(-1)^n}{n^2 + a^2}(a\cos nx - n\sin nx)\right] \quad (-\pi < x < \pi);$$

b) $$\cos ax = \frac{2}{\pi}\sin a\pi\left[\frac{1}{2a} + \sum_{n=1}^{\infty}(-1)^n\frac{a\cos nx}{a^2 - n^2}\right] \quad (-\pi \leqslant x \leqslant \pi);$$

c) $$\sin ax = \frac{2}{\pi}\sin a\pi\sum_{n=1}^{\infty}(-1)^n\frac{n\sin nx}{a^2 - n^2} \quad (-\pi < x < \pi);$$

d) $$a_0 = \frac{\pi}{2},\ a_n = \frac{\cos n\pi - 1}{n^2\pi},\ b_n = (-1)^{n-1}\frac{1}{n},$$

$$f(x) = \frac{\pi}{4} - \frac{2}{\pi}\cos x + \sin x - \frac{\sin 2x}{2} - \frac{2}{9\pi}\cos 3x + \frac{\sin 3x}{3} - \frac{\sin 4x}{4} + \cdots \quad (-\pi < x < \pi)$$

3. a) $$\cosh ax = \frac{2}{\pi}\sinh a\pi\left[\frac{1}{2a} + \sum_{n=1}^{\infty}(-1)^n\frac{a}{n^2 + a^2}\cos nx\right] \quad (-\pi \leqslant x \leqslant \pi),$$

b) $$\sinh ax = \frac{2}{\pi}\sinh a\pi\left[\sum_{n=1}^{\infty}(-1)^{n-1}\frac{n}{n^2 + a^2}\sin nx\right] \quad (-\pi < x < \pi).$$

4. a) $$\sin ax = \frac{1 - \cos a\pi}{\pi}\left[\frac{1}{a} + 2a\sum_{n=1}^{\infty}\frac{\cos 2nx}{a^2 - 4n^2}\right] + 2a\frac{1 + \cos a\pi}{\pi}\sum_{n=0}^{\infty}\frac{\cos(2n+1)x}{a^2 - (2n+1)^2} \quad (0 \leqslant x \leqslant \pi);$$

(What happens when a is an integer?)

b) $$f(x) = \frac{2h}{\pi}\left[\frac{1}{2} + \sum_{n=1}^{\infty} \frac{\sin nh}{nh} \cos nx\right] \qquad (0 \leqslant x \leqslant \pi),$$

except for the value $x = h$, where the sum equals $1/2$. (Why?)

c) $$f(x) = \frac{2h}{\pi}\left[\frac{1}{2} + \sum_{n=1}^{\infty} \left(\frac{\sin nh}{nh}\right)^2 \cos nx\right] \qquad (0 \leqslant x \leqslant \pi).$$

5. a) $$f(x) = \frac{1}{2}\sin\frac{\pi x}{l} - \frac{4}{\pi}\sum_{n=1}^{\infty} (-1)^n \frac{n}{4n^2 - 1} \sin\frac{2\pi n x}{l} \qquad (0 \leqslant x \leqslant l),$$

except for the value $x = l/2$, where the sum equals $1/2$;

b) $$f(x) = -\frac{4}{\pi}\sum_{n=2}^{\infty} \frac{n\cos(n\pi/2)}{n^2 - 1} \sin\frac{n\pi x}{l}$$

$$= \frac{4}{\pi}\left(\frac{2}{3}\sin\frac{2\pi x}{l} - \frac{4}{15}\sin\frac{4\pi x}{l} + \frac{6}{35}\sin\frac{6\pi x}{l} - \cdots\right) \qquad (0 \leqslant x \leqslant l),$$

except for the value $x = l/2$, where the sum equals 0.

6. $$f(x) = \frac{4}{\pi}\left[\frac{1}{2} + \sum_{n=1}^{\infty} (-1)^{n+1} \frac{\cos(2\pi n x/l)}{4n^2 - 1}\right].$$

7. $$a_n = \frac{1}{\pi}\int_0^{2\pi} f(x)\cos nx\,dx = -\frac{1}{\pi}\int_{-\pi/n}^{2\pi-(\pi/n)} f\left(\frac{\pi}{n} + t\right)\cos nt\,dt$$

$$= -\frac{1}{\pi}\int_0^{2\pi} f\left(\frac{\pi}{n} + t\right)\cos nt\,dt.$$

Hence

$$a_n = \frac{1}{2\pi}\int_0^{2\pi}\left\{f(x) - f\left(\frac{\pi}{n} + x\right)\right\}\cos nx\,dx,$$

so that

$$|a_n| \leqslant \frac{1}{2\pi}\int_0^{2\pi}\left|f(x) - f\left(\frac{\pi}{n} + x\right)\right| |\cos nx|\,dx \leqslant \frac{c\pi^\alpha}{n^\alpha}.$$

8. a) $$\cos x = \frac{8}{\pi}\sum_{n=1}^{\infty} \frac{n\sin 2nx}{4n^2 - 1};$$

b) $$x^3 = 2\pi^2 \sum_{n=1}^{\infty} (-1)^{n+1}\frac{\sin nx}{n} + 12\sum_{n=1}^{\infty} (-1)^n \frac{\sin nx}{n^3}.$$

CHAPTER 2

3. By the Cauchy inequality

$$|a_0 + a_1x + \cdots + a_nx^n| \leqslant (a_0^2 + \cdots + a_n^2)^{1/2}(1 + x^2 + \cdots + x^{2n})^{1/2}$$

$$= \left(\frac{1 - x^{2n+2}}{1 - x^2}\right)^{1/2} \leqslant \frac{1}{\sqrt{1 - x^2}} \quad \text{if } 0 \leqslant x \leqslant 1.$$

Now integrate. The sharper estimate is obtained by direct use of the Schwarz inequality.

4. Use the Schwarz inequality with one function equal to $|1 + e^{in_1x} + \cdots + e^{in_kx}|$ and the other equal to 1.

5. Use the Schwarz inequality.

7. Suppose $g(x)$ is orthogonal to all the $\varphi_i(x)$, and let $P_n(x) = a_n\varphi_0(x) + \cdots + a_n\varphi_n(x)$ be such that

$$\int_a^b [g(x) - P_n(x)]^2\, dx < \frac{1}{n}.$$

Then $\|g\|^2 + \|P_n\|^2 < 1/n$, so that $\|g\|^2 = 0$, i.e., $g(x) \equiv 0$, since $g(x)$ is continuous.

8. a) No orthogonal function exists, since if g were such a function and if

$$c_n = \int_a^b g(x)\varphi_n(x)\, dx,$$

then $0 = c_0 + c_1 = c_0 + c_2 = c_0 + c_3 = \cdots$, i.e., $-c_0 = c_1 = c_2 = \cdots$, which violates Bessel's inequality (Sec. 6) unless all the c_i are zero; b) No orthogonal function exists, by a similar proof; c) An orthogonal function g exists. In fact, consider the continuous function

$$g(x) = \sum_{n=0}^{\infty} (-1)^n \frac{1}{2^n} \varphi_n(x).$$

9. a) If $g = a_0\varphi_0 + \cdots + a_n\varphi_n$ is zero, then

$$0 = \|g\|^2 = |a_0|^2(\varphi_0, \varphi_0) + \cdots + |a_n|^2(\varphi_n, \varphi_n),$$

i.e., $a_0 = a_1 = \cdots = a_n = 0$ [for the definition of (φ, ψ), see Prob. 10 and Ch. 2, Sec. 10]; b) A polynomial of degree n has at most n real zeros, unless it vanishes identically.

11. Use repeated integration by parts.

12. a) $$\frac{1}{2} + \sum_{n=1}^{\infty} (-1)^{n-1} \frac{(4n-1)(2n-2)!}{2^{2n}n!(n-1)!} P_{2n-1}(x);$$

b) $$\frac{1}{2} - \sum_{n=1}^{\infty} (-1)^n \frac{(4n+1)(2n-2)!}{2^{2n}(n-1)!(n+1)!} P_{2n}(x).$$

CHAPTER 3

2. a) Neither limit exists; b) $f'_+(0) = 0$, $f'(0+)$ does not exist; c) $f'_+(0) = f'(0+) = 0$.

3. This follows from (6.3) just as in the proof of Sec. 6.

4.
$$\frac{1}{2\pi}\int_0^{2\pi} |h(x)|\,dx \leqslant \frac{1}{2\pi}\int_0^{2\pi}\frac{1}{2\pi}\int_0^{2\pi} |f(x-t)g(t)|\,dt\,dx$$
$$= \frac{1}{2\pi}\int_0^{2\pi} |g(t)|\left(\frac{1}{2\pi}\int_0^{2\pi} |f(x-t)|\,dx\right)dt$$
$$= \frac{1}{2\pi}\int_0^{2\pi} |g(t)|\left(\frac{1}{2\pi}\int_0^{2\pi} |f(x)|\,dx\right)dt.$$

Moreover, we have

$$c_n = \frac{1}{2\pi}\int_0^{2\pi} e^{-inx}h(x)\,dx$$
$$= \frac{1}{2\pi}\int_0^{2\pi} g(t)\left(\frac{1}{2\pi}\int_0^{2\pi} e^{-inx}f(x-t)\,dx\right)dt$$
$$= \frac{1}{2\pi}\int_0^{2\pi} g(t)\left(\frac{1}{2\pi}\int_0^{2\pi} e^{-inu}f(u)\,du\right)e^{-int}\,dt = a_n b_n.$$

If

$$\sum_{n=1}^{\infty} |a_n|^2 < \infty, \qquad \sum_{n=1}^{\infty} |b_n|^2 < \infty,$$

then

$$\sum_{n=1}^{\infty} |a_n b_n| < \infty$$

by the Cauchy inequality (see Ch. 2, Prob. 2).

6. We have

$$\frac{1}{2}\sum_{n=1}^{\infty} \rho_n \int_a^b (1 + \cos 2nx \cos 2\theta_n - \sin 2nx \sin 2\theta_n)\,dx \leqslant M(b-a).$$

It follows from (2.9) that

$$\left|\int_a^b (\cos 2nx \cos 2\theta_n - \sin 2nx \sin 2\theta_n)\,dx\right| < \frac{b-a}{2}$$

if $n \geqslant N$, i.e.,

$$\int_a^b (1 + \cos 2nx \cos 2\theta_n - \sin nx \sin 2\theta_n)\,dx > \frac{b-a}{2}$$

if $n \geqslant N$. Therefore

$$\frac{1}{2}\sum_{n=N}^{\infty} \rho_n \frac{b-a}{2} < M(b-a), \quad \text{so that} \quad \sum_{n=N}^{\infty} \rho_n < 4M.$$

7. a) The formula

$$\frac{x}{2} + s_n(x) = \int_0^x D_n(t)\, dt$$

follows from equations (3.2) and (4.1) after some manipulation;

b) If

$$D_n^*(x) = D_n(x) - \frac{1}{2}\cos nx = \frac{\sin nx}{2\tan(x/2)},$$

then

$$\int_0^x (D_n(t) - D_n^*(t))\, dt = \frac{1}{2}\int_0^x \cos nt\, dt \to 0$$

as $n \to \infty$ (why?). Moreover

$$\int_0^x \left(\frac{\sin nt}{t} - D_n^*(t)\right) dt = \int_0^x \left(\frac{1}{t} - \frac{1}{2}\cot\frac{t}{2}\right) \sin nt\, dt \to 0$$

as $n \to \infty$. (Why?)

Thus

$$\int_0^x D_n(t)\, dt = \int_0^x \frac{\sin nt}{t}\, dt + \omega_n(x),$$

where $\omega_n(x) \to 0$ as $n \to \infty$.

c) $\int_0^\infty \frac{\sin nt}{t}\, dt = \frac{\pi}{2}$ follows from b) when $n \to \infty$.

8. Use preceding problem. The inequality can be derived by inspecting the area under the curve $y = \sin t/t$.

CHAPTER 4

1. a) $x \neq 2k\pi$; b) for all x; c) $x \neq 2k\pi$.

2. For $x \neq 2k\pi$. No.

3. a) $x \neq (2k+1)\pi$; b) $x \neq 2k\pi$; c) for all x; d) for all x.

4. a), b) for $x \neq (2k+1)\pi$; c) for $x \neq 2k\pi/3$; d) for $x \neq (2k+1)\pi/2$.

6. a) $\sin(\cos x)\cosh(\sin x)$; b) $\cos(\cos x)\sinh(\sin x)$.

7. a) $\cos(\cos x)\cosh(\sin x)$; b) $\sin(\cos x)\sinh(\sin x)$.

8. a) $(1 + \cos x)\ln\left(2\cos\frac{x}{2}\right) + \frac{x}{2}\sin x$;

b) $\frac{x}{2}(1 + \cos x) - \sin x \ln\left(2\cos\frac{x}{2}\right) \quad (-\pi < x < \pi)$.

9. a) $\dfrac{1}{2} - \dfrac{x}{2}\sin x - \dfrac{1}{4}\cos x;$

b) $\sin x \log\left(2\cos\dfrac{x}{2}\right) - \dfrac{1}{4}\sin x \qquad (-\pi < x < \pi).$

10. a) $\cos x \ln\left(2\cos\dfrac{x}{2}\right) - \dfrac{1}{4}\cos x - \dfrac{1}{2};$

b) $\dfrac{x}{2}\cos x + \dfrac{1}{4}\sin x \qquad (-\pi < x < \pi).$

11. a) $\cos px\left[\ln\left(2\sin\dfrac{x}{2}\right) + \displaystyle\sum_{n=1}^{p}\dfrac{\cos nx}{n}\right] + \sin px\left[\dfrac{\pi - x}{2} - \displaystyle\sum_{n=1}^{p}\dfrac{\sin nx}{n}\right];$

b) $\cos px\left[\dfrac{\pi - x}{2} - \displaystyle\sum_{n=1}^{p}\dfrac{\sin nx}{n}\right] + \sin px\left[\ln\left(2\sin\dfrac{x}{2}\right) + \displaystyle\sum_{n=1}^{p}\dfrac{\cos nx}{n}\right]$

$$(0 < x < 2\pi).$$

12. Use Abel's lemma (Sec. 1).

13. Imitate the proof of Abel's lemma.

14. Assume that $0 < x_0 < \pi$. Since $\cos^2\theta \leqslant |\cos\theta|$, we have

$$\sum_{n=1}^{\infty} |a_n| \cos^2 nx_0 < \infty.$$

But $2\cos^2 nx_0 = 1 + \cos 2nx_0$, and by Theorem 1 of Sec. 3,

$$\sum_{n=1}^{\infty} |a_n| \cos 2nx_0 < \infty.$$

Therefore, the series

$$\sum_{n=1}^{\infty} |a_n| < \infty.$$

CHAPTER 5

1. a) $\dfrac{\pi^4}{90}$; b) $\dfrac{\pi^4}{96}$;

c) $$\sum_{n=1}^{\infty}\frac{(-1)^{n+1}}{n^4} = \sum_{n=1}^{\infty}\frac{1}{(2n+1)^4} - \sum_{n=1}^{\infty}\frac{1}{(2n)^4}$$

$$= \pi^4\left(\frac{1}{96} - \frac{1}{2^4\cdot 90}\right) = \frac{7\pi^4}{720};$$

d) $\frac{\pi^6}{945}$; e) $\frac{\pi^6}{960}$; f) $\pi^6\left(\frac{1}{960} - \frac{1}{2^6 \cdot 945}\right)$.

2. a) $\frac{\pi^3}{12}$; b) $\frac{\pi^3}{12}$; c) $\frac{\pi^3}{4}$.

3. By making the odd extension of $f(x)$ onto the interval $[-\pi, 0]$, we obtain an odd function $F(x)$ with equal (zero) values at the end points of the interval $[-\pi, \pi]$. Moreover, $F'(x)$ is obviously an even differentiable function, $F''(x)$ is an odd function which again has equal (zero) values at the end points of the interval $[-\pi, \pi]$, and $F'''(x)$ is a continuous even function. Therefore, $F(x)$ and its first three derivatives can be extended continuously over the whole x-axis, with $F''(x)$ a smooth function. But we know that the Fourier coefficients of a continuous function and those of its derivative are related by the formulas

$$a_n = -\frac{b_n''}{n}, \quad b_n' = \frac{a_n'}{n}$$

[see (8.3)]. Therefore, in our case

$$b_n = \frac{a_n'}{n} = -\frac{b_n''}{n^2}.$$

Moreover, for a smooth function, the sum of the squares of the absolute values of the Fourier coefficients converges (Ch. 3, Sec. 10). Since

$$\sum_{n=1}^{\infty} |b_n''| = \sum_{n=1}^{\infty} n^2 |b_n|,$$

it follows that the last series converges. This in turn implies the uniform convergence of the series obtained by one or two term by term differentiations of the original series, and the uniform convergence of these series guarantees that the term by term differentiation is legitimate.

4. $$f'(x) = -\frac{1}{2} + \sum_{n=1}^{\infty}\left[(-1)^{n+1}\frac{\cos nx}{n+1} - \frac{\sin nx}{n^2}\right] \qquad [x \neq (2k+1)\pi].$$

5. $$f'(x) = \sum_{n=1}^{\infty} (-1)^n \frac{\cos nx}{\sqrt{n}+1}.$$

6. a) $$f'(x) = -\frac{1}{2} - \sum_{n=1}^{\infty} \frac{\cos nx}{n^2+n+1},$$

b) $$f'(x) = -\frac{1}{10} - \frac{1}{5}\sum_{n=1}^{\infty} \frac{\cos nx}{5n^3+1}.$$

7. $F(x) = c_1 \cos x + c_2 \sin x + \frac{1}{2} \qquad (0 \leqslant x \leqslant \pi),$

where

$$c_1 = \sum_{n=2}^{\infty} \frac{1}{n^2 - 1} + \frac{1}{2} = 2, \quad \text{since } \frac{1}{n^2 - 1} = \frac{1}{2}\left(\frac{1}{n - 1} - \frac{1}{n + 1}\right),$$

$$c_2 = -\frac{\pi}{2}.$$

8. a) (In each case $f(x)$ denotes the sum of the series.)

$$f(x) = \sum_{n=1}^{\infty} \left(\frac{1}{n} + \frac{1}{n^3 + 1}\right) \sin nx$$

$$= \frac{\pi - x}{2} + \sum_{n=1}^{\infty} \frac{\sin nx}{n^3 + 1} \qquad (0 < x < 2\pi)$$

[see formula (2) of Sec. 12.]

b) $$f(x) = \frac{\pi - x}{2} - \sum_{n=1}^{\infty} \frac{\sin nx}{n(n^4 + 1)} \qquad (0 < x < \pi).$$

c) $$f(x) = -\ln\left(2 \sin \frac{x}{2}\right) - \frac{a}{12}(3x^2 - 6\pi x + 2\pi^2)$$

$$+ a^2 \sum_{n=1}^{\infty} \frac{\cos nx}{n^2(n + a)} \qquad (0 < x < 2\pi);$$

d) $$f(x) = \frac{x}{2} - a \int_a^x \ln\left(2 \cos \frac{x}{2}\right) dx + a^2 \sum_{n=1}^{\infty} \frac{(-1)^{n+1} \sin nx}{n^2(n + a)}$$

$$(-\pi < x < \pi);$$

e) $$f(x) = \ln\left(2 \cos \frac{x}{2}\right) + \frac{3x^2 - \pi^2}{12} + a^2 \sum_{n=1}^{\infty} \frac{(-1)^{n+1} \cos nx}{n^2(n + a)}$$

$$(-\pi < x < \pi);$$

f) $$f(x) = \frac{\pi}{2} + (2a - 1) \int_0^x \ln \tan \frac{x}{2}\, dx$$

$$+ (2a - 1)^2 \sum_{n=1}^{\infty} \frac{\sin (2n + 1)x}{(2n + 1)^2(n + a)} \qquad (0 < x < \pi);$$

g) $$f(x) = -\ln \tan \frac{x}{2} + \frac{(2a - 1)(2\pi x - \pi^2)}{4}$$

$$+ (2a - 1)^2 \sum_{n=1}^{\infty} \frac{\cos (2n + 1)x}{(2n + 1)^2(n + a)} \qquad (0 < x < \pi).$$

9. Use Theorem 1 of Sec. 3.

10. Use Theorem 1 of Sec. 8 and equation (3.1).

11. To derive formula (I), use formula (3.1). Then, substitute $h = \pi/2N$ in formula (I).

12. a) Set $N = 2^k$ and apply the last part of the preceding problem, keeping only terms with index $n \geqslant N/2$. (For such terms, $\sin^2(\pi n/2N) \geqslant \frac{1}{2}$.)

b) Use the Cauchy inequality (Ch. 2, Prob. 2);

c) Sum the inequalities b) with respect to k.

CHAPTER 6

1. a) $f(x) = 0 \quad \text{for } -\pi \leqslant x \leqslant \pi,\ x \neq 0;$

b) $f(x) = \dfrac{1}{2}\cot\dfrac{x}{2} \quad \text{for } -\pi \leqslant x \leqslant \pi,\ x \neq 0.$

4. a) $\sigma = \frac{1}{4}$; b) $\sigma = \dfrac{1}{(1+p)^2}$, the series converges.

5. $$\frac{p^3\cos x + p\cos x - 2p^2}{(1 - 2p\cos x + p^2)^2}.$$

7. a) $\frac{2}{3}$; b) $\frac{1}{2}$; c) $\frac{1}{4}$; d) Not summable by arithmetic means, but Abel-summable to 0; e) Not summable by arithmetic means, but Abel-summable to $\frac{1}{4}$.

8. Use the relation $s_n = (n+1)\sigma_{n+1} - n\sigma_n$.

9. a) Use the relation $t_n = (n+1)s_n - (n+1)\sigma_{n+1}$; b) This follows from a) and the theorem of Sec. 2.

CHAPTER 7

4. a) $\Phi(\lambda) = \sqrt{2/\pi}\,\dfrac{\lambda^3}{\lambda^4+4}$; b) $\Phi(\lambda) = \sqrt{2/\pi}\,\dfrac{\sin\pi\lambda}{1-\lambda^2}$;

c) $\Phi(\lambda) = \sqrt{2/\pi}\,\dfrac{\sin\lambda - \lambda\cos\lambda}{\lambda^2}$; d) $\Phi(\lambda) = \sqrt{2/\pi}\,\dfrac{2\lambda}{(1+\lambda^2)^2}$.

6. a) $f(x) = \dfrac{2}{\pi}\,\dfrac{1-\cos\pi x}{x}$; b) $f(x) = \dfrac{2}{\pi}\,\dfrac{1}{1+x^2}$;

c) $f(x) = \dfrac{2}{\pi}\,\dfrac{2ax}{(a^2+x^2)^2}$.

7. Imitate proofs for the discrete case (Ch. 3, Prob. 4).

8. b) $$h(x) = \begin{cases} 0 & \text{for } x < 0, \\ xe^{-x} & \text{for } x \geqslant 0; \end{cases}$$

d) $$h(x) = \begin{cases} 0 \quad \text{for } x < 0, \\ \frac{1}{6}x^3 \quad \text{for } 0 \leqslant x \leqslant 1, \\ -\frac{1}{6}(x^3 - 6x + 4) \quad \text{for } 1 \leqslant x \leqslant 2, \\ 0 \quad \text{for } x > 2. \end{cases}$$

9. a) $$f(x) = \begin{cases} 0 & \text{for } x < 0, \\ 2xe^{-x} & \text{for } x \geqslant 0; \end{cases}$$

b) $$f(x) = e^{-|x|}.$$

CHAPTER 8

1. Setting $p = 2$, we reduce the equation to the form (2.2). Then, the substitution $u = x^2y$ [see (2.1)] reduces the equation to the following Bessel's equation:

$$u'' + \frac{1}{x}u' + \left(1 - \frac{4}{x^2}\right)u = 0.$$

It follows at once that

$$y = \frac{1}{x^2}(C_1J_2(x) + C_2Y_2(x)).$$

2. $$\Gamma\left(\frac{2n+1}{2}\right) = \frac{2n-1}{2}\cdot\frac{2n-3}{2}\cdots\frac{5}{2}\cdot\frac{3}{2}\Gamma\left(\frac{3}{2}\right)$$
$$= \frac{(2n-1)(2n-3)\cdots 5\cdot 3\cdot 1}{2^n}\sqrt{\pi}$$

(see Sec. 3).

3. $$J_p'(x) = \frac{B\sin(x+\beta)}{\sqrt{x}} + \frac{\rho(x)}{x\sqrt{x}},$$

where $B = \text{const}$, $\beta = \text{const}$ and $\rho(x)$ is bounded as $x \to \infty$.

4. $$J_p''(x) = \frac{A\sin(x+\omega)}{\sqrt{x}} + \frac{\tau(x)}{x\sqrt{x}},$$

where A and ω are the same constants as in the asymptotic expression for $J_p(x)$, and $\tau(x)$ is bounded as $x \to \infty$.

6. Use formula (7.1).

9. Use the preceding problem and formula (2.9) of Ch. 3.

10. We have

$$c_n = \frac{2}{J_{p+1}^2(\lambda_n)}\int_0^1 x^{-p+1}J_p(\lambda_n x)\,dx,$$

$$\int_0^1 x^{-p+1}J_p(\lambda_n x)dx = \frac{1}{\lambda_n^{-p+2}}\int_0^{\lambda_n} t^{-p+1}J_p(t)\,dt.$$

Replacing p by $p - 1$ in formula (7.2), we obtain

$$\frac{d}{dt}[t^{-p+1}J_{p-1}(t)] = -t^{-p+1}J_p(t).$$

Therefore

$$\int_0^{\lambda_n} t^{-p+1}J_p(t)\,dt = [t^{-p+1}J_{p-1}(t)]_{t=0}^{t=\lambda_n} = \lambda_n^{-p+1}J_{p-1}(\lambda_n) - \frac{1}{2^{p-1}\Gamma(p)}$$

[see (4.3)], and hence

$$c_n = \frac{2}{J_{p+1}^2(\lambda_n)}\left[\frac{J_{p-1}(\lambda_n)}{\lambda} - \frac{1}{\lambda_n^{-p+2}2^{p-1}\Gamma(p)}\right].$$

11. $$c_n = \frac{2\lambda_n J_{p+1}(\lambda_n)}{\lambda_n^2 J_p'^2(\lambda_n) + (\lambda_n^2 - p^2)J_p^2(\lambda_n)}.$$

The series converges to x^p if $p \geqslant -\frac{1}{2}$, $p > H$.

12. $$x^3 = 16\sum_{n=1}^{\infty}\frac{J_3(\lambda_n x/2)}{\lambda_n J_4(\lambda_n)}$$

for $0 \leqslant x < 2$.

CHAPTER 9

In Problems 2 to 6, $a^2 = T/\rho$.

2. $$u(x,t) = \frac{2h}{\pi^2}\frac{l^2}{c(l-c)}\sum_{n=1}^{\infty}\frac{\sin(n\pi c/l)}{n^2}\sin\frac{n\pi x}{l}\cos\frac{n\pi at}{l}.$$

3. $$u(x,t) = \frac{32h}{\pi^3}\sum_{n=0}^{\infty}\frac{(-1)^n}{(2n+1)^3}\cos\frac{2n+1}{2l}\pi x\cdot\cos\frac{2n+1}{2l}\pi at.$$

4. $$u(x,t) = \frac{2P}{\pi a\rho}\sum_{n=0}^{\infty}\frac{\cos(2n+1)\pi x/2l}{2n+1}\sin\frac{2n+1}{2l}\pi at.$$

5. $$u(x,t) = \frac{2A}{\pi}\sum_{n=1}^{\infty}\left(\cos\frac{n\pi x_1}{l} - \cos\frac{n\pi x_2}{l}\right)\sin\frac{n\pi x}{l}\cdot\frac{\omega\sin\omega_n t - \omega_n\sin\omega t}{n\omega_n(\omega^2 - \omega_n^2)},$$

where $\omega_n = n\pi a/l$.

6. $$u(x,t) = \frac{2A}{\rho l}\sum_{n=1}^{\infty}\sin\frac{n\pi c}{l}\sin\frac{n\pi x}{l}\cdot\frac{\omega\sin\omega_n t - \omega_n\sin\omega t}{\omega_n(\omega^2 - \omega_n^2)}.$$

In Problems 7 to 9, $a^2 = E/\rho$.

7. $$u(x,t) = \frac{8F}{\pi^2 Es}\sum_{n=0}^{\infty}\frac{(-1)^n}{(2n+1)^2}\sin\frac{2n+1}{2l}\pi x\cdot\cos\frac{2n+1}{2l}\pi at.$$

8. $$u(x, t) = \frac{4P}{\pi as\rho} \sum_{n=0}^{\infty} \frac{(-1)^n}{2n+1} \sin \frac{2n+1}{2l} \pi x \cdot \sin \frac{2n+1}{2l} \pi at.$$

9. $$u(x, t) = \frac{2A}{\rho ls} \sum_{n=0}^{\infty} (-1)^n \frac{\omega_{2n+1} \sin \omega t - \omega \sin \omega_{2n+1} t}{\omega_{2n+1}(\omega_{2n+1}^2 - \omega^2)} \sin \frac{2n+1}{2l} \pi x,$$

where $\omega_n = \frac{n\pi a}{2l}$.

10. $u_{13}, u_{31}, u_{13} - u_{31}, u_{13} + u_{31}$. The nodal ring has the equation

$$\cos 2\pi x + \cos 2\pi y + 1 = 0,$$

and is not quite circular.

11. $$u(r, t) = \frac{2P}{\pi\varepsilon\rho c} \sum_{n=1}^{\infty} \frac{J_1(\mu_n \varepsilon/l)}{\mu_n^2 J_1^2(\mu_n)} J_0\left(\mu_n \frac{r}{l}\right) \sin \frac{\mu_n ct}{l},$$

where μ_n is the nth positive root of the equation $J_0(\mu) = 0$ and $c^2 = T/\rho$.

12. $$u(r, t) = \frac{2A}{\rho} \sum_{n=1}^{\infty} \frac{\omega \sin \omega_n t - \omega_n \sin \omega t}{\omega_n(\omega^2 - \omega_n^2)} \frac{J_0(\mu_n r/l)}{\mu_n J_1(\mu_n)},$$

where μ_n is the nth positive root of the equation $J_0(\mu) = 0$ and $\omega_n = \mu_n c/l$.

13. $$u(x, t) = \frac{4T_0}{\pi} \sum_{n=0}^{\infty} \frac{(-1)^n}{2n+1} \exp\left\{-\left(\frac{2n+1}{2l}\pi\right)^2 \tau\right\} \cos \frac{2n+1}{2l} \pi x.$$

where $\tau = \frac{Kt}{c\rho}$.

14. $$u(x, t) = \frac{1}{l} \sum_{n=1}^{\infty} \frac{\cos(\alpha_n x/l)}{1 + (\sin 2\alpha_n/2\alpha_n)} \exp\{-(\alpha_n^2 \tau/l^2)\} \int_{-l}^{l} f(\xi) \cos \frac{\alpha_n \xi}{l}\, d\xi$$
$$+ \frac{1}{l} \sum_{n=1}^{\infty} \frac{\sin(\beta_n x/l)}{1 - (\sin 2\beta_n/2\beta_n)} \exp\{-(\beta_n^2 \tau/l^2)\} \int_{-l}^{l} f(\xi) \sin \frac{\beta_n \xi}{l}\, d\xi,$$

where $\tau = (Kt/c\rho)$, α_n is the nth positive root of the equation $\tan x = lH/x$, and β_n is the nth positive root of the equation $\tan x = -x/lH$.

15. $$u(x, y) = \frac{2}{a} \sum_{n=1}^{\infty} \frac{\sinh(n\pi y/a)}{\sinh(n\pi b/a)} \sin \frac{n\pi x}{a} \int_0^a f(\xi) \sin \frac{n\pi\xi}{a}\, d\xi,$$

where $\sinh x$ is the hyperbolic sine (cf. Prob. 3 of Ch. 1). In the special case

$$u(x, y) = \frac{4T_0}{\pi} \sum_{n=0}^{\infty} \frac{\sinh[(2n+1)\pi y/a]}{\sinh[(2n+1)\pi b/a]} \frac{\sin[(2n+1)\pi x/a]}{2n+1}.$$

BIBLIOGRAPHY

Bochner, S., *Lectures on Fourier Integrals*, translated by M. Tenenbaum and H. Pollard, Princeton University Press, Princeton (1959).

Bochner, S. and K. Chandrasekharan, *Fourier Transforms*, Princeton University Press, Princeton (1949).

Byerly, W. A., *An Elementary Treatise on Fourier's Series and Spherical, Cylindrical, and Ellipsoidal Harmonics*, Ginn and Co., Boston (1893).

Carslaw, H. S., *Introduction to the Theory of Fourier's Series and Integral*, Macmillan & Co., Ltd., London (1930), reprinted by Dover Publications, Inc., New York.

Carslaw, H. S. and J. C. Jaeger, *Conduction of Heat in Solids*, second edition, Oxford University Press, New York (1959).

Churchill, R. V., *Fourier Series and Boundary Value Problems*, McGraw-Hill Book Co., Inc., New York (1941).

Franklin, P., *Fourier Methods*, McGraw-Hill Book Co., Inc., New York (1949), reprinted by Dover Publications, Inc., New York.

Goldberg, R. R., *Fourier Transforms*, Cambridge University Press, New York (1961).

Jackson, D., *Fourier Series and Orthogonal Polynomials*, Carus Mathematical Monograph No. 6, Math. Assoc. America (1941).

Jeffrey, R. L., *Trigonometric Series*, University of Toronto Press, Toronto (1956).

McLachlan, N. W., *Bessel Functions For Engineers*, second edition, Oxford University Press, New York (1955).

Rogosinski, W.W., *Fourier Series*, translated by H. Cohn and F. Steinhardt, second edition, Chelsea Publishing Co., New York (1959).

Sneddon, I. A., *Fourier Transforms*, McGraw-Hill Book Co., Inc., New York (1951).

Titchmarsh, E. C., *Introduction to the Theory of Fourier Integrals*, second edition, Oxford University Press, New York (1948).

Titchmarsh, E. C., *Eigenfunction Expansions Associated with Second-Order Differential Equations*, Oxford University Press, New York, Part I (1946), Part II (1958).

Watson, G. N., *A Treatise on the Theory of Bessel Functions*, second edition, Cambridge University Press, New York (1945).

Wiener, N., *The Fourier Integral and Certain of Its Applications*, Cambridge University Press, New York (1933), reprinted by Dover Publications, Inc., New York.

Zygmund, A., *Trigonometric Series*, in two volumes, Cambridge University Press, New York (1959).

INDEX

F

G

H

I

J

L

M

N

O

P

R

S

T

U

V

W